UNDERSTANDING

Human Anatomy
and
Physiology

UNDERSTANDING

Human Anatomy
and
Physiology

Sylvia S. Mader

WCB **Wm. C. Brown Publishers**

Book Team

Editor *Kevin Kane*
Developmental Editor *Carol Mills*
Production Editor *Gloria G. Schiesl*
Designer *K. Wayne Harms*
Art Editor *Jess Schaal*
Photo Editor *Mary Roussel*
Permissions Editor *Vicki Krug*
Visuals Processor *Andreâ Lopez-Meyer*

WCB **Wm. C. Brown Publishers**

President *G. Franklin Lewis*
Vice President, Publisher *George Wm. Bergquist*
Vice President, Publisher *Thomas E. Doran*
Vice President, Operations and Production *Beverly Kolz*
National Sales Manager *Virginia S. Moffat*
Advertising Manager *Ann M. Knepper*
Marketing Manager *Craig S. Marty*
Editor in Chief *Edward G. Jaffe*
Managing Editor, Production *Colleen A. Yonda*
Production Editorial Manager *Julie A. Kennedy*
Production Editorial Manager *Ann Fuerste*
Publishing Services Manager *Karen J. Slaght*
Manager of Visuals and Design *Faye M. Schilling*

Cover illustration by Carlyn Iverson

The credits section for this book begins on page 340, and is considered an extension of the copyright page.

Library of Congress Catalog Card Number: 90–80267

ISBN 0–697–13544–6 (cloth)
ISBN 0–697–07856–6 (paper).

Printed in the United States of America by Wm. C. Brown Publishers, 2460 Kerper Boulevard, Dubuque, IA 52001

10 9 8 7 6 5 4 3 2

Contents

v

Part Two

Support and Movement

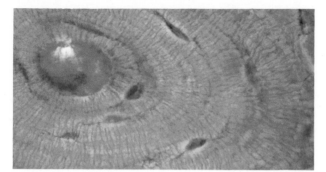

Part Three

Integration and Coordination

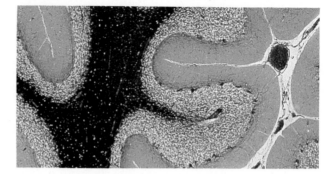

Part Four

Processing and Transporting

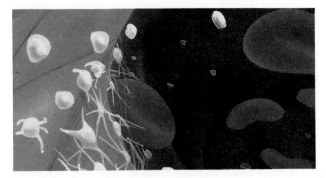

10 Digestion 180

11 Blood 196

12 Circulation 210

13 Lymphatic System 230

14 Respiration 244

Preface

Understanding Human Anatomy and Physiology is written for students who are taking a one-semester course in anatomy and physiology. It covers all the basic information that is needed to have a general understanding of the structure and function of the human body.

The writing style and depth of presentation are appropriate for students who have had little background in science and who are just beginning to pursue a career in an allied health field. Each chapter presents the topic clearly, simply, and distinctly so that students will feel capable of mastering the learning objectives of that chapter.

The text's logical organization will be especially helpful to instructors and students.

Organization

Understanding Medical Terminology

An understanding of medical terminology is critical to students of anatomy and physiology. It will help them in their study of this subject and will contribute to success in their chosen careers. Understanding Medical Terminology was written by Emily Boegli, an expert in the field. This section gives an overview of the basics of medical terminology and introduces students to the correct pronunciation of medical terms.

In addition to this introduction to medical terminology, every chapter of the text ends with a Medical Terminology Reinforcement Exercise, which reviews medical terms used in that chapter. These exercises reinforce the principles covered in Understanding Medical Terminology. Answers to these exercises are included in the Instructor's Manual.

Part I: Human Organization

The first chapter in this part explains the organization of the human body and the terms that are used to describe the location of body parts. It introduces the various organ systems and the concept of homeostasis, an equilibrium that is maintained by these systems. The other chapters of this part describe the cell, body tissues and membranes, and the integumentary system. An entire chapter is devoted to the anatomy and physiology of the skin.

Part II: Support and Movement

Two very important chapters are found in this part. They concern the skeletal system and the muscular system, which support and protect the body and allow its parts to move. Instructors will appreciate the logical organization of these chapters and the manner in which the basics are presented for easy understanding and learning.

Part III: Integration and Coordination

In this part, there are separate chapters devoted to the nervous system, senses, and endocrine system. The central and peripheral nervous systems as well as the mechanics of the nerve impulse are described. Again, the level of presentation is appropriate to the beginning student. In the chapter on senses, both general and specific receptors are described. Special emphasis is given to the eye and the ear. The presentation of the endocrine system stresses medical conditions associated with the malfunctioning of endocrine glands. Included is an expanded section on diabetes mellitus.

Part IV: Processing and Transporting

This part contains chapters on digestion, circulation, and blood. The lymphatic, respiratory, and excretory systems are also described. The chapter on the lymphatic system includes a discussion of immunity.

Since many students more easily grasp anatomy, the anatomy of the system is covered before introducing the physiology, which is kept at an appropriate level. To increase student interest, these chapters include references to pathological conditions.

Part V: Reproduction and Development

In this part, there are chapters on reproduction, development, and genetics. The reproduction chapter not only includes the male and female anatomy and physiology, it also includes a discussion of birth control measures and infertility. The development chapter stresses development before birth and developmental changes after birth. It ends with a section on aging. The genetics chapter presents some common human genetic disorders.

Special Features of the Text

Understanding Human Anatomy and Physiology includes a number of aids that will help students study and successfully learn anatomy and physiology.

Acetate Overlay

Inside the front cover of *Understanding Human Anatomy and Physiology* is an acetate transparency with outlines of the human torso and the nine regions of the abdomen. The acetate can be overlaid on the reference plates (see pp. 53–60) as a visual aid for students learning the placement of organs within each region.

Illustrations

One of the outstanding features of this text is the full color illustration program. The drawings have been selectively chosen and designed to help students easily visualize structures and processes. The illustrations have been placed near their related textual discussion.

Readability

The writing style is appropriate for the beginning student. The author is well known for her ability to write in a clear and concise manner, and at a level that is easy for students to understand.

Chapter Outlines

Each chapter begins with a **chapter outline** which allows students to see the chapter organization at a glance and to identify the major topics quickly. The chapter outlines include the first- and second-level headings for each chapter.

Learning Objectives

Each chapter has a list of **learning objectives** that tells students what they should be able to do after studying the chapter. Their study of the chapter is complete when they can satisfy these objectives.

Internal Summary Statements

Summary statements are placed at strategic locations throughout the chapter to immediately reinforce the material that has just been discussed. The summary statements will aid student retention of the chapter's main points.

Tables

Each chapter has several tables that are placed near their related textual discussion. The tables clarify complex ideas and summarize sections of the narrative. Once students have achieved an understanding of the subject matter, these tables become valuable review tools.

Boldfaced Words

New terms appear in boldface print as they are introduced within the text and are immediately defined in context. Boldface terms are in the text glossary, where a phonetic pronunciation is given.

Boxed Readings

Several chapters have interesting boxed readings which expand on the core information presented in each chapter. These are entitled **"More about . . ."** and include additional information on such topics as fractures and pain.

Chapter Summaries

Chapter summaries offer a concise review of material in each chapter. Students may read the chapter summaries to preview the topics of importance, and they may also use the summaries to refresh their memories after they have a firm grasp of the information presented in each chapter.

Chapter Questions and Exercises

The **study questions** at the end of each chapter allow students to test their grasp of the material in the chapter. The **objective questions** allow students to quiz themselves with matching and short fill-in-the-blank objective questions. Answers to the objective questions are found in the Instructor's Manual.

The **medical terminology reinforcement exercises** at the end of each chapter give students the opportunity to apply the information that was presented in the section on medical terminology at the beginning of the book. Pronunciation practice is also provided in these exercises. Answers to the Medical Terminology Reinforcement Exercises appear in the Instructor's Manual.

Appendices

The appendices contain optional information for student referral. Appendix A, **Chemistry,** concerns a topic that some instructors may wish to include in their regular curriculum. Appendix B, **Further Readings,** is a list of references for those students who would like more information about a particular topic.

Glossary

The text glossary defines terms that are necessary to the study of anatomy and physiology. Students can use the glossary to practice pronunciations and to review the definitions of the boldfaced terms in the text.

Index

A complete index helps students find topics quickly and easily.

Additional Aids

Paperback and Casebound Texts

This text is available in both paperback and casebound versions.

Instructor's Manual—Test Item File

The *Instructor's Manual* was written by Mitzi Bryant, an experienced instructor of anatomy and physiology. This manual is designed to assist instructors as they plan and prepare for classes using *Understanding Human Anatomy and Physiology*. Possible course organizations including alternate sequencing of chapters is provided. Each chapter includes an outline and chapter synopsis, concepts to emphasize, suggested student activities, and a listing of audiovisual materials. Arthur Cohen, a long-time instructor of beginning students, wrote the objective test questions and several essay questions for each chapter.

Study Workbook

The author was pleased to have Jay Templin, who is very experienced in the field of anatomy and physiology, write the Study Workbook. Each text chapter has a corresponding study workbook chapter offering many exercises to help students fulfill the learning objectives of the chapter. Answers to the exercises in the Study Workbook are provided at the end of each chapter.

Transparencies

Fifty full-color transparencies are available with *Understanding Human Anatomy and Physiology*. Chosen by the author, they were judged to be the illustrations that instructors would most like to have as an adjunct to their lectures.

WCB TestPak and WCB GradePak

WCB TestPak, a computerized testing service, provides instructors with either a mail-in/call-in testing program or the complete test item file on diskette for use with the Apple® and IBM® PC computers. **WCB** TestPak requires no programming experience. **WCB** GradePak, a component of TestPak, is a computerized grade management system for instructors. This program tracks student performance on examinations and assignments. It will compute each student's percentage and corresponding letter grade, as well as the class average.

Also Available from WCB . . .

Anatomy and Physiology Laboratory Textbook, Short Version, Fourth Edition, by Harold J. Benson, *Professor Emeritus,* Stanley E. Gunstream, Arthur Talaro, and Kathleen P. Talaro, *all of Pasadena City College.* This excellent lab text includes 64 exercises and a full-color histology atlas. It's the perfect complement for a one- or two-semester combined anatomy and physiology course where no primary dissection animal is used. This text is accompanied by an Instructor's Handbook and 40mm Histology Slides.

The Coloring Review Guide to Human Anatomy by Hogin McMurtrie, *McMurtrie Design,* and James Krall Rikel, *Pierce College.* Offering nearly 800 outstanding illustrations, this distinctive new learning tool combines clear explanations, a unique design, clearly-labeled illustrations, and concise summary charts. Opening vertically, this coil-bound coloring guide lies flat, allowing equal coloring ease for the right- or left-handed student. To further help the student, the authors begin each chapter with an overview of the system to be studied and then progress to the specifics of that system.

Case Histories in Human Physiology by Donna Van Wynsberghe, *University of Wisconsin,* Milwaukee and Gregory M. Cooley, *Medical College of Wisconsin.* This unique anatomy and physiology supplement presents 25 case histories allowing students to "experience" real-life applications of human physiology. Perforated question and answer pages allow students to diagnose problems and hand in for evaluation. An answer key is available for instructor use.

Acknowledgments

The personnel at Wm. C. Brown Publishers have lent their talents to the production of *Understanding Human Anatomy and Physiology*. I want to especially thank Kevin Kane, biology editor, and Carol Mills, developmental editor, for guiding the book from its inception to its completion. Gloria Schiesl was the production editor who coordinated the efforts of many. Wayne Harms designed the book not only in an attractive manner, but in a way that facilitates student learning.

Several instructors read the entire manuscript and offered suggestions for improvements. The author is extremely thankful to each of the following:

Judy A. Donaldson
North Seattle Community College

Nicolette P. Baumgartner
Grand Valley State University

Emily H. Boegli
Savannah Tech

Mitzi L. Bryant
St. Louis Board of Education

Patricia Turner
Howard Community College

Catherine Relihan
Region IV School of Practical Nursing

Stephen G. Lebsack
Linn Benton Community College

Understanding Medical Terminology

Upon completion of this section, you should be able to:

1. Discuss the importance of medical terminology and how it can be incorporated into the study of the human body.
2. Differentiate between a prefix, suffix, word root, and compound word.
3. Link word parts to form medical terms.
4. Differentiate between singular and plural endings of medical terms.
5. Practice pronunciation of medical words.
6. Dissect (cut apart) compound medical words into parts to analyze the meaning.
7. Recognize the more commonly used prefixes, suffixes, and root words used in medical terminology.

Introduction to Medical Terminology

As students of medical science, we are the inheritors of a vast fortune of knowledge. This fortune, amassed by giants of eighteenth- and nineteenth-century scholarship, was nurtured largely in the atmospheres of universities in which Latin and Greek were the languages of lecture and writing. Scientists then strove to define a universal language in which to communicate their findings. Latin and Greek, studied throughout Europe, became the languages of choice for scholars whose native tongue was English, German, French, Spanish, and so on, because they all read Latin and Greek. So, many seminal works in medicine were first penned in Latin and their vocabularies remain to this day.

Anatomy and physiology were born in the eighteenth century in the midst of a glut of quacks, frauds, charlatans, myths, and superstitions. Honest scholars sought proofs to banish practices that should have been questioned by reason and proved wrong by experience. These scholars were among the first to connect disease with the failure of function or structure of body tissue; thus the race to name and define all anatomical structures began.

Problems arose, inevitably, with the discovery of heretofore unknown tissue. Names were virtually created from parts of existing words by combining parts until they approximated an acceptable description. Medical terminology is simply a catalog of parts which allows us to take apart and reassemble the special language of medicine. The study of medical terminology is easier than it first seems.

Medical words have three basic parts—prefix, root word, and suffix. The prefix comes before a root word and alters the meaning. Prefixes will be seen time and again. For example, *hyper* means over or above. Hyper/kinetic means overactive, hyper/esthesia is overly sensitive, hyper/tension is high blood pressure, and hyper/trophy is overdevelopment.

A suffix is attached to the end of a root word and changes the meaning of the word. For example, the suffix *-itis* means inflammation. Inflammation can occur at almost any part of the body, so *-itis* can be added to root words to make hundreds of words. Dermat/itis is inflammation of the skin, rhin/itis is inflammation of the nose, gastr/itis is inflammation of the stomach, and so on.

A root word is the main part of the word. Once the root word is known for each part of the anatomy, the prefixes and suffixes can be used to analyze and/or build many medical words. The root word for heart is *cardi*. Here are a few terms in which *cardi* appears: cardi/algia means pain in the heart, cardi/omegaly means enlarged heart, brady/cardia means slow heart, and peri/cardio/centesis means puncture to aspirate fluid from around the heart.

Many medical words have, in addition to a prefix and/or a suffix, more than one word part. These are called compound words and can be analyzed by breaking them into parts. For example, hysterosalpingo-oophorectomy is made up of three root words and a suffix. *Hyster* is the root word for uterus, *salping* is the root word for tube, *oophor* is the root word for ovary, and *-ectomy* is the suffix for cut out. Now you know that hysterosalpingo-oophorectomy means the surgical excision of the uterus, tube, and ovary.

To facilitate pronunciation, word parts need to be linked together. The linkage for word parts is *o* and may be referred to as a combining form. For example, linking the root *cardi* with the suffix *-pathy* would produce a word that would be difficult to pronounce; therefore an *o* is used to link the root word with the suffix. The complete word is written cardiopathy and pronounced kar″de-op′ah-the, and the combining form is cardi/o.

When a word is only a root or ends with a root, the word ending depends on whether the word is a noun or an adjective. For example, duodenum (noun) is a part of the small intestine. Duodenal (adjective) is related to the duodenum (e.g., duodenal ulcer).

Accurate spelling of each word part is essential:

1. Changing one letter may change the word part. For example, *ileum* is a part of the small intestine, whereas *ilium* is a pelvic bone.
2. Finding a word in the dictionary requires a knowledge of spelling—at least of the beginning of the word. For example, *pneumonia* and *psychology* have a silent *p*, *rhinitis* (inflammation of the nose) has a silent *h*, and *eupnea* (easy breathing) has an initial silent *e*.

Plural Endings

In many English words, the plurals are formed by adding *s* or *es*; but in Greek and Latin, the plural may be designated by changing the ending.

Singular Ending	Plural Ending	Examples
a	ae	aorta—aortae
ax	aces	thorax—thoraces
en	ena	lumen—lumena
ex, ix	ices	cortex—cortices appendix—appendices
is	es	testis—testes
on, um	a	phenomenon—phenomena medium—media
ur	ora	femur—femora
us	i	bronchus—bronchi
x	ces	calyx—calyces
y	ies	anomaly—anomalies
ma	mata	adenoma—adenomata

If a word ends in *s* and the vowel in the last syllable is short, the word is singular. If the word ends in *s* and the vowel in the last syllable is long, the word is plural.

Any word ending in a consonant is singular (e.g., um, us, at).

In Latin, *a* is neuter and is a singular ending (e.g., aorta).

Al, *as*, *ir*, and *i* are plural endings.

Pronunciation

1. Words are made up of syllables.
2. Syllables are made up of letters—consonants and vowels (a,e,i,o,u).
3. Vowels have a long sound (pronounced by saying its name) when:
   a. a syllable ends with an unmarked vowel. For example, glomeruli (glo mer′ u li).
   b. When it has a macron (‾) over it. For example, gastroscopy (gas tros′ kopē).
4. A short sound (ah, eh, ih, oo, uh) is used when:
   a. there is an unmarked vowel in a syllable ending with a consonant. For example, anginal (an′-je nal).
   b. When the vowel constitutes a syllable or ends a syllable, and is indicated by a breve (˘). For example, effect (ĕ-fekt′).
5. The syllable *ah* is used for the sound *a* in open, unaccented syllables: For example, abortion (ah-bor′ shun) or amenorrhea (ah-men″ o-re′ ah).
6. The primary accent in a word is indicated by a single accent.
7. The secondary accent is indicated by a light face, double accent. For example, duodenostomy (du″od-e-nos′to me).
8. The accent on medical terms is generally on the third from the last syllable.

Practice pronouncing the following words:

1. hematemesis (hem″ah-tem′ĕsis)—vomiting blood
2. hysterosalpingo-oophorectomy (his″ter-o-sal-ping″go o″of-o-rek′ to me)
3. phrenohepatic (fren″o-hĕ-pat′ik)—pertaining to the diaphragm and liver
4. gastropathy (gas″trop′ ah-the)—disease of the stomach
5. metatarsus (met″ah-tar′ sus)—part of the foot between the tarsus (ankle) and toes

More Commonly Used Prefixes

Prefix	Meaning	Example
a-, an-, in-	without, negative	a/men/orrhea—without a monthly flow
ab-	from, away from	ab/normal—away from normal
ad-, ac-, as-, at-	to, toward	ad/duct—carry toward
aniso-	unequal	an/iso/cyt/osis—abnormal condition of unequal cells
ante-, pre-	before	anterior—front; pre/natal—before birth
anti-, ant-, ob-	against	anti/pyre/tic—agent used against fever
bi-	two	bi/lateral—two sides;
bio-	life	bio/logy—study of life
brachy-	short	brachy/dactyl/ism—short fingers and toes
brady-	slow	brady/cardia—slow heart rate
cent-	hundred	centi/meter—one one-hundredth of a meter
circum-	around	circum/cis/ion—to cut around
co-, com-, con-	with, together	con/genital—born with
contra-	against	contra/indicated—against indication
de-	away from	de/hydrate—loss of water
dextr-	right	dextr/o/cardia—heart displaced to right
dia-	through	dia/rrhea—flow through
dis-	apart	dis/sect—to cut apart
dys	bad, difficult	dys/pnea—difficult breathing
e-, ex-	out, out from	ex/cise—to cut out
ect-, exo-, extra-	outside	extra/corporeal—outside the body
en-	in, on	en/capsulated—in a capsule
end-	within	endo/scopy—visualization within
epi-	upon	epi/dermis—upon the skin
eu-	good	eu/phonic—good sound
hemi-, semi-	half	hemi/gastr/ectomy—surgical removal of half of the stomach
hyper-	over, above	hyper/kinetic—overactive
hypo-	under, below	hypo/glossal—under the tongue
infra-	beneath	infra/mammary—beneath the breast
inter-	between	inter/cellular—between the cells
intra-	within	intra/cranial—within the cranium
kil-	thousand	kilo/gram—1,000 grams
macr-	large	macro/cyte—large cell
mal-	bad	mal/nutrition—bad nourishment
mes-	middle	mes/entery—middle of intestine
meta-	after, beyond	meta/carpals—beyond the carpals (wrist)
micr-	small	micro/cephal/ic—having a small head
milli-	one-thousandth	milli/liter—one one-thousandth of a liter
multi-	many	multi/para—one who has many children
neo-	new	neo/plasm—new growth
olig-	scanty, few	olig/uria—scanty amount of urine
onc-	tumor	onc/ology—study of tumors
per-	through	per/cutaneous—through the skin
peri-	around	peri/tonsillar—around the tonsil
poly-	much, many	poly/cystic—many cysts

More Commonly Used Prefixes—*continued*

Prefix	Meaning	Example
post-	after	post/mortem—after death
pre-	before	pre/natal—before birth
presby-	old	presby/opia—old vision
primi-	first	primi/gravida—first pregnancy
pro-	before	pro/gnosis—foreknowledge, predict outcome
re-	back, again	re/generate—produce, develop again
retr-	behind	retro/sternal—behind the sternum
sub-	under	sub/lingual—below the tongue
super-, supra-	above	superior—above
syn-, sym-	with, together	syn/ergism—working together
tachy-	fast	tachy/phasia—fast speech

Frequently Used Suffixes

Suffix	Meaning	Example
-algia	pain	dent/algia—pain in the tooth
-atresia	without an opening	proct/atresia—rectum without an opening
-cele	hernia	omphalo/cele—umbilical hernia
-centesis	puncture to aspirate fluid	arthro/centesis—puncture to aspirate fluid from a joint
-cept	take, receive	re/cept/or—something that receives again
-cide	kill	bacteri/cidal—able to kill bacteria
-cis	cut	circum/cis/ion—cutting around
-cyte	cell	erythro/cyte—red cell
-denia	pain	cephalo/denia—pain in the head
-desis	fusion	arthro/desis—fusion of a joint
-ectasia	expansion	cor/ectasis—expanding/dilating pupil
-ectomy	cut out, excise	nephr/ectomy—surgically remove kidney
-edema	swelling	cephal/edema—swelling of head
-emesis	vomiting	hyper/emesis—excessive vomiting
-emia	blood	hyper/glyc/emia—elevated blood sugar
-gnosis	knowledge	dia/gnosis—knowledge through examination (determining cause of disease)
-gram	record	myelo/gram—X ray of the spinal cord
-graphy	making a record	angio/graphy—making a record of vessels
-iasis	condition	chole/lith/iasis—condition of gallstones
-ist	one who	opto/metr/ist—one who measures vision
-itis	inflammation	aden/itis—inflammation of a gland
-lepsy	seizures	narco/lepsy—seizures of numbness
-logist	one who specializes	ophthalmo/log/ist—one specializing in eyes
-logy	study of	bio/logy—study of life
-lysis, -lytic	break down, dissolve	teno/lysis—destruction of tendons
-malacia	abnormal softening	osteo/malacia—abnormal softening of bone
-mania	madness	pyro/mania—irresistible urge to set fires
-megaly	enlargement	spleno/megaly—enlargement of spleen
-meter	measure	thermo/meter—instrument to measure temperature
-oid	resembling	muc/oid—resembling mucus
-oma	tumor	neur/oma—nerve tumor
-opia	vision	ambly/opia—dim vision

Frequently Used Suffixes—*continued*

Suffix	Meaning	Example
-osis	abnormal condition	nephr/osis—abnormal condition of kidney
-osme	smell	an/osmia—inability to smell
-ostomy	create an opening	col/ostomy—create an opening in the colon
-otia	ear	macr/otia—large ear
-pathy	disease	encephalo/pathy—disease of the brain
-penia	deficiency, poor	leuko/cyto/penia—deficiency of white cells
-pepsia	digestion	dys/pepsia—bad digestion
-pexy	surgical fixation	nephro/pexy—surgical fixation of kidney
-phasia	speak, say	a/phasia—without ability to speak
-philia	love, attraction	chromo/philic—attracted to color
-phobia	abnormal fear	agora/phobia—abnormal fear of crowds
-plasia	formation	hyper/plasia—excessive formation
-plasm	substance	proto/plasm—original substance
-plasty	make, shape	rhino/plasty—to shape the nose
-plegia	paralysis	hemi/plegia—paralysis of one half of body
-pnea	breath	tachy/pnea—fast breathing
-ptosis	prolapse, dropping	hystero/ptosis—prolapse of uterus
-scope	instrument for viewing	oto/scope—instrument to look in ears
-scopy	visualization	laryngo/scopy—visualization of larynx
-some, -soma	body	lyso/some—body which lysis/dissolves
-spasm	twitching	blepharo/spasm—twitching of eyelid
-stasis	stop, control	hemo/stasis—control bleeding
-therapy	treatment	hydro/therapy—treatment with water
-tome	instrument to cut	osteo/tome—instrument to cut bone
-tomy	to cut	laparo/tomy—to cut into the abdomen
-tripsy	crushing	nephro/litho/tripsy—crushing stone in kidney
-trophy	development	hyper/trophy—overdevelopment
-rrhagia	burst forth	metro/rrhagia—hemorrhage from uterus
-rrhaphy	suture, sew	hernio/rrhaphy—suture a hernia
-rrhea	flow, discharge	oto/rrhea—discharge from ear
-rrhexis	rupture	spleno/rrhexis—rupture of the spleen
-uria	urine	hemat/uria—blood in the urine

Frequently Used Root Words*

Roots	Meaning	Example
acr-	extremity, peak	acro/megaly—enlarged extremities acro/phobia—abnormal fear of heights
aden-	gland	adeno/pathy—disease of a gland
aer-	air	aero/phagia—swallowing air
angi-	vessel	angi/oma—tumor of a vessel
arthr-	joint	arthr/algia—pain in the joint
blast-	bud, growing thing	neuro/blast—growing nerve cell
blephar-	eyelid	blepharo/ptosis—drooping of eyelid
brachi-	arm	brachial—pertaining to the arm
carcin-	cancer	adeno/carcin/oma—cancerous tumor of a gland

*Words that are the same as anatomical terms used in English have been omitted (i.e., pancrease, tonsil, and so on).

Roots	Meaning	Example
cardi-	heart	myo/cardi/tis—inflammation of heart muscle
carp-	wrist	flexor carpi—muscle to bend wrist
caud-	tail	caudal—pertaining to tail
cephal-	head	cephalo/dynia—pain in the head
cervic-	neck	cervic/itis—inflammation of the neck of uterus
cheil-	lip	cheilo/plasty—shaping the lip
cheir-, chir-	hand	chiro/megaly—large hands
chol-	bile, gall	chole/cyst/ectomy—surgical removal of the gallbladder
chondr-	cartilage	chondro/malacia—softening of cartilage
chrom-	color	poly/chromatic—having many colors
chron-	time	syn/chron/ous—occuring at the same time
col-	colon	mega/colon—enlarged colon
colp-	vagina	colp/orrhaphy—suture of vagina
cost-	rib	inter/costal—between the ribs
crani-	skull	crani/otomy—incision into the skull
cry-	cold	cryo/philic—cold loving
crypt-	hidden	crypt/orchid/ism—hidden (undescended) testicle
cutan-, cut-	skin	sub/cutaneous—below the skin
cyan-	blue	acro/cyan/osis—abnormal condition of blueness of extremities
cyst-	bladder	cysto/cele—bladder hernia
cyt-	cell	thrombo/cyte—clotting cell (platelet)
dacry-	tear	dacryo/rrhea—flow of tears
dactyl-	fingers, toes	poly/dactyl/ism—too many fingers and toes
dent-, odont-	tooth	peri/odontal—around the teeth
		dent/algia—toothache
derm-, dermat-	skin	intra/dermal—within the skin
dextr-	right	dextro/cardia—heart displaced to the right
dips-	thirst	poly/dipsia—excessive thirst
dors-	back	dorsal—pertaining to the back
duct-	carry	ovi/duct—tube to carry ova (eggs)
encephal-	brain	encephalo/cele—hernia of the brain
enter-	intestine	gastro/enter/itis—inflammation of stomach and intestine
erg-	work	en/ergy—working within
erythr-	red	erythro/cyto/penia—deficiency of red cells
esthe-	sensation	an/esthe/tic—agent to eliminate sensation
esthen-	weakness	my/esthenia—muscle weakness
febr-	fever	a/febrile—without a fever
flex-	bend	dorsi/flex—bend backward
gastr-	stomach	gastro/scopy—visualization of the stomach
gen-	produce	patho/genic—agent which produces disease
gingiv-	gums	gingiv/ectomy—removal of gums
gloss-	tongue	hypo/glossal—under the tongue
glyc-, glu-	sugar	hypo/glyc/emia—low blood sugar
gnath-	jaw	micro/gnath/ism—small jaw
grav-	heavy	secundi/gravida—second pregnancy

Roots	Meaning	Example
gynec-	female	gyneco/logy—study of female conditions
hem-, hemat-	blood	hemat/emesis—vomiting blood
hepat-	liver	hepato/megaly—enlarged liver
heter-	different	hetero/genous—different origins
hidr-	perspiration	hidro/rrhea—flow of perspiration
hist-	tissue	histo/logy—study of tissue
home-, hom-	same	homeo/stasis—stay same, equilibrium
hydr-, hydra-	water	de/hydra/tion—process of losing water
iatr-	physician	iatro/genic—produced by the physician
irid-	iris	irid/ectomy—surgical removal of iris
is-	equal	iso/tonic—equal in tone
kary-	nut, nucleus	mega/karyo/cyte—cell with large nucleus
kerat-	cornea	kerato/plasty—repair of cornea
kin-	move	kinesio/logy—study of movement
lacrim-	tear	lacrima/tion—crying
lact-, galact-	milk	lacto/genic—milk producing
lapar-	abdomen	laparo/rrhaphy—suture of the abdomen
laryng-	larynx	laryngo/scopy—visualization of larynx
later-	side	bi/lateral—two sides
leuk-, leuc-	white	leuko/rrhea—white discharge
lingu-	tongue	sub/lingual—under the tongue
lip-	fat	lip/oma—tumor of fat
lith-	stone	litho/tripsy—crushing a stone
mast-, mamm-	breast	mast/itis—inflammation of the breast mammo/gram—X ray of breast
melan-	black	melan/oma—black tumor
men-	monthly, menses	dys/meno/rrhea—difficult monthly flow
metr-	uterus	endo/metr/ium—lining of uterus
morph-	shape, form	poly/morphic—pertaining to many shapes
my-	muscle	myo/metr/itis—inflammation of muscle of uterus
myc-	fungus	onycho/myc/osis—fungus condition of the nails
myel-	marrow, spinal cord	myelo/gram—X-ray/record of spinal cord
myring-	eardrum	myringo/tomy—opening into eardrum
nas-	nose	naso/pharyng/eal—pertaining to nose and throat
necr-	dead	necr/opsy—examining dead bodies; autopsy
nephr-, ren-	kidney	hydro/nephr/osis—abnormal condition of water in the kidney
neur-	nerve	neur/algia—nerve pain
noct-, nyct-	night	noct/uria—voiding at night
null-	none	nulli/gravida—woman who has had no pregnancies

Roots	Meaning	Example
oo-	ova, egg	oo/genesis—producing eggs
ocul-	eye	mon/ocular—pertaining to one eye
omphal-	umbilicus	omphalo/rrhea—discharge from the naval
onych-	nail	onycho/crypt/osis—condition of hidden nail (ingrown)
oophor-	ovary	oophoro/cyst/ectomy—removal of cyst from ovary
ophthalm-	eye	ex/ophthalmos—condition of protruding eyes
or-	mouth	oro/pharyngeal—pertaining to mouth and throat
orchid-	testis	orchid/ectomy—removal of testis
orexis-	appetite	an/orexis—absence of appetite
orth-	straight	orth/odont/ist—one who straightens teeth
oste-, oss-	bone	osteo/chondr/oma—tumor of bone and cartilage
ot-, aur-	ear	ot/itis—inflammation of the ear post/auricular—behind the ear
para-	to bear	primi/para—to bear first child
path-	disease	patho/physio/logy—study of effects of disease on body functioning
pect-	chest	pectoralis—chest muscle
ped-	child	ped/iatrician—one who specializes in children
peps-	digest	dys/pepsia—bad digestion
phag-	swallow, eat	a/phagia—inability to swallow
pharmac-	drug	pharmaco/logy—study of drugs
pharyng-	throat	pharyng/itis—inflammation of the throat
phas-	speak, say	tachy/phasia—speaking fast
phleb-	vein	phlebo/thromb/osis—abnormal condition of clot in vein
phon-	voice	a/phonic—absence of voice
phren-	diaphragm	phreno/hepatic—pertaining to the diaphragm and liver
pil-, trich-	hair	tricho/glossia—hairy tongue
pneum-	air, breath	pneumo/thorax—air in the chest
pneumon-	lung	pneumon/ectomy—surgical removal of the lung
pod-	foot	pod/iatrist—one who specializes in foot problems
proct-	rectum	procto/scopy—visualization of the rectum
pseud-	false	pseudo/cyesis—false pregnancy
psych-	mind	psycho/somatic—pertaining to the mind and the body
pulmo(n)-	lung	cardio/pulmonary—pertaining to heart and lungs
py-	pus	pyo/rrhea—flow of pus

Frequently Used Root Words—*continued*

Roots	Meaning	Example
pyel-	kidney pelvis	pyelo/nephr/itis—inflammation of the kidney pelvis
pyr-	fire, fever	anti/pyretic—agent used against fever
quadri-	four	quadri/plegia—paralysis of all four extremities
rhin-	nose	rhino/plasty—revision of the nose
salping-	tube	salping/itis—inflammation of the Fallopian tube
sanguin-	blood	ex/sanguina/tion—process of bleeding out (bleed to death)
scler-	hard	arterio/scler/osis—condition of hardening of arteries
sect-	cut	dis/section—cutting apart
sept-	contamination	anti/septic—agent used against contamination
sial-	saliva	poly/sialia—excessive salivation
sten-	narrow, constricted	pyloric stenosis—narrowing of pylorus
stomat-	mouth	stomat/itis—inflammation of the mouth
strict-	draw tight	vaso/con/strict/or—agent that compresses vessels
tax-	order, arrange	a/taxic—uncoordinated
ten-	tendon	teno/rrhaphy—suture a tendon
therm-	heat	hyper/thermia—raising body heat
thorac-	chest	thoraco/centesis—puncture to aspirate fluid from chest
thromb-	clot	thrombo/cyte—clotting cell
tox-	poison	tox/emia—poison in the blood
trache-	windpipe	tracheo/malacia—softening of tracheal cartilages
trachel-	neck	trachel/orrhaphy—suture of cervix (neck of uterus)
traumat-	wound	traumat/ology—study of trauma
tri-	three	tri/geminal—having three beginnings
troph-	turn	ec/tropion—turned out
ur-	urine	ur/emia—urine constituents in the blood
vas-	vessel	vaso/constriction—narrowing of a vessel
vert-	turn	retro/vert/ed—turned backward
vesic-	bladder	vesico/cele—hernia of the bladder
viscer-	internal organs	e/viscera/tion—process of viscera protruding from abdominal wall
vita	life	vital—necessary for life

Now you are ready to apply your knowledge! Think of the word parts when you encounter new words in the text and try to analyze the meaning. At the end of each chapter, you will be given an opportunity to reinforce your knowledge by pronouncing the words and dissecting them into parts to arrive at a meaning. You can also begin to build medical words and use them in your everyday conversation. You will be amazed at how rapidly your vocabulary will grow, and how your study of the human body will become easier and more enjoyable.

Human Organization

An organ such as the heart contains different types of tissues and cells. These are fat cells found in adipose tissue that accumulates on the outer surface of the heart.

1

Body Organization

Chapter Outline

Anatomy and Physiology

Organization of Body Parts

Anatomical Terms

   Relative Positions of Body Parts

   Planes and Sections of the Body

   Regions of the Body

   Cavities of the Body

Organ Systems

   Integumentary System

   Support and Movement

   Integration and Coordination

   Processing and Transporting

   Reproduction and Development

Homeostasis

Learning Objectives

After you have studied this chapter, you should be able to:

1. Define anatomy and physiology, and explain how they are related.

2. Describe each level of organization in reference to an example.

3. Use the terms that describe relative positions, body sections, and body regions.

4. List the cavities of the body and state their locations.

5. State the organs located in each of the body cavities.

6. List the organ systems of the body and state the major organs associated with each.

7. Describe, in general, the functions of each organ system.

8. Define homeostatis and explain, in general, how it is maintained.

Anatomy and Physiology

Anatomy is the study of the structure of body parts. For example in figure 1.1, we can observe that the stomach is a J-shaped pouchlike organ in the abdominal cavity that lies near the liver and large intestine. And, we can go on to observe the structure of the wall of the stomach, which has thick folds called rugae. These folds tend to disappear as the stomach expands to increase its capacity. And there is no need to stop there, because with the aid of the microscope we can observe that the rugae themselves have much smaller folds.

Physiology is the study of the function of body parts. For example, the stomach functions in temporarily storing food received from the esophagus, secreting digestive juices, and passing on the partially digested food to the small intestine.

There is a close connection between anatomy and physiology. It is said that the structure suits the function. The pouchlike shape of the stomach and the rugae are suitable to its function of storing food. And, the microscopic structure of the stomach wall is suitable to its secreting digestive juices as we shall see later on in the text.

Anatomy is the study of the structure of the body parts and physiology is the study of the function of these parts. Structure is suited to the function of a part.

Organization of Body Parts

You can tell from our discussion above that it's possible to study the structure of body parts at different *levels of organization* (fig. 1.2). First of all, any substance, including any body part, is composed of chemicals that are made up of submicroscopic particles called **atoms.** The atoms most frequently found in the body are carbon, hydrogen, oxygen, and nitrogen, but you may be more familiar with the necessity of calcium to build strong bones or iron to prevent anemia.

Atoms join together to form **molecules** and these molecules can join with others to form macromolecules. For example, the small molecules called amino acids join together to form the very large molecules called proteins. Muscles contain a significant amount of protein; therefore, meat is a rich source of this basic nutrient.

The macromolecules called proteins and fats contribute to the makeup of the **cell,** the basic unit of all living things. Within cells, there are **organelles,** tiny structures that carry on the functions of the cell. For example, there is an organelle, called the nucleus, that is especially concerned with reproduction of the cell; and another, called the mitochondrion, that supplies the cell with energy. Muscle cells use much energy when they contract, for example.

Cells are the smallest living portion of any living thing. Human beings are multicellular animals, and it is at the cellular level that we can best understand health and disease. Cells are found in tissues and tissues make up organs. A **tissue** is composed of similar types of cells and performs a specific function. An **organ** is composed of several types of tissues and performs a particular function within an **organ system.** The stomach, for instance, is a part of the digestive system. It has a specific role in this system, in which the overall function is to supply us with the nutrients the body needs for growth and repair. The other systems of the body (fig. 1.9) also have specific functions.

Finally, all of the systems together make up the **organism**—in this case, the human being. A human being is quite complex, but it is possible to break down this complexity and study it at different levels, each level being simpler than the one studied previously. Still, each level is organized and is constructed in a particular way.

The body has levels of organization that progress from atoms to molecules, cells, tissues, organs, body systems and, finally, to the organism.

Anatomical Terms

It is a custom to use certain terms to describe the relationship of body parts, imaginary planes and sections of the body, various regions, and cavities of the body. It is essential to become accustomed to these terms before your study of anatomy and physiology begins. When these terms are used, it is assumed that the body is in the *anatomical position*. In this position, the body is standing erect, the face is forward, and the arms are at the sides with the palms and toes directed forward.

Figure 1.1 Anatomy of the stomach. *a.* Drawing shows that the stomach is a **J**-shaped pouched organ in the digestive system. *b.* Scanning electron micrograph shows that the stomach wall has folds called rugae (Ru). (The micrograph also shows the detailed anatomy of the wall: Mu = mucosa, MM = muscularis mucosa, Su = submucosa, ML = muscularis, Se = serosa. The arrows point to openings of gastric glands.) *c.* Photomicrograph shows that the rugae themselves have folds.

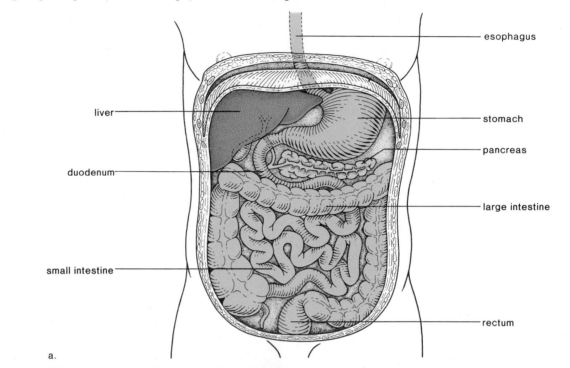

esophagus

liver

stomach

pancreas

duodenum

large intestine

small intestine

rectum

a.

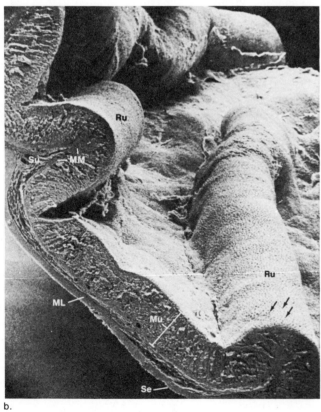

b.

c.

Figure 1.2   Levels of organization of the human body. Each level is more complex than the previous level.

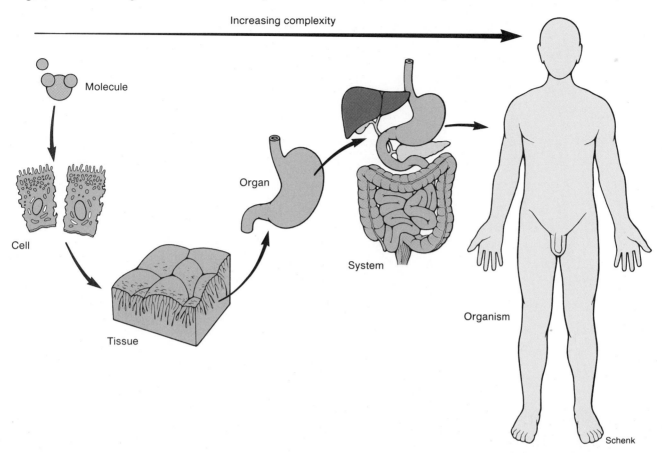

Relative Positions of Body Parts

The terms used in figure 1.3 describe the location of a part in relation to another part of the body:

Superior means that a part is above, or closer to the head than, another part. **Inferior** means that the part is below, or closer to the feet. For example, the stomach is superior to the small intestine, but inferior to the esophagus.

Anterior (or ventral) means that the body part is located toward the front. **Posterior** (or dorsal) means that the part is toward the back. The windpipe (trachea) is anterior to the esophagus while the esophagus is posterior to the windpipe.

Medial means that a part is nearer than another part to an imaginary midline of the body. **Lateral** means that the part is further than another from this midline. The nose is medial to the eyes; the ears are lateral to the eyes.

Proximal means that an appendage part is closer to a point of attachment or closer to the trunk than another part. **Distal** means that an appendage part is further from a point of attachment or further from the trunk than another part. The upper arm is proximal to the elbow and the lower arm is distal to the elbow.

Superficial means a body part is located near the surface. **Deep** means that the body part is located away from the surface. Superficial blood vessels are closer to the skin than those that lie deep in the abdominal cavity.

Visceral pertains to internal organs or the covering of internal organs. **Parietal** refers to the wall of a body cavity. The visceral pleura is a membrane that adheres to the surface of the lungs while the parietal pleura is a membrane that lines the pleural cavities within the thoracic cavity.

Figure 1.3 Terms such as these are used to describe the relative positions of body parts.

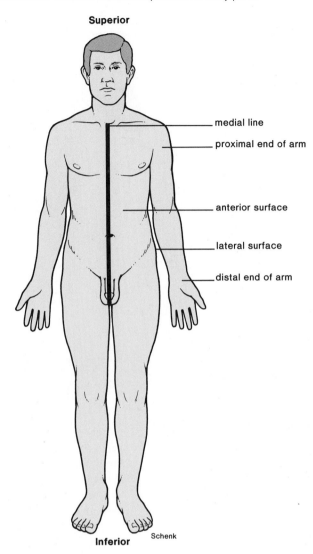

Superior

medial line

proximal end of arm

anterior surface

lateral surface

distal end of arm

Inferior   Schenk

The terms superior/inferior, medial/lateral, anterior/posterior, proximal/distal, superficial/deep, and visceral/parietal are used to describe the relative position of body parts.

Planes and Sections of the Body

In order to observe the internal body parts, it is necessary to section (cut open) the body in some way. It is customary to describe these sections in terms of imaginary planes that divide the body (fig. 1.4).

A **sagittal** (vertical) **plane** is a lengthwise cut that divides the body into right and left portions. If the section passes exactly through the midline it is called a midsagittal cut.

A **transverse** (horizontal) **plane** is a cut that divides the body horizontally to give a cross section. A transverse cut divides the body into superior and inferior portions.

A **frontal plane** divides the body lengthwise into anterior and posterior portions.

The terms *longitudinal section* and *cross section* are often applied to body parts that have been removed and cut either lengthwise or straight across, respectively.

A sagittal (vertical) cut divides the body into right and left parts; a transverse (horizontal) cut is a cross section; and a frontal cut divides the body into anterior and posterior parts.

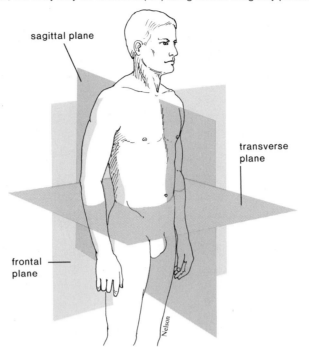

Regions of the Body

The human body can be divided into the **axial** portion, which includes the head, neck, and trunk, and the **appendicular** portion, which includes the arms and legs. The **trunk,** or torso, contains the thorax, abdomen, and pelvis. Each of these portions has been further divided into the specific regions mentioned (fig. 1.5):

Head
cephalic (head)
cranial (skull)
frontal (forehead)
occipital (back of head)
oral (mouth, buccal)
nasal (nose)
orbital (eyes)
ophthalmic (eyes)

Neck
cervical (neck)

Thorax (chest)
pectoral (chest)
mammary (breast)
axillary (armpit)
vertebral (backbone)
costal (ribs)

Abdomen
celiac (abdomen)
pelvic (lower portion of abdomen)
gluteal (buttock)
groin (depressed region of abdomen near thigh)
inguinal (groin)
lumbar (lower back)
sacral (where vertebrae terminate)
perineal (region between anus and external sex organs)

Limbs (arms and legs)
brachial (upper limb, arm)
upper arm
forearm (lower arm)
carpal (wrist)
cubital (elbow)
palmar (palm)
lower limb (leg)
femoral (thigh)
lower leg
pedal (foot)

Other
cutaneous (skin)

There are terms that identify each particular region of the body.

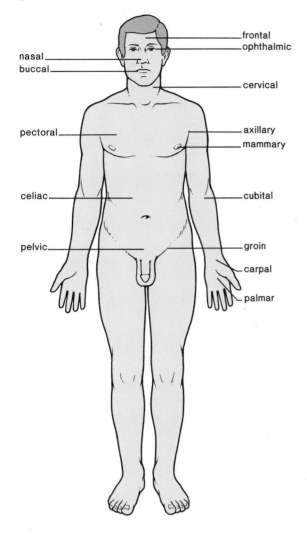

Cavities of the Body

The internal organs, called the **visceral** organs, are located within specific cavities (fig. 1.6). The two main cavities are the *dorsal cavity* and the larger *ventral cavity*. The dorsal cavity can be subdivided into two parts: the **cranial cavity** within the skull contains the brain; and the **spinal cavity,** protected by the vertebrae, contains the spinal cord.

The ventral cavity is subdivided into the **thoracic cavity** and the **abdominopelvic cavity.** The lungs, heart, esophagus, trachea, and thymus gland are located in the thoracic cavity. This cavity is separated from the abdominopelvic cavity by a horizontal muscle called the *diaphragm.*

The abdominopelvic cavity has two portions: the upper **abdominal** portion and the lower **pelvic** portion. The stomach, liver, spleen, gallbladder, and most of the small and large intestine are in the abdominal cavity.

The pelvic cavity contains the rest of the large intestine, the rectum, the urinary bladder, and the internal reproductive organs. In males, there is an external extension of the abdominal wall called the **scrotum** where the testes are found.

Since the abdominopelvic cavity is quite large, it is sometimes divided into four sections by running a transverse plane across the midsagittal plane at the point of the navel (fig. 1.7). The abdominal cavity can also be divided into a greater number of sections as is done in figure 1.8.

The human body has two major cavities called the dorsal and ventral cavities. Each of these is subdivided into smaller cavities, and the visceral organs can be associated with one of these subcavities.

Figure 1.6 Major cavities of the body. The ventral cavity is divided into the thoracic and abdominopelvic cavity. The latter is further divided into the abdominal and pelvic cavity. The dorsal cavity is divided into the cranial and spinal cavity.

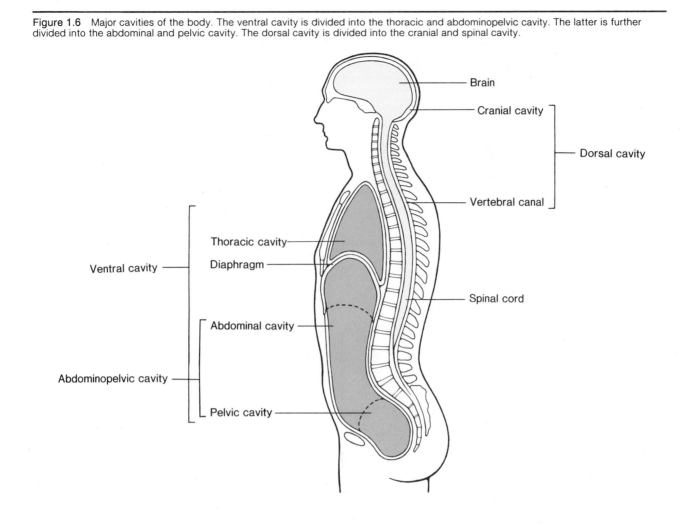

Organ Systems

The organ systems (fig. 1.9) of the body have been divided into the following categories in this text. Consult the reference figures on pages 53–60 to further serve as an aid to learning the organ systems and their placement.

Integumentary System

The **integumentary system,** discussed in chapter 4, includes the skin and accessory organs such as the hair, nails, sweat glands, and sebaceous glands. The skin protects underlying tissues, helps regulate body temperature, contains sense organs, and even synthesizes certain chemicals that effect the rest of the body.

The integumentary system, which includes the skin, not only protects the body, but has other functions as well.

Support and Movement

The **skeletal system** and the **muscular system** give the body support, and are involved in the ability of the body and its parts to move.

The skeletal system consists of bones of the skeleton and associated cartilages, as well as the ligaments that bind these structures together. The skeleton protects body parts. For example, the skull forms a protective encasement for the brain, as does the rib cage for the heart and lungs. Some bones produce red blood cells and all of them are a storage area for calcium and

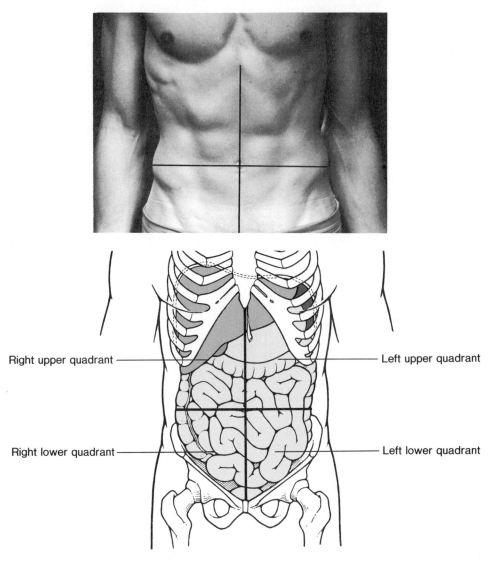

Right upper quadrant

Left upper quadrant

Right lower quadrant

Left lower quadrant

phosphorus salts. The skeleton, as a whole, serves as a place of attachment for the muscles.

The contraction of *skeletal muscles,* discussed in chapter 6, accounts for our ability to move voluntarily and to respond to outside stimuli. These muscles also maintain our posture and are responsible for the production of body heat. *Cardiac* and *smooth muscle* are called involuntary muscles because they contract automatically. Cardiac muscle makes up the heart, and smooth muscle is found within the walls of the internal organs.

The skeletal system contains the bones and the muscular system contains the three types of muscles. The primary function of these systems is support and movement, but they have other functions as well.

Integration and Coordination

The **nervous system,** discussed in chapter 7, consists of the brain, spinal cord, and associated nerves. The nerves conduct nerve impulses from the sense organs to the brain and spinal cord. They also conduct nerve impulses from the brain and spinal cord to the muscles.

The *sense receptors,* discussed in chapter 8, are the organs that inform us about the outside environment. This information is then processed by the brain and spinal cord and the individual responds to environmental stimuli through the muscular system.

The **endocrine system,** discussed in chapter 9, consists of the hormonal glands which secrete chemicals that serve as messengers between body parts. Both the nervous and hormonal systems help maintain a relatively constant internal environment by coordinating the

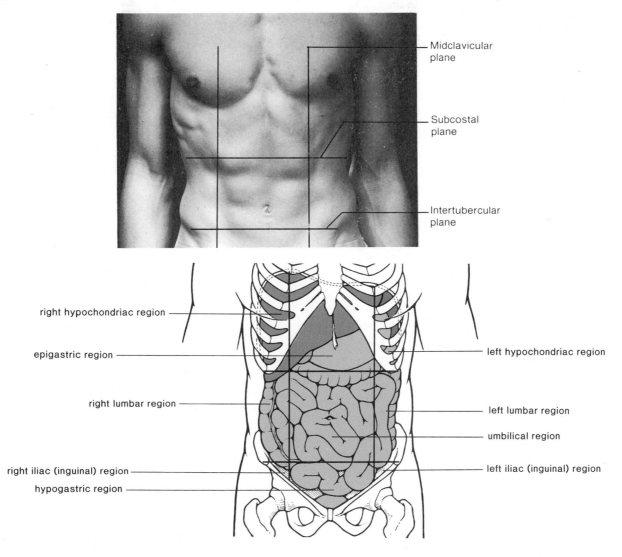

Figure 1.8 To be more precise than figure 1.7, the abdominal region can be further subdivided into these regions.

Midclavicular plane

Subcostal plane

Intertubercular plane

right hypochondriac region

epigastric region

right lumbar region

right iliac (inguinal) region

hypogastric region

left hypochondriac region

left lumbar region

umbilical region

left iliac (inguinal) region

functions of the body's other systems. The nervous system acts quickly, but provides short-lived regulation; the hormonal system acts more slowly, but provides a more sustained regulation of body parts. The endocrine system also helps maintain the proper functions of the reproductive organs in males and females.

The nervous system contains the brain, spinal cord, and nerves. Because the nervous system is in communication with both the sense organs and muscles, it allows us to respond to outside stimuli. The endocrine system contains the hormonal glands. Both the nervous and endocrine systems regulate the activities of the other systems.

Processing and Transporting

Each of these systems also functions in a way that helps maintain the normal internal conditions of the human body. The **digestive system** (fig. 1.1) discussed in chapter 10, consists of the mouth, esophagus, stomach, small intestine, and large intestine (colon) along with the associated organs: teeth, tongue, salivary glands, liver, gallbladder, and pancreas. This system receives food and digests it into the nutrient molecules, which can enter the cells of the body.

The **circulatory system,** discussed in chapter 12, consists of the heart and the blood vessels that carry blood through the body. Blood transports nutrients and

Figure 1.9 Organ systems of the body.

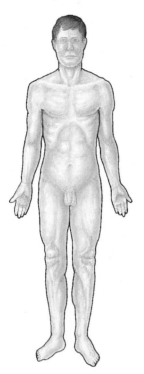

Integumentary system
Function: external support
and protection of body

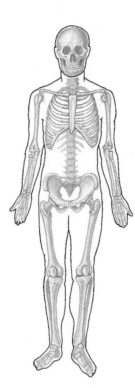

Skeletal system
Function: internal support and
flexible framework for body
movement; production of
blood cells

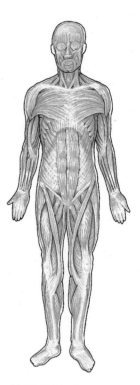

Muscular system
Function: body movement;
production of body heat

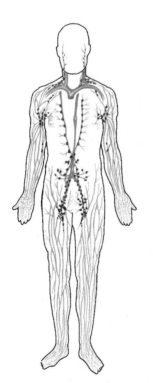

Lymphatic system
Function: body immunity;
absorption of fats; drainage
of tissue fluid

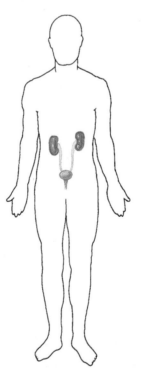

Urinary system
Function: filtration of blood;
maintenance of volume and
chemical composition
of the blood

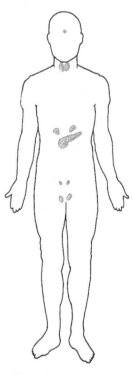

Endocrine system
Function: secretion of
hormones for
chemical regulation

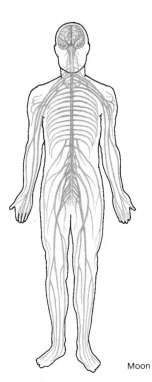

Nervous system
Function: regulation of
all body activities:
learning and memory

Moon

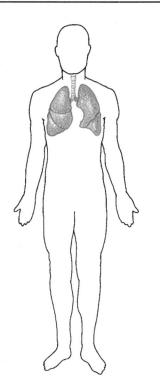

Respiratory system
Function: gaseous exchange
between external environment
and blood

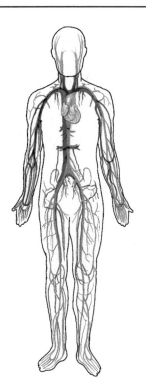

Circulatory system
Function: transport of life-
sustaining materials to body
cells; removal of metabolic
wastes from cells

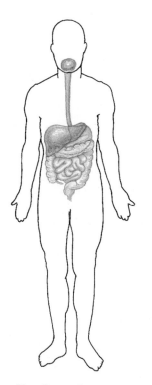

Digestive system
Function: breakdown and
absorption of food materials

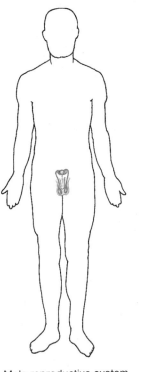

Male reproductive system
Function: production of male
sex cells (sperm); transfer of
sperm to reproductive system
of female

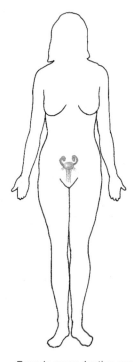

Female reproductive system
Function: production of
female sex cells (ova);
receptacle of sperm from
male; site for fertilization
of ovum, implantation, and
development of embryo
and fetus; delivery of fetus

oxygen to the cells, and takes away their waste molecules to be excreted from the body. The blood also contains cells that are produced by the **lymphatic system,** discussed in chapter 13. The lymphatic system protects us from disease.

The **respiratory system,** discussed in chapter 14, consists of the lungs and the tubes that take air to and from the lungs. The respiratory system brings oxygen into the lungs and takes carbon dioxide out of the lungs.

The **urinary system,** discussed in chapter 15, contains the kidneys and the urinary bladder. This system rids the body of nitrogenous wastes and helps regulate the fluid level and chemical content of the blood.

The digestive system (mouth, esophagus, stomach, and small and large intestine); the circulatory system (heart and vessels); immune system; respiratory system (lungs and conducting tubes); and the urinary system (kidneys and bladder) all have specific functions in processing and transporting to maintain the normal conditions of the body.

Reproduction and Development

The **reproductive system,** discussed in chapter 16, has different organs in the male and female. The *male reproductive system* consists of the testes, other glands, and various ducts that conduct semen to the penis, which serves as the copulatory organ of males. The *female reproductive system* contains the ovaries, oviducts, uterus, vagina, and external genitalia. It not only produces sex cells and receives those of the male, the female reproductive system also nourishes and protects the unborn until the time of birth.

The reproductive system in males (testes, ducts, and penis) and in females (ovaries, oviducts, uterus, and vagina) carries out those functions that give humans the ability to procreate.

Homeostasis

Homeostasis means that the internal environment remains relatively constant regardless of the conditions in the external environment. In humans, for example:

1. Blood glucose concentration remains at about 0.1%.
2. The pH of the blood is always near 7.4 (see p. 202).

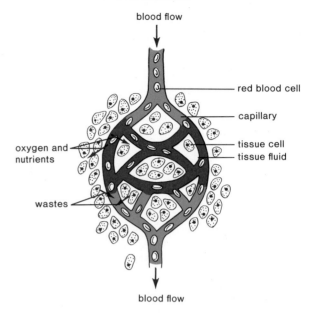

Figure 1.10   The internal environment of the body is the blood and tissue fluid. Tissue cells are surrounded by tissue fluid, which is continually refreshed when nutrient molecules exit and waste molecules enter the bloodstream.

blood flow

red blood cell

capillary

oxygen and nutrients

tissue cell
tissue fluid

wastes

blood flow

3. Blood pressure in the brachial artery averages near 120/80.
4. Blood temperature averages around 37° C (98.6° F).

The ability of the body to keep the internal environment within a certain range allows humans to live in a variety of habitats, such as the arctic regions, deserts, or the tropics.

The internal environment includes a tissue fluid that bathes all of the tissues of the body. The composition of tissue fluid must remain constant if cells are to remain alive and healthy. Tissue fluid is created when water (H_2O), oxygen (O_2), and nutrient molecules leave a capillary (the smallest of the blood vessels), and it is purified when water, carbon dioxide (CO_2), and other waste molecules enter a capillary from the fluid (fig. 1.10). Tissue fluid remains constant only as long as blood composition remains constant. Although we are accustomed to using the word *environment* to mean the external environment of the body, it is important to realize that it is the internal environment of tissues that is ultimately responsible for our health and well-being.

The internal environment of the body consists of blood and tissue fluid, which bathes the cells.

Most systems of the body contribute to maintaining a constant internal environment. The digestive system takes in and digests food, providing nutrient molecules, which enter the blood and replace the nutrients that are constantly being used up by the body cells. The respiratory system adds oxygen to the blood and removes carbon dioxide. The amount of oxygen taken in and carbon dioxide given off can be increased to meet bodily needs. The chief regulators of blood composition, however, are the liver and the kidneys. They monitor the chemical composition of blood and alter it as required. Immediately after glucose enters the blood, it can be removed by the liver for storage as glycogen. Later, glycogen can be broken down to replace the glucose used by the body cells; in this way, the glucose composition of the blood remains constant. The hormone insulin, secreted by the pancreas, regulates glycogen storage. The liver also removes toxic chemicals, such as ingested alcohol and drugs, and nitrogenous wastes given off by the cells. These are converted to molecules that can be excreted by the kidneys, organs that are also under hormonal control.

All of the systems of the body contribute to homeostasis; that is, maintaining the relative constancy of the internal environment.

Although homeostasis is, to a degree, controlled by hormones, it is ultimately controlled by the nervous system. The brain contains centers that regulate such factors as temperature and blood pressure. Maintaining proper temperature and blood pressure levels

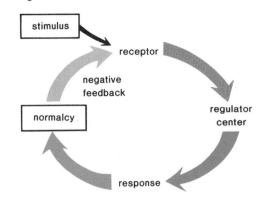

Figure 1.11 Diagram illustrating how homeostasis is maintained. A receptor (sense organ) signals a regulatory center that directs a response. Once normalcy is achieved, the receptor is no longer stimulated.

requires sensors that detect unacceptable levels and signal a control center. If a correction is required, the center then directs an adaptive response (fig. 1.11). Once normalcy is obtained, the receptor no longer stimulates the center. This is called control by **negative feedback,** because the control center brings about a response that is opposite to present conditions. For example, a fall in body temperature is followed by a rise in body temperature. This type of homeostatic regulation results in fluctuation between two levels, as illustrated for temperature control in figure 4.7.

Control of normalcy by negative feedback is a self-regulatory mechanism that results in slight fluctuations within narrow limits.

Summary

I. Anatomy and Physiology. Anatomy is the study of the structure of the body parts and physiology is the study of the function of these parts. Structure is suited to the function of a part.
II. Organization of Body Parts. The body has levels of organization that progress from atoms to molecules to cells to tissues to organs to body systems to the organism.
III. Anatomical Terms. Various types of terms are used to help describe the location of the organs of the body. It is assumed that the body is in the anatomical position. In this position, the body is standing erect, the face is forward, and the arms are at the sides with the palms and toes directed forward.
   A. The terms superior/inferior, medial/lateral, anterior/posterior, proximal/distal, superficial/deep, and visceral/parietal are used to describe the relative position of body parts.
   B. The body or its parts may be cut along certain imaginary planes. A sagittal cut divides the body into right and left parts, a transverse cut divides the body into superior and inferior parts, and a frontal cut divides the body into anterior and posterior parts.
   C. The body can be divided into axial and appendicular regions, and there are terms that indicate even more specific regions of the body. For example, brachial refers to the arm, and pedal refers to the foot.
   D. The human body has two major cavities called the dorsal and ventral cavities. Each of these is subdivided into smaller cavities and the visceral organs can be associated with one of these subcavities.

Summary—continued

IV. Organ Systems. There are a number of systems that function to maintain the normal conditions of the body. These have been categorized as follows:

  A. The integumentary system, which includes the skin, not only protects the body, but has other functions as well.

  B. Support and movement. The skeletal system contains the bones and the muscular system contains the three types of muscles. The primary function of these systems is support and movement, but they have other functions as well.

  C. Integration and coordination. The nervous system contains the brain, spinal cord, and nerves. Because the nervous system is in communication with both the sense organs and muscles, it allows us to respond to outside stimuli. The endocrine system contains the hormonal glands. Both the nervous and endocrine systems regulate the activities of the other systems.

  D. Processing and transporting. The digestive system (mouth, esophagus, stomach, small and large intestines, and accessory organs); the circulatory system (heart and vessels); immune system; respiratory system (lungs and conducting tubes); and the urinary system (kidneys and bladder) all have specific functions in processing and transporting to maintain the normal conditions of the body.

  E. Reproduction and development. The reproductive system in males (testes, other glands, ducts, and penis) and in females (ovaries, oviducts, uterus, vagina, and external genitalia) carries out those functions that give humans the ability to have children.

V. Homeostasis. Homeostasis is the relative constancy of the internal environment, which is blood and tissue fluid.

  A. All the systems of the body contribute to homeostasis. Some, like the respiratory and digestive systems, remove and/or add substances to blood.

  B. The nervous and endocrine systems regulate the activities of other systems. Negative feedback is a self-regulatory mechanism that enables the nervous and endocrine systems to maintain bodily conditions within normal limits.

Study Questions

1. Distinguish the study of anatomy from the study of physiology.
2. Give an example to show the relationship between the form and function of body parts.
3. List the levels of organization within the human body in reference to a specific organ.
4. Distinguish between the axial and appendicular portions of the body. State at least two anatomical terms that pertain to the head, thorax, abdomen, and extremities.
5. What effects do a midsagittal cut, a transverse cut, and a frontal cut have on the body?
6. Distinguish between the dorsal and ventral body cavities, and name the smaller cavities that occur within each.
7. Name the four quadrants of the abdominopelvic cavity. Use the following terms to divide the abdominopelvic cavity into nine regions: epigastric, umbilical, hypogastric, hypochondriac, lumbar, and iliac.
8. Name the major organ systems and describe the general functions of each.
9. List the major organs found within each organ system.
10. Define homeostasis and explain its importance.

Objective Questions

I. For questions 1–5, match the terms in the key to the items below.

  Key: a. superior
       b. inferior
       c. anterior
       d. posterior
       e. medial
       f. lateral
       g. proximal
       h. distal

  1. the esophagus in relation to the stomach
  2. the ears in relation to the nose
  3. the shoulder in relation to the hand
  4. the intestines in relation to the vertebrae
  5. the rectum in relation to the mouth

II. For questions 6–12, match the terms in the key to the items below.

  Key: a. buccal
       b. occipital
       c. gluteal
       d. cutaneous
       e. palmar
       f. cervical
       g. axillary

  6. buttocks
  7. palm
  8. back of head
  9. mouth
  10. skin
  11. armpit
  12. neck

III. For questions 13–22, match the items in the key to the organs below.

  Key: a. cranial cavity
       b. spinal cavity
       c. thoracic cavity
       d. abdominal cavity
       e. pelvic cavity

13. stomach
14. heart
15. urinary bladder
16. brain
17. liver
18. trachea
19. esophagus
20. rectum
21. small intestine
22. pancreas

IV. For questions 23–29, match a system in the key to the organs below.

Key:
a. digestive system
b. urinary system
c. respiratory system
d. circulatory system
e. reproductive system
f. nervous system
g. endocrine system

23. lungs
24. heart
25. ovaries
26. brain
27. stomach
28. kidneys
29. thyroid gland

V. Fill in the blanks.

30. The level of organization above the tissue level is the _____ level.

31. The imaginary plane that passes through the midline of the body is called the _____ plane.

32. All the organ systems of the body together function to maintain _____, relative constancy of the internal environment.

Medical Terminology Reinforcement Exercise

Pronounce, dissect, and fill in the blank to give a brief meaning to the following terms:

1. Suprapubic (soo″prah-pu′bik) means _____ the pubis.

2. Infraorbital (in-frah″or′bi-tal) means _____ the eye orbit.

3. Gastrectomy (gas-trek′to-me) means excision of the _____ .

4. Celiotomy (se″le-ot′o-me) means incision (cut into) the _____ .

5. Macrocephalous (mak″ro-sef′ah-lus) means large _____ .

6. Transthoracic (trans″tho-ras′ik) means across the _____ .

7. Bilateral (bi-lat′er-al) means two or both _____ .

8. Ophthalmoscope (of-thal′mo-skōp) is an instrument to view inside the _____ .

9. Dorsalgia (dor-sal′je-ah) means pain in the _____ .

10. Endocrinology (en″do-kri-nol′o-je) is the _____ of the endocrine system (secretions within).

2

Cell Structure and Function

Chapter Outline

The Generalized Cell
    Cell Membrane
    Nucleus
    Membranous Canals and Vacuoles
    Mitochondria
    Other Organelles
Movement of Molecules across the
  Cell Membrane
    Diffusion
    Osmosis
    Filtration
    Active Transport
    Phagocytosis and Pinocytosis
Cell Division

Learning Objectives

After you have studied this chapter, you should be able to:

1. Label the parts of a generalized cell.
2. Describe the structure and function of the cell membrane.
3. Describe the nucleus and its parts.
4. Describe the structure, function, and relationship of the endoplasmic reticulum, Golgi apparatus, and lysosomes.
5. Describe the structure and function of mitochondria.
6. Describe the composition of the cytoskeleton.
7. State how centrioles are related to cilia and flagella.
8. Describe how substances move across the cell membrane; and distinguish between diffusion, osmosis, filtration, and active transport.
9. Give an overview for cell division, explaining the mechanism by which the chromosome number stays constant.

The Generalized Cell

The cells of your body perform specific functions and, therefore, their structure varies greatly. Even so, because all cells have the same basic organization, it is possible for us to begin our study of cell structure by examining a generalized cell. Our knowledge of the generalized animal cell depicted in figure 2.1 was obtained by using the light microscope and the electron microscope. The *light microscope,* which utilizes light to view the object, does not show much detail, but the *electron microscope,* which uses electrons to view the object, allows cell biologists to discern cell structure in great detail. Table 2.1 contrasts these two types of microscopes.

Electron micrographs, photographs obtained by use of the electron microscope, have helped us develop an understanding of cell structure.

Figure 2.1 shows that a human cell is surrounded by an outer membrane, or **cell membrane,** within which is found the **cytoplasm,** the substance of the cell outside the nucleus, in which there are various **organelles,** small bodies with specific structures and functions. Table 2.2 summarizes the organelles that will be studied.

Cell Membrane

All cells have a **cell,** or plasma, **membrane** (fig. 2.2) that serves as a boundary to regulate the entrance and exit of molecules into and out of the cell. The membrane

---

Figure 2.1   Cell structure. This generalized cell is based on electron micrographs.

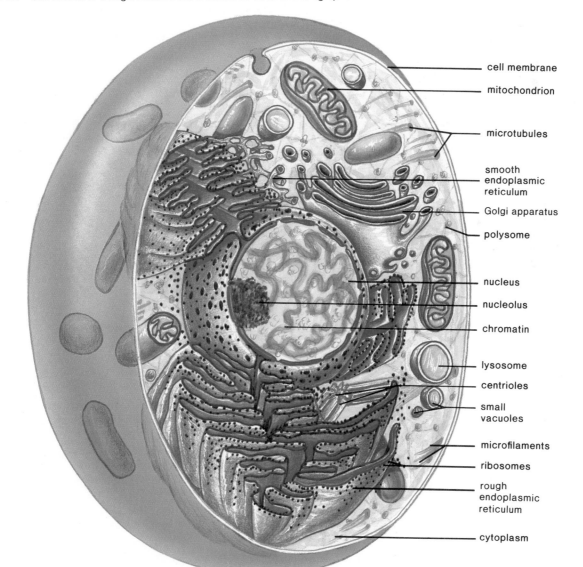

cell membrane

mitochondrion

microtubules

smooth endoplasmic reticulum

Golgi apparatus

polysome

nucleus

nucleolus

chromatin

lysosome

centrioles

small vacuoles

microfilaments

ribosomes

rough endoplasmic reticulum

cytoplasm

model accepted by most investigators today is the *fluid-mosaic model* whose main structural component is a double layer of lipid (fat) molecules that has the consistency of a light oil. Attached to and/or embedded in the membrane are protein molecules whose composition and concentration vary according to the particular cell.

Certain proteins located within the membrane serve as receptors for chemical molecules, such as hormones, that influence the metabolism of the cell. Others are carriers that transport molecules across the membrane. Both the outer lipids and proteins bear markers that probably function to identify the cell. Cells that are not recognized by the body are likely to be rejected, as sometimes happens when individuals receive organ transplants. On the other hand, cell recognition processes may be faulty when cancerous cells are permitted to grow and reproduce.

The cell membrane, the outer boundary of the cell, consists of a double layer of lipid molecules in which protein molecules form a mosaic pattern.

Nucleus

The **nucleus,** the largest organelle found within the cell, is enclosed by a double-layered nuclear envelope that is actually continuous with the endoplasmic reticulum, which will be discussed in the following paragraphs. As illustrated in figure 2.3, there are pores, or openings,

Table 2.1 A Comparison of a Light and an Electron Microscope

Light	Electron
Glass lenses	Electromagnetic lenses
Illumination by light	Illumination due to beam of electrons
Resolution[a] ≃ 0.1 μm	Resolution ≃ 1nm
Magnifies to 2,000×	Magnifies to 100,000×
Cost: up to thousands of dollars	Cost: up to hundreds of thousands of dollars
Specimen may be living or dead	Specimen must be dead

[a]Resolution is the ability to distinguish two points as being separate.

Table 2.2 Organelles (Simplified)

Name	Structure	Function
Cell membrane	Bilayer of phospholipid and globular proteins	Regulates passage of molecules into and out of cell
Nucleus	Nuclear envelope surrounds chromatin, nucleolus, and nucleoplasm	Control center of cell
Nucleolus	Concentrated area of RNA in the nucleus	Ribosome formation
Chromatin (chromosomes)	Composed of DNA and protein	Contains hereditary information
Endoplasmic reticulum	Folds of membrane forming flattened channels and tubular canals	Transport by means of vesicles
Rough	Studded with ribosomes	Protein synthesis
Smooth	Having no ribosomes	Lipid synthesis
Ribosome	RNA and protein in two subunits	Protein synthesis
Golgi apparatus	Stack of membranous saccules	Packaging and secretion
Vacuole and vesicle	Membranous sacs	Containers of material
Lysosome	Membranous container of hydrolytic enzymes	Intracellular digestion
Mitochondrion	Inner membrane (cristae) within outer membrane	Aerobic cellular respiration
Microfilament	Actin and myosin proteins	Movement of organelles and shape of cell
Microtubule	Tubulin protein	Movement of cell
Centriole	9 + 0 pattern of microtubules	Directs microtubule organization; associated with cell division
Cilium and flagellum	9 + 2 pattern of microtubules	Moves particles along surface; movement of cell

Figure 2.2 Fluid-mosaic model of a membrane indicates a bilayer of lipid molecules with an irregular distribution of protein molecules.

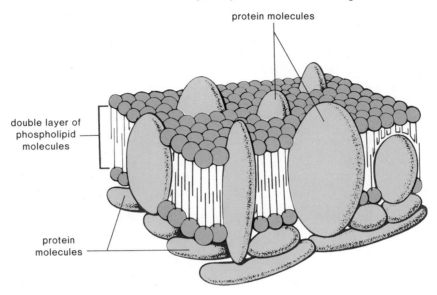

protein molecules

double layer of
phospholipid
molecules

protein
molecules

in this membrane through which large molecules pass from the nucleoplasm, the fluid portion of the nucleus, to the cytoplasm.

The nucleus is of primary importance in the cell because it is the control center that oversees the metabolic functioning of the cell and ultimately determines the cell's characteristics. Within the nucleus, there are masses of threads called *chromatin,* so named because they take up stains and become colored. Chromatin is indistinct in the nondividing cell, but it condenses to rodlike structures called **chromosomes** at the time of cell division. Chemical analysis shows that chromatin and, thus, chromosomes contain the chemical DNA (deoxyribonucleic acid).

It has been known for some time that the chromosomes contain the genes and now we know that the genes are composed of DNA. Each gene codes for the production of a specific type of protein in the cytoplasm. Many of these proteins are **enzymes,** organic catalysts that speed up chemical reactions. **Metabolism** is all the chemical reactions that occur in a cell and, therefore, we see that DNA controls the cell because it controls the metabolism of the cell.

RNA (ribonucleic acid) is a chemical that is found in the **nucleolus,** a nearly circular body in the nucleus (fig. 2.3). This type of RNA eventually becomes a part of the ribosomes in the cytoplasm.

The nucleus contains chromatin, which condenses into chromosomes just prior to cell division. The genes, composed of DNA, are on the chromosomes and they code for the production of proteins in the cytoplasm. The RNA of the nucleolus is found in the ribosomes.

Figure 2.3 An electron micrograph of a nucleus (*N*) with a clearly defined nucleolus (*Nu*) and irregular patches of chromatin scattered throughout the nucleoplasm. The nuclear envelope (*NE*) contains pores indicated by the arrows. This nucleus is surrounded by endoplasmic reticulum (*ER*), and its size may be compared to the mitochondrion (*M*) that appears to the left.

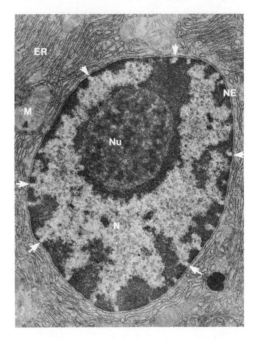

Membranous Canals and Vacuoles

Endoplasmic reticulum, the Golgi apparatus, vacuoles, and lysosomes (fig. 2.1) are structurally and functionally related membranous structures. Ribosomes are not composed of membrane, but are included under this heading because they are often associated with the endoplasmic reticulum.

Figure 2.4 Rough endoplasmic reticulum. *a.* Electron micrograph showing that in some cells the cytoplasm is packed with this organelle. *b.* A three-dimensional drawing gives a better idea of the organelle's actual shape.

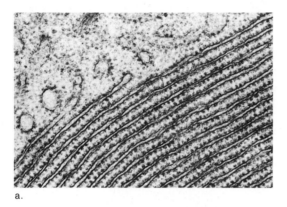

a.

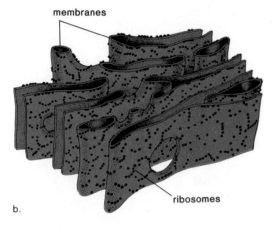

membranes

ribosomes

b.

Figure 2.5 Golgi apparatus function. The Golgi apparatus receives vesicles from the endoplasmic reticulum and thereafter forms at least two types of vesicles, lysosomes and secretory vesicles. Lysosomes contain hydrolytic enzymes that can break down large molecules. Sometimes lysosomes join with vesicles, bringing large molecules into the cell. Thereafter, any nondigested residue is voided at the cell membrane. The secretory vesicles formed at the Golgi apparatus also discharge their contents at the cell membrane.

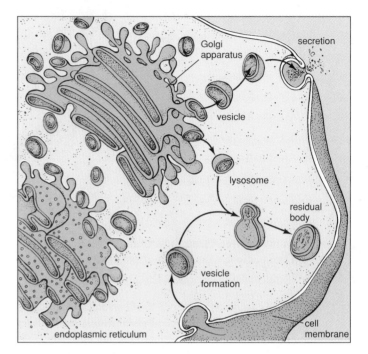

Golgi apparatus

secretion

vesicle

lysosome

residual body

vesicle formation

endoplasmic reticulum

cell membrane

Endoplasmic Reticulum

The **endoplasmic reticulum** (**ER**) forms a membranous system of tubular canals that begins at the nuclear envelope and branches throughout the cytoplasm. Small granules, called **ribosomes,** are attached to some portions of the endoplasmic reticulum. If they are present, the reticulum is called **rough ER** (fig. 2.4); and if they are not present, it is called **smooth ER.** Smooth endoplasmic reticulum contains enzymes that make lipids such as steroid hormones in certain cells. Also, it is known that the administration of drugs increases the amount of smooth ER in the liver. It would seem then that the reticulum has enzymes that detoxify drugs.

Ribosomes As mentioned previously, ribosomes contain RNA from the nucleolus. Once ribosomes are fully assembled within the cytoplasm, they function in the process of protein synthesis. *Synthesis* refers to the joining together of small molecules to make larger ones. When proteins are synthesized, amino acids are joined together in an order dictated by a gene, and it is the specific order of the amino acids that makes one protein different from another.

Sometimes ribosomes lie free in the cytoplasm and form aggregates called *polysomes.* All the ribosomes

of a polysome are making multiple copies of the same protein. The ribosomes attached to rough ER are making proteins that are exported (secreted) from the cell. Such proteins are stored temporarily in the channels of the reticulum. Small portions of the endoplasmic reticulum then break away to form membrane-enclosed vesicles (small membranous sacs) that migrate to the Golgi apparatus, where the product is received and repackaged before being secreted.

The endoplasmic reticulum is a membranous system of tubular canals that may be smooth or rough. Smooth ER functions in steroid metabolism and possibly detoxification of drugs. Rough ER functions in protein synthesis.

Golgi Apparatus

The **Golgi apparatus** (fig. 2.5) is named for the person who first discovered its presence in cells. It is composed of a stack of about a half-dozen or more saccules that look like flattened vacuoles. At the edges of the saccules are rounded vacuoles and vesicles.

The Golgi apparatus is especially well developed in cells that secrete (export) a product; for example, in the pancreatic cells that make digestive enzymes or the

Figure 2.6 Electron micrograph showing two types of lysosomes. The darker one is a residual body containing nondigested material; the lighter one is an autodigestive vacuole containing a mitochondrion.

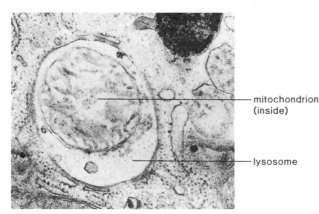

mitochondrion (inside)

lysosome

bronchial cells that produce mucus. When the Golgi apparatus packages a product for export, the product is enclosed within a vesicle that moves toward the cell membrane where it discharges its contents.

The Golgi apparatus processes and packages products that will be secreted at the cell membrane. It also produces lysosomes.

Lysosomes

A **lysosome** is a special type of vesicle (fig. 2.6), most likely formed by the Golgi apparatus. All lysosomes are concerned with intracellular digestion and contain *digestive enzymes*. Following formation, the lysosome may fuse with a vesicle that contains a substance to be digested. The products of digestion enter the cytoplasm, and only nondigested residue is retained.

Occasionally, a person is unable to manufacture an enzyme normally found within the lysosome. In these cases, the lysosome fills to capacity with macromolecules that cannot be broken down. The cells may become so filled with lysosomes of this type that it brings about the death of the individual. The best known of these lysosomal storage disorders is Tay Sachs disease.

Lysosomes also carry out *autodigestion,* or the disposal of worn-out or damaged cell components, such as mitochondria (fig. 2.6), which have a short life span in the cell. This is an essential part of the normal process of cytoplasmic maintenance and turnover. Turnover here means that the cell is constantly breaking down and remaking its parts.

Lysosomes contain hydrolytic enzymes that function in digestion of substances taken in at the cell membrane and also of the cell parts themselves.

Figure 2.7 Mitochondrion structure. *a.* Electron micrograph of a mitochondrion surrounded by rough endoplasmic reticulum. Note the shelflike cristae formed by the inner membrane. *b.* A diagrammatic drawing shows outer and inner structure more clearly.

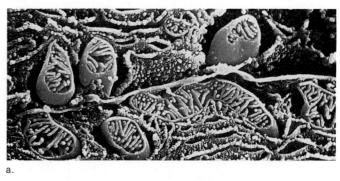

a.

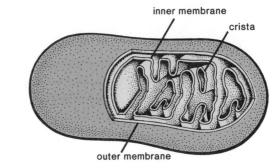

inner membrane

crista

outer membrane

b.

Mitochondria

A **mitochondrion** is a rather complex organelle that produces ATP (adenosine triphosphate), the chemical energy needed by cells. Every cell needs a certain amount of ATP energy to synthesize molecules, but many cells need ATP to carry out their specialized functions. For example, muscle cells need it for muscle contraction and nerve cells need it for the conduction of nerve impulses.

Mitochondria are often referred to as the powerhouses of the cell because, just as a powerhouse burns fuel to produce electricity, the mitochondria burn glucose products to produce ATP molecules. In the process, mitochondria use up oxygen and give off carbon dioxide and water. The oxygen you breathe in enters cells and then the mitochondria; the carbon dioxide you breathe out is released by the mitochondria. Since gas exchange is involved, it is said that mitochondria carry on **aerobic cellular respiration.** A shorthand way to indicate the chemical transformation associated with aerobic cellular respiration is:

carbohydrate + oxygen → carbon dioxide + water + ATP energy

Each mitochondrion is composed of two membranes, an outer membrane and an inner membrane (fig. 2.7). The inner membrane is convoluted into shelflike projections called *cristae*. The molecules that aid in the production of energy are located in an assembly-line

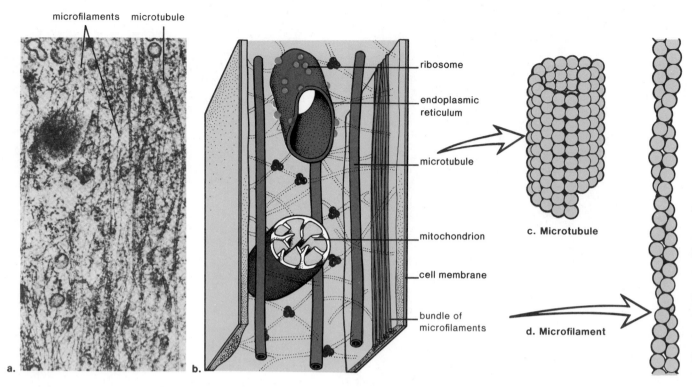

Figure 2.8 Cytoskeleton of cell. Notice that the various organelles are suspended in a cytoplasm that includes microtubules and microfilaments. *a.* Electron micrograph. *b.* Drawing of microtubules and microfilaments. *c.* Detailed structure of a microtubule. *d.* Detailed structure of a microfilament.

microfilaments microtubule

ribosome

endoplasmic reticulum

microtubule

mitochondrion

cell membrane

bundle of microfilaments

c. Microtubule

d. Microfilament

a.

b.

fashion on these membranous shelves. The membrane is divided into functional units, and a very small area of each crista contains one respiratory unit. The inner membrane lends itself to this type of arrangement and, therefore, we see that structure aids function.

Mitochondria are the sites of aerobic cellular respiration, a process that provides ATP energy molecules to the cell.

Other Organelles

Cytoskeleton

Several types of filamentous protein structures form a **cytoskeleton** (fig. 2.8) that helps maintain the cell's shape, anchors the organelles, or allows them to move as appropriate. The cytoskeleton includes microfilaments and microtubules. **Microfilaments** are long, extremely thin fibers that usually occur in bundles or other groupings. Microfilaments have been isolated from a number of cells. When analyzed chemically, their composition is similar to that of actin or myosin, the two proteins responsible for muscle contraction.

Microtubules are shaped like thin cylinders and are several times larger than microfilaments. Each cylinder contains thirteen rows of tubulin, a globular pro-

tein, arranged in a helical fashion. Aside from existing independently in the cytoplasm, microtubules are also found in certain organelles, such as cilia, flagella, and centrioles.

Remarkably, both microfilaments and microtubules assemble and disassemble within the cell. When they are assembled, the protein molecules are bonded together and when they are disassembled, the protein molecules are not attached to one another. When microfilaments and microtubules are assembled, the cell has a particular shape and when they disassemble, the cell can change shape (fig. 2.9).

The cytoskeleton contains microfilaments and microtubules. Microfilaments and microtubules maintain the shape of the cell and also direct the movement of cell parts.

Centrioles

Centrioles are short cylinders that contain nine microtubule triplets arranged in a circle. There is always one pair (fig. 2.1) lying at right angles to one another near the nucleus. Before a cell divides, the centrioles duplicate and the members of each pair are also at right angles to one another.

Figure 2.9 A scanning electron micrograph of an individual cell in a tissue culture. Notice the fingerlike projections on the "ruffle," which marks the leading edge of the cell. Microfilaments and microtubules are most likely present in this ruffle.

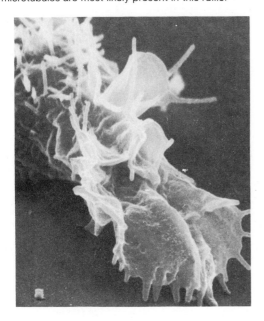

Figure 2.10 Sperm cells use long whiplike flagella to move about.

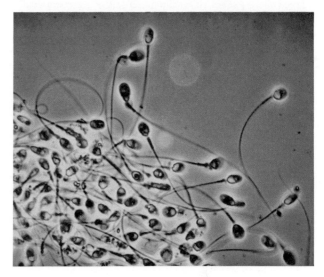

Centrioles are believed to give rise to basal bodies that direct the formation of cilia and flagella. Centrioles may also be involved in other cellular processes that use microtubules, such as the movement of material throughout the cell, or the appearance and disappearance of the spindle apparatus (p. 38). Their exact role in these processes is uncertain, however.

Cilia and Flagella

Cilia and flagella are hairlike projections of cells that can move either in an undulating fashion, like a whip, or stiffly, like an oar. Cells that have these organelles are capable of movement. For example, sperm cells, carrying genetic material to the egg, move by means of flagella (fig. 2.10). The cells that line our upper respiratory tract are ciliated. The cilia sweep debris trapped within mucus back up the throat and this action helps keep the lungs clean.

Cilia are much shorter than flagella, but even so, they both are constructed similarly. They are membrane-bound cylinders enclosing a matrix area. In the matrix are nine microtubule doublets arranged in a circle around two central microtubules. Each cilium and flagellum has a basal body that resembles a centriole. Therefore, it is reasoned, as mentioned previously, that centrioles might give rise to basal bodies.

Centrioles that lie near the nucleus may be involved in the production of the spindle apparatus and are believed to give rise to the basal bodies of cilia and flagella.

Movement of Molecules across the Cell Membrane

The cell membrane allows only certain molecules to enter and exit the cytoplasm freely; therefore, the cell membrane is said to be *selectively permeable*. Some of the molecules that can cross the membrane diffuse across.

Diffusion

Diffusion is the movement of molecules from the area of higher concentration to the area of lower concentration. To illustrate diffusion, imagine opening a perfume bottle in the corner of a room (fig. 2.11). The smell of the perfume soon penetrates the room, because the molecules that make up the perfume have drifted to all parts of the room. Another example is putting a tablet of dye into water. The water eventually takes on the color of the dye as the tablet dissolves.

Figure 2.11 Diffusion occurs when *a.* a perfume bottle is opened and the scent fills a room, because the molecules have moved away from the bottle, and *b.* a tablet of dye is placed in a beaker and water becomes colored, because the molecules have moved away from the original area of the tablet.

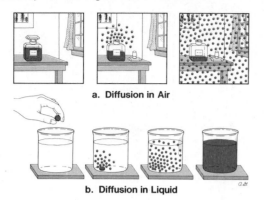

a. Diffusion in Air

b. Diffusion in Liquid

In the body, oxygen enters the blood from the alveoli (air sacs) of the lungs (fig. 2.12) by diffusion. During kidney dialysis, substances diffuse across a membrane from the area of greater concentration to the area of lesser concentration. No energy need be expended by cells whenever substances can simply diffuse across the cell membrane.

Osmosis

Osmosis is the diffusion of water across a cell membrane. It occurs whenever there is an unequal concentration of water on either side of a selectively permeable membrane. Normally, body fluids are *isotonic* to cells (fig. 2.13)—there is an equal concentration of substances (solutes) and water (solvent) on either side of the cell membrane and cells maintain their usual size and shape. For this reason, most intravenous solutions are also isotonic to cells.

If red blood cells are placed in a *hypotonic* solution that has a higher concentration of water (lower concentration of solute) than do the cells, water will enter the cells and they will swell to bursting. Bursting of red blood cells is called **hemolysis.** On the other hand, if red blood cells are placed in a *hypertonic* solution that has a lower concentration of water (higher concentration of solute) than do the cells, water will leave the cells and they will shrink. The shrinking of red blood cells is called **crenation.**

Some substances can simply diffuse across a cell membrane. The diffusion of water is called osmosis. In an isotonic solution, cells neither gain nor lose water. In a hypotonic solution, they swell, and in a hypertonic solution, they shrink.

Figure 2.12 Oxygen (dots) diffuses into the capillaries of the lungs because there is a greater concentration of oxygen in the alveoli than in the capillaries.

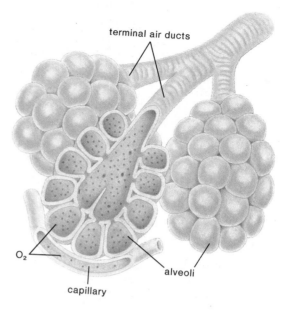

terminal air ducts

O_2

capillary

alveoli

Filtration

A capillary wall is one-cell thick and, therefore, we would expect for substances to diffuse across these cells from the area of higher to lower concentration. Indeed, small molecules (water plus small solutes) do diffuse across the capillary wall. However, any blood pressure present helps matters along, and pushes water and the dissolved solutes out of the capillary. This process is called **filtration.**

Filtration is easily observed in the laboratory when a solution is poured past filter paper into a flask. Large substances stay behind, but small molecules and water pass through. Filtration of water and substances in the region of capillaries is largely responsible for the formation of tissue fluid, the fluid that surrounds the cells. It is also at work in the kidneys when water and small molecules move from the blood to the inside of the kidney tubules.

During filtration, diffusion of substances out of a blood vessel is aided by blood pressure.

Active Transport

In **active transport,** substances accumulate either inside or outside the cell in the region of *higher* concentration. For example, iodine collects in the cells of the thyroid gland; sugar is completely absorbed from the gut by the

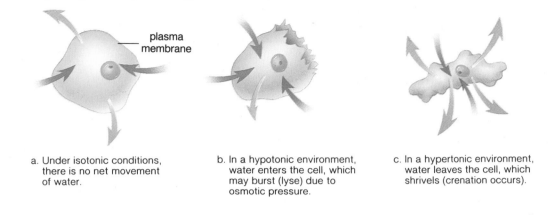

Figure 2.13 Osmosis. The arrows indicate the net movement of water. In an isotonic solution, a cell neither gains nor loses water; in a hypotonic solution, a cell gains water; and in a hypertonic solution, a cell loses water.

plasma membrane

a. Under isotonic conditions, there is no net movement of water.

b. In a hypotonic environment, water enters the cell, which may burst (lyse) due to osmotic pressure.

c. In a hypertonic environment, water leaves the cell, which shrivels (crenation occurs).

Figure 2.14 Active transport is apparent when a molecule crosses the cell membrane toward the area of greater concentration. An expenditure of ATP energy is required, presumably to allow a protein carrier to transport molecules across the cell membrane.

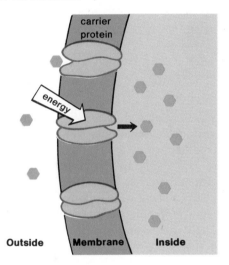

carrier protein

energy

Outside   Membrane   Inside

Figure 2.15 Phagocytosis and pinocytosis both depend on vesicle formation to bring a substance into the cell. During phagocytosis, the material may be as large as a small cell; during pinocytosis, the substance is suspended or dissolved in water.

cell membrane

substance

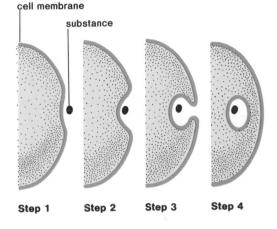

Step 1   Step 2   Step 3   Step 4

cells lining the digestive tract; and sodium is sometimes almost completely withdrawn from urine by cells lining the kidney tubules.

Both protein carriers and an expenditure of energy (fig. 2.14) are needed to transport substances from an area of lower to an area of higher concentration. A **carrier** is a cell membrane protein that specializes in combining with and transporting substances across the cell membrane. ATP energy is needed to cause a carrier to combine with the substance to be transported. Therefore, it is not surprising that cells primarily involved in active transport, such as kidney cells, have a large number of mitochondria near the membrane where active transport is occurring.

During active transport, which requires cell membrane carriers and ATP energy, substances move against a concentration gradient and accumulate in the area of higher concentration.

Phagocytosis and Pinocytosis

At times, substances are taken into cells by vesicle formation (fig. 2.15). When the material taken in is quite large, the process is called phagocytosis (cell eating). **Phagocytosis** is common to amoeboid-type cells, such as macrophages, which are the body's scavengers, because they engulf worn-out red blood cells and other

Figure 2.16 Mitosis overview. Following duplication, each chromosome in the mother cell contains two chromatids. During mitotic division, the chromatids separate so that daughter cells have the same number and kinds of chromosomes as the mother cell.

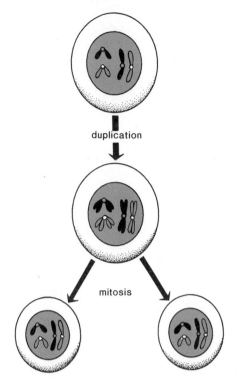

duplication

mitosis

types of debris. Cells commonly take in, by vesicle formation, material that is small enough to be dissolved or suspended in water. In that case, the process is called **pinocytosis** (cell drinking).

Substances enter and exit cells by either diffusion, active transport, or vesicle formation.

Cell Division

During ordinary cell division, called mitosis, the chromosome number stays constant. The original cell, called the mother cell, has forty-six chromosomes, and the two daughter, or resulting cells also have forty-six chromosomes.

Figure 2.16 shows a diagram of a mother cell that for simplicity's sake has only two pairs of chromosomes. Notice that following duplication of the chromosomes, each one has two identical portions called chromatids. During mitosis, the chromatids separate and one chromatid from each duplicated chromosome goes to each daughter cell. Therefore, each daughter cell will have the same number and kinds of chromo-

somes as the mother cell. Before the daughter cells can divide again, the chromosomes must duplicate to have two chromatids.

Figure 2.17 diagrams the process of mitosis as it occurs in somatic cells. The process actually requires several stages during which the nuclear envelope disappears and a spindle apparatus with spindle fibers forms. The chromosomes are attached to the spindle fibers by structures called centromeres and the chromosomes move to the center of the mother cell before the centromere splits and the chromatids separate. The separated chromatids, now called daughter chromosomes, move away from each other into the newly forming daughter cells.

Mitosis is the type of cell division required for growth and repair of body cells. The process of mitosis assures that each cell in the body has the same number and kinds of chromosomes and genes because the genes are on the chromosomes.

Ordinarily, a cell only divides about fifty times and then with maturity, a cell divides no more. Cancer cells never mature and, instead, they continue to divide indefinitely. We are only now beginning to learn that cancer cells either produce or respond to growth factors in a way that causes them to keep on dividing.

Figure 2.17   Mitosis, in detail. During *interphase,* the nuclear envelope is intact and the nucleolus is clearly visible in the nucleus. The chromosomes (in chromatin form) are in the process of duplicating as are the centrioles, organelles associated with the formation of the spindle apparatus. During *prophase,* the nuclear envelope and nucleolus are disappearing and the spindle fibers are appearing as the centrioles move apart. The chromosomes are distinct and duplicated. During *metaphase,* the chromosomes are lined up at the middle of the spindle apparatus. During *anaphase,* the chromatids separate and the resulting daughter chromosomes move apart. During *telophase,* the nuclear envelope and nucleolus are reappearing and the spindle fibers are disappearing. Furrowing of the cell membrane divides the cytoplasm between two daughter cells.

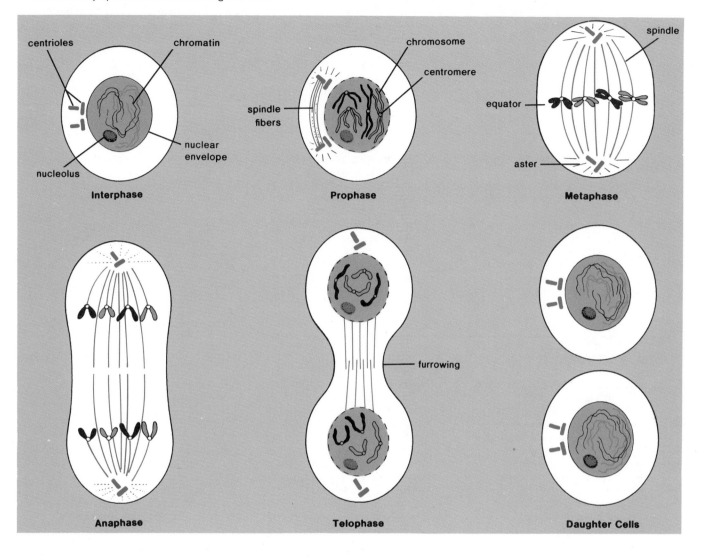

Summary

I. The Generalized Cell. Cells differ in shape and function, but even so, it is possible to describe a generalized cell.
  A. Electron micrographs, photographs obtained by use of the electron microscope, have helped biologists develop an understanding of cell structure.
  B. The cell membrane, the outer boundary of the cell, consists of a double layer of lipid (fat) molecules in which protein molecules form a mosaic pattern.
  C. The nucleus contains chromatin, which condenses into chromosomes just prior to cell division. The genes, composed of DNA, are on the chromosomes and they code for the production of proteins in the cytoplasm. The RNA of the nucleolus is found in the ribosomes.
  D. The endoplasmic reticulum is a membranous system of tubular canals that may be smooth or rough. Smooth ER functions in steroid hormone metabolism and detoxification of drugs. Rough ER functions in protein synthesis.
  E. The Golgi apparatus processes and packages products that will be secreted at the cell membrane. It also produces lysosomes.
  F. Lysosomes contain hydrolytic enzymes that function in digestion of substances taken in at the cell membrane and

also of the cell parts themselves.

G. Mitochondria are the sites of aerobic cellular respiration, a process that provides ATP energy molecules to the cell.

H. The cytoskeleton contains microfilaments and microtubules. Microfilaments and microtubules maintain the shape of the cell and also direct the movement of cell parts.

I. Centrioles that lie near the nucleus may be involved in the production of the spindle apparatus and are believed to give rise to the basal bodies of cilia and flagella.

II. Movement of Molecules across the Cell Membrane. Certain molecules move across the membrane passively (no energy required) by diffusion, filtration, or active transport. Other types of molecules are transported actively

(energy required) across the membrane by active transport or vesicle formation.

A. The cell membrane is selectively permeable. A few types of small molecules can diffuse through the cell membrane from the area of higher concentration to the area of lower concentration.

B. Osmosis is the diffusion of water across a cell membrane. When a cell is placed in a hypotonic solution, the cell gains water. When a cell is placed in a hypertonic solution, the cell loses water and the cytoplasm shrinks.

C. During filtration, diffusion of small molecules out of a blood vessel is aided by blood pressure.

D. During active transport, substances are transported across the cell membrane by protein carriers, and energy is

required. The molecules may be going toward the region of higher concentration.

E. Substances enter cells by vesicle formation. During phagocytosis (cell eating), the substance may be as large as a smaller cell; during pinocytosis, the substance is suspended or dissolved in water.

III. Cell Division. During mitosis, each newly formed cell receives a copy of each kind of chromosome. Later, the cytoplasm divides by furrowing.

A. Mitosis ensures that each cell in the body is genetically identical. At the time of division, a chromosome consists of chromatids. When these separate, each newly forming cell receives the same number and kinds of chromosomes as the original cell.

Study Questions

1. Describe the fluid-mosaic model of membrane structure.
2. Describe the nucleus and its contents, including the terms DNA and RNA in your description.
3. Describe the structure and function of endoplasmic reticulum. Include the terms rough and smooth ER, and ribosomes in your description.
4. Describe the structure and function of the Golgi apparatus.

Mention vacuoles and lysosomes in your description.
5. Describe the structure and function of mitochondria. Mention the energy molecule, ATP, in your description.
6. Describe the composition of the cytoskeleton, and give a function for microfilaments and microtubules.
7. Describe the structure and function of centrioles, cilia, and flagella.

8. What are the mechanisms by which substances enter and exit cells? Contrast diffusion with active transport of molecules across the cell membrane.
9. Define osmosis and discuss the effects of placing red blood cells in an isotonic, hypotonic, and hypertonic solution.
10. Describe mitosis, and discuss the function of mitosis in humans.

Objective Questions

I. For questions 1–5, match the organelles in the key to their functions.

Key: a. mitochondria
b. nucleus
c. Golgi apparatus
d. rough ER
e. centrioles

1. packaging and secretion
2. cell division
3. powerhouses of the cell
4. protein synthesis
5. control center for cell

II. Fill in the blanks.

6. The fluid-mosaic model of membrane structure says that _____ molecules drift about within a _____ bilayer.
7. Rough ER has _____ , but smooth ER does not.
8. Microfilaments and microtubules are a part of the _____ , the framework of the cell that provides its shape and regulates movement of organelles.

9. Basal bodies that organize the microtubules within cilia and flagella are derived from

_____ .

10. Water will enter a cell when it is placed in a _____ solution.
11. Active transport requires a protein _____ and _____ energy.
12. At the conclusion of mitosis, each newly formed cell in humans contains _____ chromosomes.

Medical Terminology Reinforcement Exercise

Pronounce, dissect, and analyze the following terms:

1. phagocytosis (fag″o-si-to′sis)
2. hemolysis (he″mol′i-sis)
3. isotonic (i″so-ton′ik)
4. cytology (si-tol′o-je)
5. cytometer (si-tom′e-ter)
6. nucleoplasm (nu′kle-o-plazm″)
7. lysosome (li′so-sōm)
8. pancytopenia (pan″si-to-pe′ne-ah)
9. cytogenic (si-to-jen′ik)
10. erythrocyte (ĕ-rith′ro-sit)

3

Body Tissues and Membranes

Chapter Outline

Body Tissues
    Epithelial Tissue
    Connective Tissue
    Muscular Tissue
    Nervous Tissue

Body Membranes and Organs
    Body Membranes
    Organs

Learning Objectives

After you have studied this chapter, you should be able to:

1. Describe the general characteristics and functions of epithelial tissues.
2. Name the major types of epithelial tissues and relate each one to a particular organ.
3. Describe the general characteristics and functions of connective tissues.
4. Name the major types of connective tissues and relate each one to a particular organ.
5. Describe the general characteristics and functions of muscular tissues.
6. Name the major types of muscular tissues and relate each one to a particular organ.
7. Describe the general characteristics and functions of nerve tissue.
8. Name the different types of membranes and relate each type to a particular location in the body.
9. Give the names and locations for the serous membranes found in the ventral cavity of humans.

Body Tissues

The tissues of the human body can be categorized into four major types: *epithelial tissue,* which covers body surfaces and lines body cavities; *connective tissue,* which binds and supports body parts; *muscular tissue,* which is specialized for movement; and *nervous tissue,* which responds to stimuli and transmits impulses from one body part to another.

Epithelial Tissue

Epithelial tissue, also called epithelium, forms a continuous layer, or sheet, over the entire body surface and most of the body's inner cavities. On the external sur-face, it protects the body from drying out, injury, and bacterial invasion. On internal surfaces, this tissue may be specialized for other functions in addition to protection; for example, it secretes mucus along the digestive tract; it sweeps up impurities from the lungs by means of hairlike extensions called cilia; and it efficiently absorbs molecules from kidney tubules because of fine cellular extensions called microvilli.

There are three types of epithelial tissue. **Squamous epithelium** (fig. 3.1a) is composed of flat cells and is found lining the lungs and blood vessels. **Cuboidal epithelium** (fig. 3.1b) contains cube-shaped cells and is found lining the kidney tubules. In **columnar epithelium** (fig. 3.1c), the cells resemble pillars or columns, and nuclei are usually located near the bottom of each

Figure 3.1   Major types of epithelial tissues. *a.* Photomicrograph of simple squamous epithelium. *b.* Simple cuboidal epithelium: (*left*) photomicrograph, (*right*) drawing. *c.* Simple columnar epithelium: (*left*) photomicrograph, (*right*) drawing. Arrow on left points to a goblet cell.

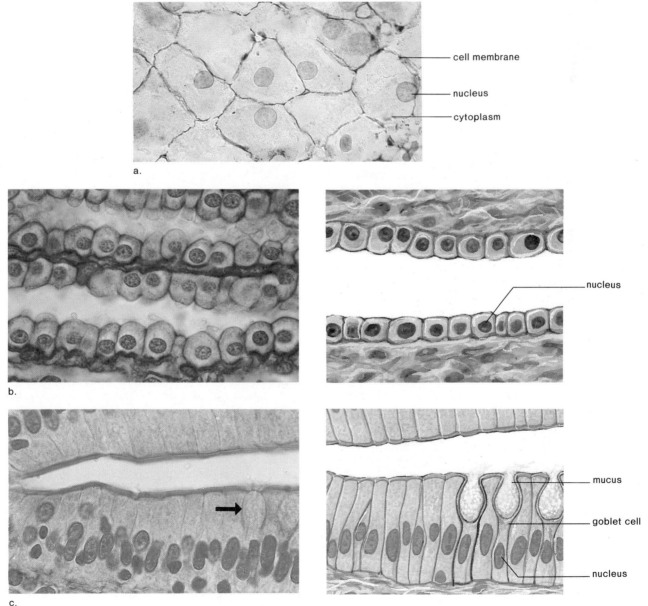

Figure 3.2 Scanning electron micrograph of ciliated columnar epithelium that lines the oviducts. The beating of the cilia propels the egg toward the uterus. It has also been suggested that the return motion of the cilia sets up currents that help move the sperm toward the egg. (Ci = cilia; Mv = microvilli).

Kessel, R. G., and Kardon, R. H.: *Tissues and Organs: A Text-Atlas of Scanning Electron Microscopy.* © 1979 by W. H. Freeman and Co.

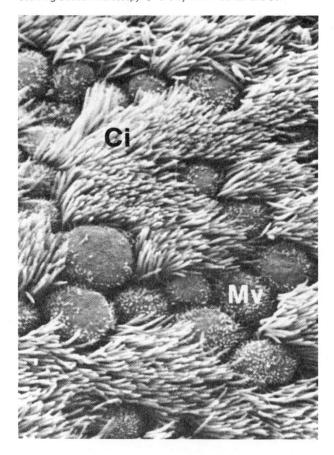

Figure 3.3 Pseudostratified ciliated columnar epithelium from the lining of the windpipe (trachea). When you cough, material trapped in the mucus secreted by goblet cells is moved upward to the throat, where it can be swallowed. *a.* Photomicrograph. *b.* Drawing.

a.

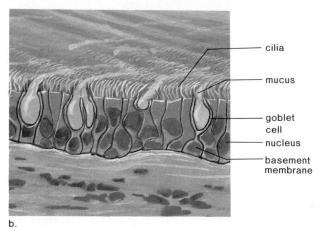

b.

— cilia
— mucus
— goblet cell
— nucleus
— basement membrane

cell. This epithelium is found lining the digestive tract. An epithelium may have microvilli or cilia as appropriate for its particular function. For example, the oviducts are lined by ciliated columnar cells whose beat propels the egg toward the uterus or womb (fig. 3.2).

An epithelium may also be simple or stratified. Simple means that the cells occur in a single layer. **Stratified** means to exist as layers piled one over the other. The nose, mouth, esophagus, anal canal, and vagina are all lined by stratified squamous epithelium. As we shall see, the outer layer of skin is also stratified squamous epithelium, but here the cells have been reinforced by keratin, a protein that strengthens cells. One type of epithelium is **pseudostratified.** It appears to be layered; however, true layers do not exist because each cell touches the base line. The lining of the windpipe, or trachea, is called *pseudostratified ciliated columnar epithelium* (fig. 3.3).

An epithelial sometimes secretes a product, in which case it is described as glandular. A gland can be a single epithelial cell, as in the case of the mucus-secreting goblet cells found within the columnar epithelium lining

the digestive tract (fig. 3.1c), or a gland can contain numerous cells. Glands that secrete their products into ducts are called **exocrine glands,** and those that secrete directly into the bloodstream are called **endocrine glands.**

Epithelial tissue is classified according to the shape of the cell. There can be a single layer or many layers of cells, and the layer lining a cavity can be ciliated and/or secretory (table 3.1).

Connective Tissue

Connective tissue (table 3.2) binds structures together, provides support and protection, fills spaces, and stores fat. As a rule, connective tissue cells are widely separated by a noncellular **matrix.** The matrix may have fibers of two types. White fibers contain collagen, a substance that gives them flexibility and strength. Yellow fibers contain elastin, a substance that is not as strong as collagen but is more elastic.

Table 3.1 Epithelial Tissue

Type	Function	Location
Simple squamous	Filtration, diffusion, osmosis	Walls of capillaries; linings of blood vessels
Simple cuboidal	Secretion, absorption	Surface of ovaries; linings of kidney tubules
Simple columnar	Protection, secretion, absorption	Linings of uterus; tubes of the digestive tract
Pseudostratified columnar	Protection, secretion, movement of mucus and sex cells	Linings of respiratory passages; various tubes of the reproductive systems
Stratified squamous	Protection	Outer layers of skin; oral cavity, esophagus, vagina, and anal canal

Table 3.2 Connective Tissue

Type	Function	Location
Loose connective	Binds organs together	Beneath the skin; beneath most epithelial layers
Adipose	Insulation; storage of fat	Beneath the skin; around the kidneys
Fibrous connective	Binds organs together	Tendons; ligaments
Hyaline cartilage	Support; protection	Ends of bones; nose; rings in walls of respiratory passages
Elastic cartilage	Support; protection	External ear; part of the larynx
Fibrocartilage	Support; protection	Between bony parts of backbone and knee
Bone	Support; protection	Bones of skeleton

Figure 3.4 Loose connective tissue has plenty of space between components. This type of tissue is found surrounding and between the organs.

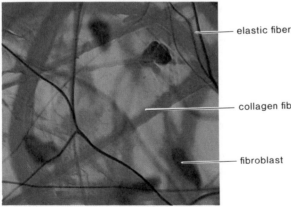

— elastic fiber

— collagen fiber

— fibroblast

Figure 3.5 Adipose tissue cells look like white "ghosts" because they are filled with fat.

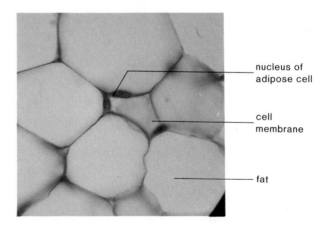

— nucleus of adipose cell

— cell membrane

— fat

Loose Connective Tissue

Loose connective tissue binds structures together (fig. 3.4). The cells of this tissue are mainly fibroblasts, large star-shaped cells that produce extracellular fibers. In loose connective tissue, the fibroblasts are located some distance from one another and are separated by a jellylike intercellular material that contains many white and yellow fibers. The white fibers occur in bundles and are strong and flexible. The yellow fibers form networks that are highly elastic—when stretched they return to their original length. Loose connective tissue commonly lies beneath an epithelium. In certain instances, the epithelium and its underlying connective tissue forms a body membrane (p. 49).

Adipose tissue (fig. 3.5) is a type of loose connective tissue in which the fibroblasts enlarge and store fat, and the intercellular matrix is reduced.

Figure 3.6 Hyaline cartilage cells, located in lacunae, are separated by a flexible matrix rich in protein and fibers. This type of cartilage forms the embryonic skeleton, which is later replaced by bone.

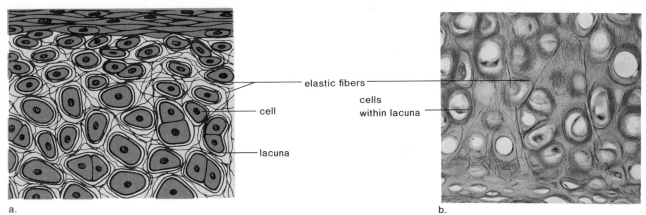

a.

b.

Fibrous Connective Tissue

Fibrous connective tissue contains large numbers of collagenous fibers that are closely packed together. This type of tissue has more specific functions than does loose connective tissue. For example, fibrous connective tissue is found in **tendons,** which connect muscles to bones, and **ligaments,** which connect bones to other bones at joints. Tendons and ligaments take a long time to heal following an injury because their blood supply is relatively poor.

Loose and fibrous connective tissues, which bind body parts together, differ according to the type and abundance of fibers in the matrix.

Cartilage

In **cartilage,** the cells lie in small chambers called **lacunae,** separated by a matrix that is solid yet flexible. Unfortunately, because this tissue lacks a direct blood supply, it heals very slowly. There are three types of cartilage according to the type of fiber in the matrix.

Hyaline cartilage (fig. 3.6), the most common type, contains only very fine collagenous fibers. The matrix has a milk glass appearance. This type of cartilage is found in the nose, at the ends of the long bones and ribs, and in the supporting rings of the windpipe. The fetal skeleton is also made of this type of cartilage. Later, the cartilaginous fetal skeleton is replaced by bone.

Elastic cartilage has more elastic fibers than hyaline cartilage. For this reason, it is more flexible and is found, for example, in the framework of the outer ear.

Fibrocartilage has a matrix containing strong collagenous fibers. It is found in structures that withstand tension and pressure, such as the pads between the vertebrae in the backbone and the wedges found in the knee joint.

Figure 3.7 Compact bone is highly organized. The cells are arranged in circles about a central (Haversian) canal that contains a nutrient-bearing blood vessel.

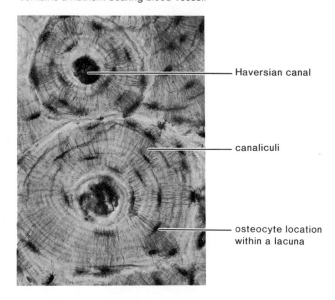

Bone

Bone (fig. 3.7) is the most rigid of the connective tissues. It consists of an extremely hard matrix of calcium salts deposited around protein fibers. The minerals give it rigidity, and the protein fibers provide elasticity and strength, much as steel rods do in reinforced concrete.

The outer portion of a long bone contains compact bone. In **compact bone,** bone cells (osteocytes) are located in lacunae that are arranged in concentric circles around tiny tubes called Haversian canals. There are nerve fibers and blood vessels in these canals. The latter bring the nutrients that allow bone to renew itself. The nutrients can reach all of the cells, because there are minute canals (canaliculi) containing thin processes of the osteocytes that connect them with one another and with the Haversian canals.

Figure 3.8 Blood is classified as connective tissue. It contains two components: plasma, the liquid portion of blood, and formed elements (red cells, white cells, and platelets).

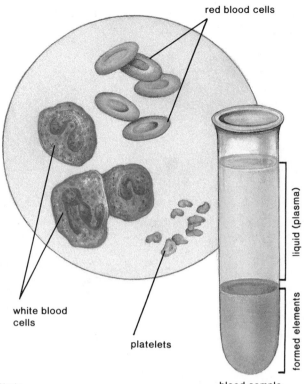

red blood cells

liquid (plasma)

formed elements

white blood cells

platelets

O'Keefe

blood sample

The ends of a long bone contain spongy bone, which has an entirely different structure. Spongy bone contains numerous bony bars and plates separated by irregular spaces. Although lighter than compact bone, **spongy bone** is still designed for strength. Just as braces are used for support in buildings, the solid portions of spongy bone follow lines of stress.

Cartilage and bone are support tissues. Cartilage is more flexible than bone because the matrix is rich in protein and not calcium salts like that of bone.

Blood

Blood (fig. 3.8) is a connective tissue in which the cells are separated by a liquid called **plasma.** Blood cells are of two types; **red** (erythrocytes), which carry oxygen, and **white** (leukocytes), which aid in fighting infection. Also present are **platelets,** which are important to the initiation of blood clotting. Platelets are not complete cells; rather, they are fragments of giant cells found in the bone marrow.

Blood is a connective tissue in which the matrix is plasma.

Muscular Tissue

Muscle fibers contain **actin** and **myosin** microfilaments, whose interaction accounts for the movements we associate with animals. There are three types of muscles: *skeletal, smooth,* and *cardiac* (table 3.3).

Skeletal muscle (fig. 3.9) is attached to the bones of the skeleton and functions to move body parts. Skeletal muscle fibers are cylindrical and run the length of a muscle. They are multinucleated, the nuclei appearing just inside the cell membrane. The fibers also have characteristic light and dark bands perpendicular to the length of the cell. These bands give the muscle a striated appearance. Skeletal muscle is under our voluntary control and has the fastest contraction of all the muscle types.

Smooth muscle (fig. 3.10) is so named because it lacks striations. The spindle-shaped cells that make up smooth muscle are not under voluntary control and are said to be *involuntary.* Smooth muscle, which is found in the viscera (intestine, stomach, and so on) and blood vessels, contracts more slowly than skeletal muscle, but can remain contracted for a longer time. The cells tend to form layers in which the thick middle portion of one cell is opposite the thin ends of adjacent cells. Consequently, the nuclei form an irregular pattern in the tissue.

Table 3.3 Muscular Tissues

Type	Fiber Appearance	Locations	Control
Skeletal	Striated	Attached to skeleton	Voluntary
Smooth	Spindle shaped	Internal organs	Involuntary
Cardiac	Striated and branched	Heart	Involuntary

Figure 3.9 Skeletal muscle is striated. *a.* Photomicrograph. *b.* Drawing.

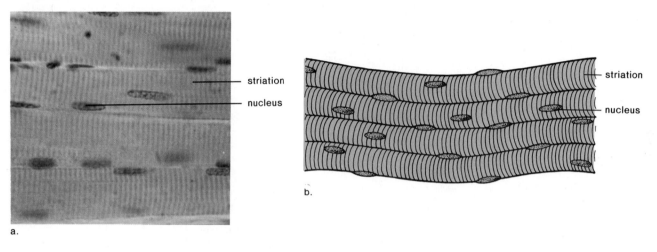

Figure 3.10 Smooth muscle is composed of spindle-shaped cells. There are no striations. *a.* Photomicrograph (arrow at nucleus). *b.* Drawing.

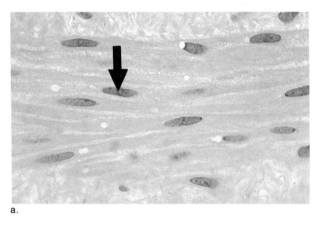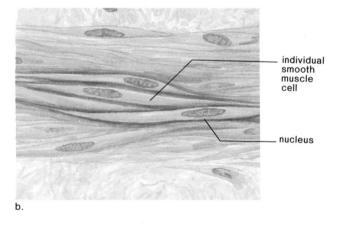

Cardiac muscle (fig. 3.11), which is found only in the heart, is responsible for the heartbeat. Cardiac muscle seems to combine features of both smooth and skeletal muscle. It has *striations* like those of skeletal muscle, but the contraction of the heart is involuntary for the most part. Cardiac muscle fibers also differ from skeletal muscle fibers in that they are branched and seemingly fused, one with the other, so that the heart appears to be composed of one large, interconnecting mass of muscle cells. Actually, however, cardiac muscle fibers are separate and individual, but they are bound end to end at *intercalated disks,* areas of folded cell membrane between the cells.

All muscle tissue contains actin and myosin microfilaments; these form a striated pattern in skeletal and cardiac muscle, but not in smooth muscle.

Nervous Tissue

The brain and spinal cord contain conducting cells called neurons. A **neuron** (fig. 3.12) is a specialized cell that has three parts: (1) *dendrites* conduct impulses (send a message) to the cell body; (2) the *cell body*

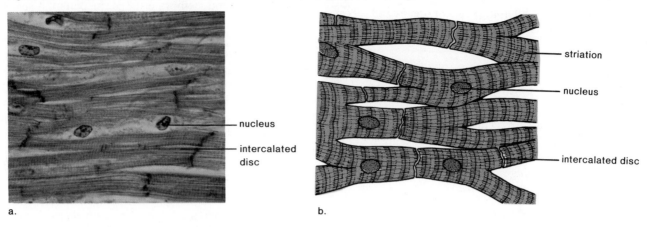

a.

b.

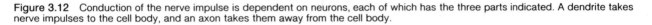

Figure 3.12 Conduction of the nerve impulse is dependent on neurons, each of which has the three parts indicated. A dendrite takes nerve impulses to the cell body, and an axon takes them away from the cell body.

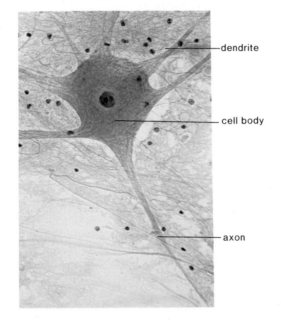

contains most of the cytoplasm and the nucleus of the neuron; and (3) the *axon* conducts impulses away from the cell body.

When axons and dendrites are long, they are called *nerve fibers.* Outside the brain and spinal cord, nerve fibers are bound together by connective tissue to form **nerves.** Nerves conduct impulses from sense organs to the spinal cord and brain, where the phenomenon called sensation occurs; they also conduct nerve impulses away from the spinal cord and brain to the muscles, causing them to contract.

In addition to neurons, nervous tissue contains **glial cells.** These cells maintain the tissue by giving support and protection to neurons. They also provide nutrients to neurons and help keep the tissue free of debris.

Nerve cells, called neurons, have processes (fibers) called axons and dendrites. Outside the brain and cord, these fibers are found in nerves.

Body Membranes and Organs

More than one type of tissue is found in membranes and organs. Membranes line the internal spaces of organs and tubes that open to the outside, and they also line the body cavities discussed in chapter 1. Organs contain several tissues, each performing a function consistent with the overall function of the organ.

Body Membranes

Membranes line the internal spaces of organs and tubes that open to the outside, and they also line the body cavities discussed in chapter 1.

Meninges

The **meninges** are membranes found within the dorsal cavity (p. 19). They are composed only of connective tissue, and they serve as a protective covering for the brain and spinal cord. *Meningitis* is a life-threatening infection of the meninges.

Synovial Membranes

Synovial membranes line freely movable joint cavities. They are composed of connective tissues. These membranes secrete synovial fluid into the joint cavity and this fluid lubricates the ends of the bones so that they can move freely. In *rheumatoid arthritis,* the synovial membrane becomes inflamed and grows thicker. Fibrous tissue then invades the joint and may eventually become bony so that the bones of the joint are no longer capable of moving.

Mucous Membranes

Mucous membranes line the interior of the organs and tubes of the digestive, respiratory, urinary, and reproductive systems. These membranes consist of an epithelium overlying a layer of connective tissue. The epithelium contains goblet cells that secrete mucus.

The mucus secreted by mucous membrane ordinarily protects the wall from invasion by bacteria and viruses; hence, more mucus is secreted when we have a cold and have to "blow our nose." In addition, mucus usually protects the wall of the stomach and small intestine from digestive juices, but this protection breaks down when a person develops an *ulcer.*

Serous Membranes

Serous membranes line the interior of the thoracic and abdominal cavities, and they also cover the organs within these cavities. These membranes consist of a layer of simple squamous epithelium overlying a layer of connective tissue. They secrete a watery fluid that keeps the membranes lubricated. These membranes support the internal organs, and tend to compartmentalize the large thoracic and abdominal cavities. This helps to hinder the spread of any infection.

In the thorax, the **pleural membranes** line the chest cavity and then double back to cover the lungs. The potential space between the pleural membranes is called the pleural cavity. A well-known infection of these membranes is called *pleurisy.* The heart is covered by a double layer of serous membrane called the **pericardium** (pericardial membranes) and the potential space between them is called the pericardial cavity.

In the abdomen, the interior wall and organs are lined by **peritoneum** (peritoneal membranes). Between the internal organs, the peritoneum sometimes comes together to form a double layer called **mesentery** (fig. 3.13). The omentums are double-layered peritoneum that cover certain abdominal organs. The greater omentum covers the intestines.

Peritonitis is a much dreaded infection of the peritoneum. Such an infection is quite likely if an inflamed appendix should burst before it is removed.

Meninges are membranes that cover the brain and spinal cord; synovial membranes line certain joint cavities; mucous membranes line organs and tubes that exit to the outside; and serous membranes line the thoracic and abdominal cavities, and cover the organs within these cavities.

Organs

While it may seem that some organs in the body are composed largely of one type of cell, most are a composite of many different types of tissues. For example, the wall of the small intestine (fig. 3.14) is composed of these layers:

Mucosa This tissue is made up of columnar epithelium that lines the central cavity and contains glandular epithelial cells that secrete mucus. This layer is arranged in deep folds, called **villi,** to increase the absorptive surface of the intestine. Within the crevices of the villi are **digestive glands** that secrete digestive juices.

Submucosa The digestive glands extend down into this layer, which is a broad band of loose connective tissue containing blood vessels and nerves.

Muscularis Two layers of smooth muscle make up this section. The inner is a circular layer of cells that go around the gut; the outer is a longitudinal layer that lies in the same direction as the gut.

Serosa The serosa is a very thin outermost layer of squamous epithelium.

Organs usually contain several different types of tissues; the wall of the small intestine contains columnar epithelium (mucosa); loose connective tissue (submucosa); smooth muscle (muscularis); and squamous epithelium (serosa).

Figure 3.13 This diagram illustrates how the peritoneum lines the abdominal cavity and its organs. Mesentery occurs wherever there is a double layer of peritoneum.

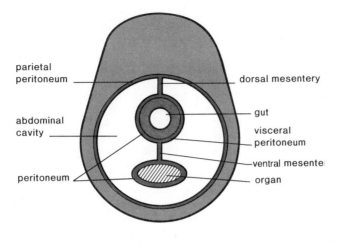

parietal peritoneum

dorsal mesentery

abdominal cavity

gut

visceral peritoneum

ventral mesente

peritoneum

organ

Figure 3.14 The intestinal wall. *a.* Major layers of intestinal wall. For an explanation of mesentery, see page 50. *b.* Enlargement of a section of the wall, showing how the mucosa forms deep folds called villi. These absorb nutrient molecules resulting from food digestion.

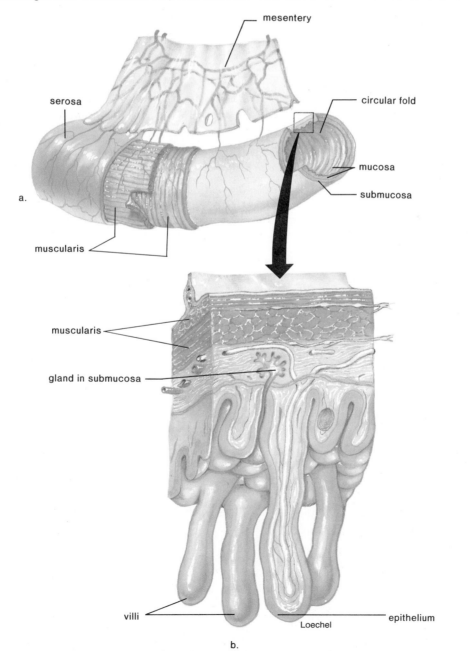

mesentery

serosa

circular fold

mucosa

submucosa

a.

muscularis

muscularis

gland in submucosa

villi

epithelium

Loechel

b.

Summary

I. Body Tissues. Body tissues are categorized into the four types noted.
   A. Epithelial tissue is classified according to the shape of the cell, which may be squamous, cuboidal, or columnar. The cells may be stratified and/or ciliated, and the tissue may be secretory.
   B. Connective tissue. Loose and fibrous connective tissues, which bind body parts together, differ according to the type and abundance of fibers in the matrix.
      Cartilage and bone are support tissues. Cartilage is more flexible than bone because the matrix is rich in protein and not calcium salts, like that of bone.
      Blood is a connective tissue in which the matrix is plasma.
   C. All muscle tissue contains actin and myosin microfilaments; these form a striated pattern in skeletal and cardiac muscle, but not in smooth muscle.
   D. Nerve cells, called neurons, have processes (fibers) called axons and dendrites. Outside the brain and cord, these fibers are found in nerves.
II. Body Membranes and Organs
   A. Body Membranes. Meninges are membranes that line the brain and spinal cord; synovial membranes line certain joint cavities; mucous membranes line organs and tubes that exit to the outside; and serous membranes line the thoracic and abdominal cavities, and the organs within these cavities.
   B. Organs. Organs usually contain several different types of tissues: the wall of the small intestine contains columnar epithelium (mucosa); loose connective tissue (submucosa); smooth muscle (muscularis); and squamous epithelium (serosa).

Study Questions

1. What is a tissue?
2. Name the four major groups of tissues.
3. What are the functions of epithelial tissue? Name the different kinds, and give a location for each.
4. What are the functions of connective tissue? Name the different kinds, and give a location for each.
5. Contrast the structure of cartilage with that of bone using the words lacunae and Haversian canal in your description.
6. Describe the composition of blood and give a function for each type of cell.
7. What are the functions of muscular tissue? Name the different kinds, and give a location for each.
8. Nervous tissue contains what types of cells? Which organs in the body are made up of nervous tissue?
9. Name the different types of body membranes and associate each type with a particular location in the body.
10. Name the different types of serous membranes, and associate each type with a particular organ or organs.

Objective Questions

I. Fill in the blanks.
   1. Most organs contain several different types of _____ .
   2. Mucous membrane contains _____ tissue overlying _____ tissue.
   3. Pseudostratified ciliated columnar epithelium contains cells that appear to be _____ , have projections called _____ , and are _____ in shape.
   4. Both cartilage and blood are classified as _____ tissue.
   5. Connective tissue cells are widely separated by a _____ that contains _____ .

II. For questions 6–9, match the organs in the key to the epithelial tissues below.
   Key: a. kidney tubules
        b. small intestine
        c. inside mouth
        d. trachea (windpipe)
   6. squamous
   7. cuboidal
   8. columnar
   9. pseudostratified ciliated columnar

III. For questions 10–12, match the muscle tissues in the key to the descriptions below.
   Key: a. skeletal muscle
        b. visceral muscle
        c. cardiac muscle
   10. striated and branched, involuntary
   11. striated and voluntary
   12. smooth and involuntary

Medical Terminology Reinforcement Exercise

Pronounce, dissect, and analyze the following terms:

1. epithelioma (ep″ĭ the″le-o′mah)
2. fibrodysplasia (fi″bro-dis-pla′se-ah)
3. meningoencephalopathy (mĕ-ning″go-en-sef″al-lop′ah-the)
4. mesenteric (mes′en-ter-ik)
5. pericardiocentesis (per″i-kar″de-o-sen-te′sis)
6. peritonitis (per″i-to-ni′tis)
7. intrapleural (in″tra-ploor′al)
8. neurofibromatosis (nu″ro-fi″bro″mah-to′sis)
9. submucosa (sub″mu-ko′sah)
10. polyarthritis (pol″e-ar-thri′tis)

Reference Figures

The Human Organism

Figure 1 Anterior view of the human torso with the superficial muscles exposed.

Figure 2 The torso with the deep muscles exposed.

Figure 3 The torso with the anterior abdominal wall removed to expose the abdominal viscera.

Figure 4 The torso with the anterior thoracic wall removed to expose the thoracic viscera.

Figure 5 The torso as viewed with the thoracic viscera sectioned in a coronal plane, and the abdominal viscera as viewed with most of the small intestine removed.

Figure 6 The torso as viewed with the heart, liver, stomach, and portions of the small and large intestines removed.

Figure 7 The torso with the anterior thoracic and abdominal walls removed, along with the viscera, to expose the posterior walls and body cavities.

Figure 1 A human torso, with a view of the anterior surface on one side and the superficial muscles exposed on the other side (*m.* stands for *muscle,* and *v.* stands for *vein*).

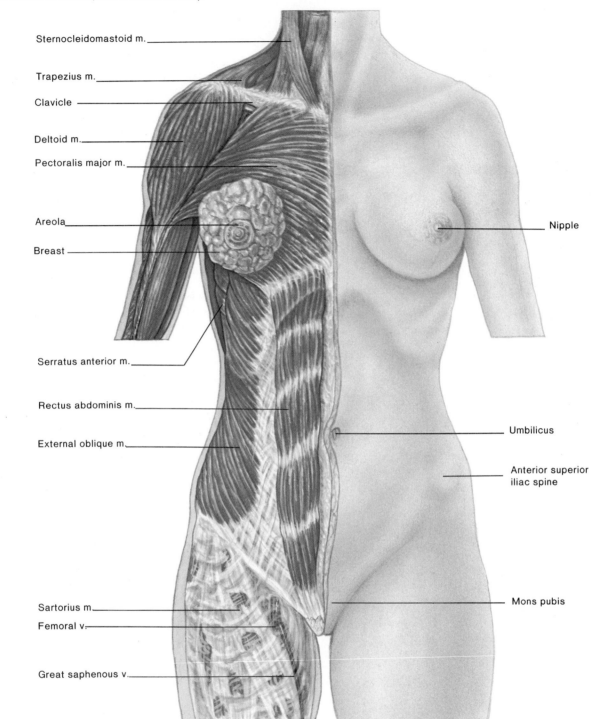

Sternocleidomastoid m.

Trapezius m.

Clavicle

Deltoid m.

Pectoralis major m.

Areola

Breast

Serratus anterior m.

Rectus abdominis m.

External oblique m.

Sartorius m.

Femoral v.

Great saphenous v.

Nipple

Umbilicus

Anterior superior
iliac spine

Mons pubis

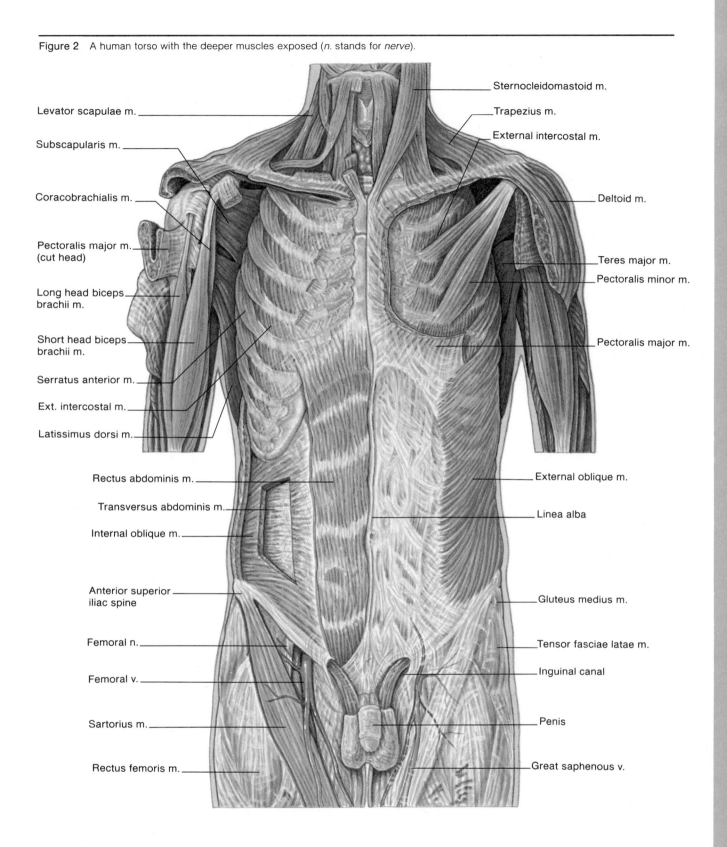

Figure 2   A human torso with the deeper muscles exposed (*n.* stands for *nerve*).

Levator scapulae m.

Subscapularis m.

Coracobrachialis m.

Pectoralis major m. (cut head)

Long head biceps brachii m.

Short head biceps brachii m.

Serratus anterior m.

Ext. intercostal m.

Latissimus dorsi m.

Rectus abdominis m.

Transversus abdominis m.

Internal oblique m.

Anterior superior iliac spine

Femoral n.

Femoral v.

Sartorius m.

Rectus femoris m.

Sternocleidomastoid m.

Trapezius m.

External intercostal m.

Deltoid m.

Teres major m.

Pectoralis minor m.

Pectoralis major m.

External oblique m.

Linea alba

Gluteus medius m.

Tensor fasciae latae m.

Inguinal canal

Penis

Great saphenous v.

Figure 3 A human torso with the deep muscles removed and the abdominal viscera exposed (*a.* stands for *artery*).

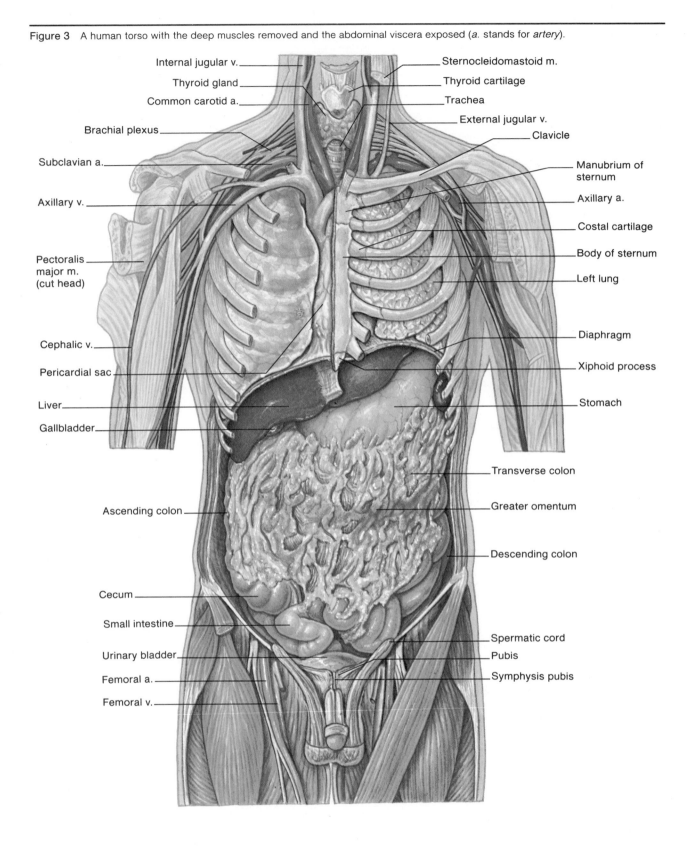

Internal jugular v.

Thyroid gland

Common carotid a.

Brachial plexus

Subclavian a.

Axillary v.

Pectoralis major m. (cut head)

Cephalic v.

Pericardial sac

Liver

Gallbladder

Ascending colon

Cecum

Small intestine

Urinary bladder

Femoral a.

Femoral v.

Sternocleidomastoid m.

Thyroid cartilage

Trachea

External jugular v.

Clavicle

Manubrium of sternum

Axillary a.

Costal cartilage

Body of sternum

Left lung

Diaphragm

Xiphoid process

Stomach

Transverse colon

Greater omentum

Descending colon

Spermatic cord

Pubis

Symphysis pubis

Figure 4   A human torso with the thoracic viscera exposed.

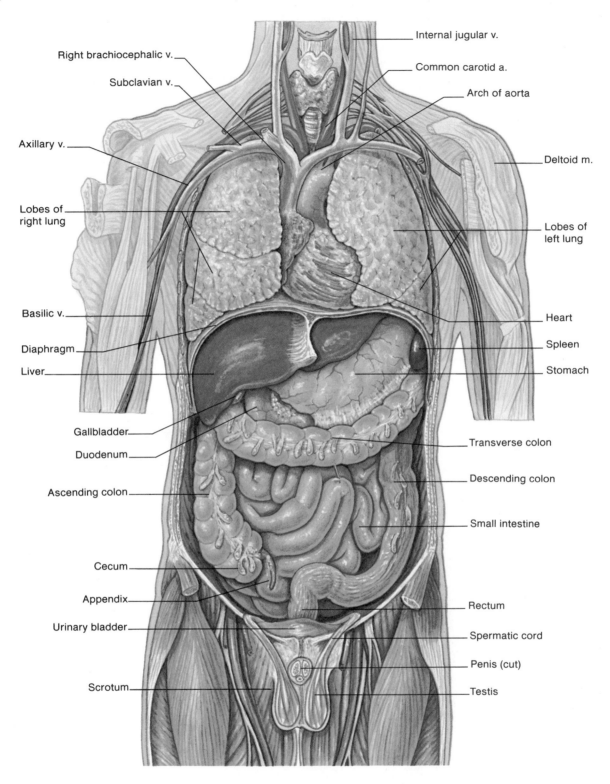

Figure 5 A human torso with the lungs, heart, and small intestine sectioned.

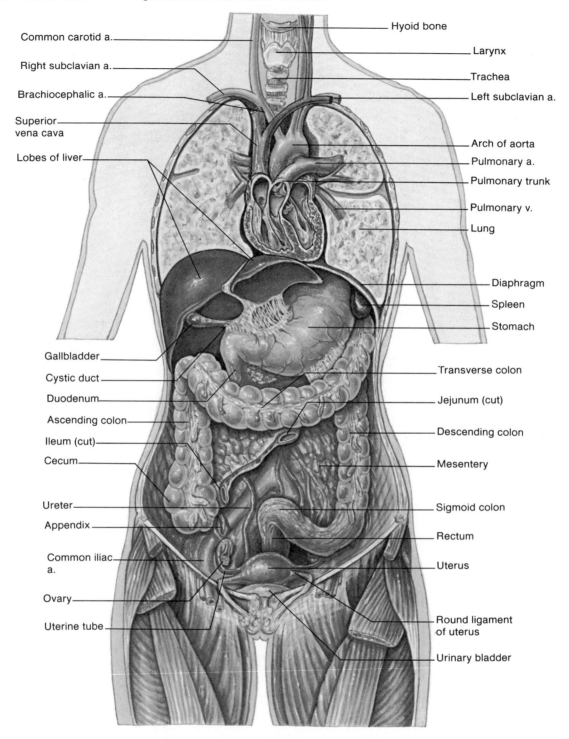

Common carotid a.

Right subclavian a.

Brachiocephalic a.

Superior vena cava

Lobes of liver

Gallbladder

Cystic duct

Duodenum

Ascending colon

Ileum (cut)

Cecum

Ureter

Appendix

Common iliac a.

Ovary

Uterine tube

Hyoid bone

Larynx

Trachea

Left subclavian a.

Arch of aorta

Pulmonary a.

Pulmonary trunk

Pulmonary v.

Lung

Diaphragm

Spleen

Stomach

Transverse colon

Jejunum (cut)

Descending colon

Mesentery

Sigmoid colon

Rectum

Uterus

Round ligament of uterus

Urinary bladder

Figure 6 A human torso with the heart, stomach, liver, and parts of the intestines and lungs removed.

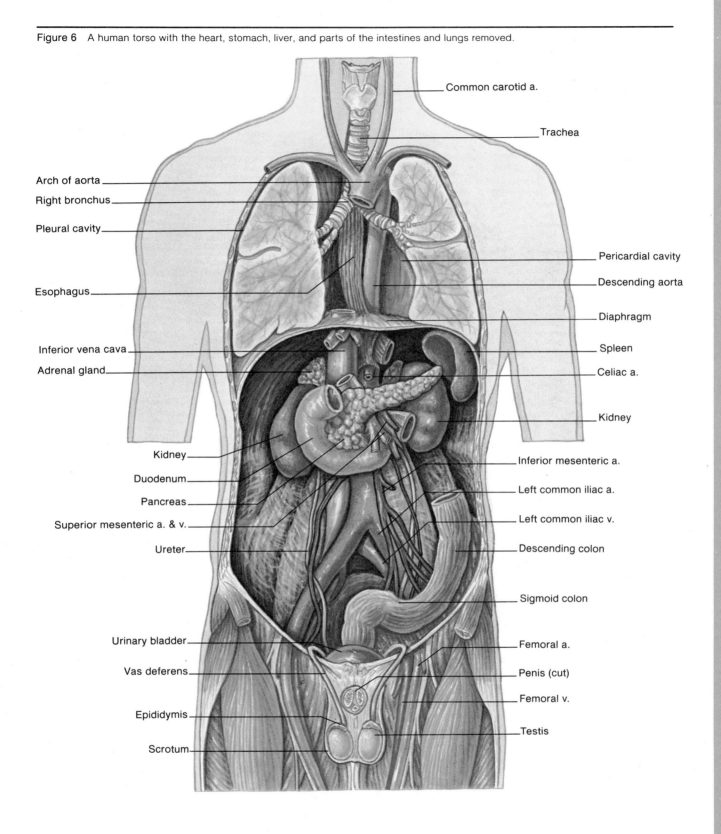

Common carotid a.

Trachea

Arch of aorta

Right bronchus

Pleural cavity

Esophagus

Pericardial cavity

Descending aorta

Diaphragm

Inferior vena cava

Adrenal gland

Spleen

Celiac a.

Kidney

Kidney

Duodenum

Pancreas

Superior mesenteric a. & v.

Ureter

Inferior mesenteric a.

Left common iliac a.

Left common iliac v.

Descending colon

Sigmoid colon

Urinary bladder

Vas deferens

Femoral a.

Penis (cut)

Femoral v.

Epididymis

Scrotum

Testis

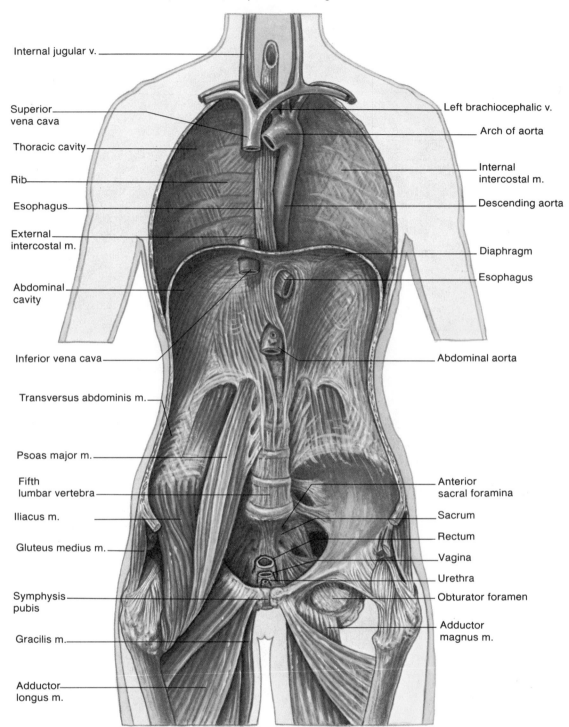

Internal jugular v.

Superior vena cava

Thoracic cavity

Rib

Esophagus

External intercostal m.

Abdominal cavity

Inferior vena cava

Transversus abdominis m.

Psoas major m.

Fifth lumbar vertebra

Iliacus m.

Gluteus medius m.

Symphysis pubis

Gracilis m.

Adductor longus m.

Left brachiocephalic v.

Arch of aorta

Internal intercostal m.

Descending aorta

Diaphragm

Esophagus

Abdominal aorta

Anterior sacral foramina

Sacrum

Rectum

Vagina

Urethra

Obturator foramen

Adductor magnus m.

4

Skin

Chapter Outline

Structure of the Skin
    Epidermis
    Dermis
    Subcutaneous Layer
Accessory Structures of the Skin
    Hair and Nails
    Glands
Functions of the Skin
    Protection
    Sensory Reception and
      Communication
    Synthesis of Vitamin D
    Regulation of Body Temperature
Disorders
    Cancer of the Skin
    Wound Healing
    Burns

Learning Objectives

After you have studied this chapter, you should be able to:

1. Name the three layers of skin and describe their structure.
2. Describe the structure of hair and nails, and their growth.
3. Name three glands of the skin, and describe their structure and function.
4. List four functions of the skin and discuss each one.
5. Name the three types of skin cancer and state their cause.
6. Describe the steps by which a skin wound heals.
7. Name and describe the three types of burns.

Structure of the Skin

The skin covers the entire exterior of the human body and has a surface area of about 1.8 square meters (20.83 square feet). Usually the skin is only loosely attached to underlying muscle tissue, but where there are no muscles, the skin attaches directly to bone. For example, there are *flexion creases* where the skin attaches directly to the joints of the fingers.

The skin is sometimes called the **integument** and since the skin has several accessory organs, it is also possible to speak of the **integumentary system.**

The skin (fig. 4.1) is usually considered to have three layers: the epidermis, the dermis, and the subcutaneous layer. The epidermis and dermis together, however, form the so-called **cutaneous membrane.** This explains how the term subcutaneous arose.

Epidermis

The **epidermis** is the outer and thinner layer of the skin. It is made up of *stratified squamous epithelium,* which are continually produced by a bottom layer of cells that lies next to the dermis. This is appropriately termed a *germinal layer* and the cells are called *basal cells* because they lie right next to the dermis. Here, constant cell division produces many new cells that rise to the surface of the epidermis in about six to eight weeks. As the cells push away from the dermis, they get further and further away from the blood vessels in the dermis. This means that they are not being supplied with nutrients and oxygen, and eventually they die and are sloughed off.

Specialized cells in the epidermis called **melanocytes** produce **melanin,** the pigment responsible for skin color in darker skinned persons. The pigment carotene is also present in epidermal cells, but this color is only apparent when melanocytes lack melanin. When one sunbathes, the melanocytes become more active, producing melanin in an attempt to protect the skin from the damaging effects of ultraviolet (UV) radiation in sunlight.

Cells gradually become flattened and hardened as they are pushed toward the surface of the epidermis. Hardening is caused by keratinization, cellular production of a fibrous, waterproof protein called **keratin.** Over much of the body, keratinization is minimal, but the palm of the hand and the sole of the foot normally have a particularly thick outer layer of dead, keratinized cells arranged in spiral and concentric patterns. These patterns are due to the configuration of the upper dermal layer. They are unique to each person and are known as **fingerprints** and footprints (fig. 4.2). Continual pressure can cause the epidermis to thicken irregularly into calluses and corns.

The epidermis, the outer layer of skin, is a stratified squamous epithelium. New cells continually produced in the innermost layer of the epidermis push outward and become dead keratinized cells that are sloughed off.

Dermis

The **dermis** is a layer of dense connective tissue that is deeper and thicker than the epidermis. It contains elastic and collagen fibers. The *collagen* fibers form bundles that run parallel to each other and the skin surface. As a person ages and is exposed to the sun, the number of fibers become less and those remaining have characteristics that make the skin less supple and cause wrinkling to occur. Some of the newer methods of attempting to do away with wrinkles try to restore the amount of collagen in the skin.

The dermis contains blood vessels that nourish the skin. Blood rushes into these vessels when a person "blushes" and is reduced in them when a person turns "blue." Sometimes blood flow to a particular area is restricted in bedridden patients and consequently they develop **bedsores** (decubitis ulcers) (fig. 4.3). These can be prevented by changing the position of the patient frequently and massaging the skin to stimulate blood flow.

There are also numerous nerve fibers in the dermis that go to and from accessory structures to be mentioned shortly.

The dermis is composed of dense connective tissue and lies beneath the epidermis. It contains blood vessels and nerve fibers.

Subcutaneous Layer

The **subcutaneous layer,** which lies below the dermis, is composed of loose connective tissue including adipose tissue. Adipose tissue helps insulate the body from either gaining heat from the outside or losing heat from the inside. A well-developed subcutaneous layer gives a rounded appearance to the body. Excessive development accompanies obesity.

The subcutaneous layer of skin is made up of loose connective and adipose tissue that insulates the body.

Accessory Structures of the Skin

These structures are of epidermal origin even though some are largely found in the dermis.

Figure 4.1 Skin anatomy. Skin is composed of two layers: the epidermis and the dermis. The subcutaneous layer is also considered to be a part of skin. Most cells in the epidermal layer are no longer living. Skin cancer brought on by UV radiation from the sun starts in the lower epidermal cells because they are mitotically active. The dermis contains the various structures depicted.

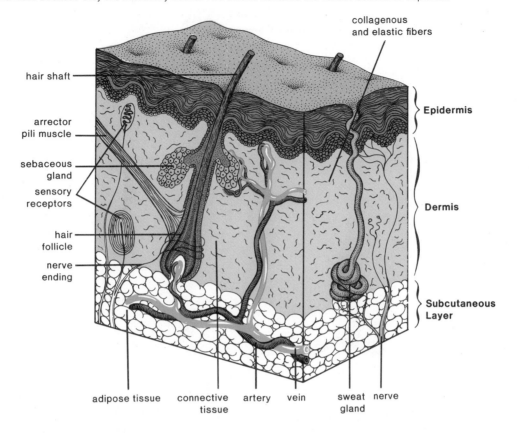

Figure 4.2 Basic dermatoglyphic patterns in the digits: *a.* arch, *b.* whorl, *c.* loop, and *d.* combination.

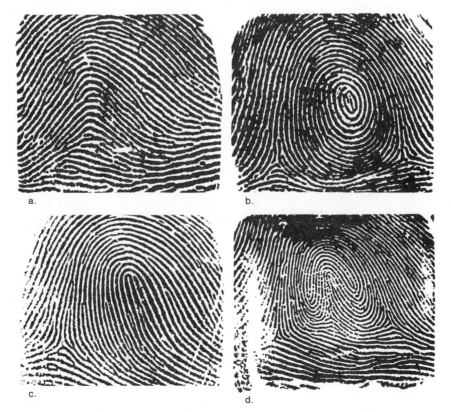

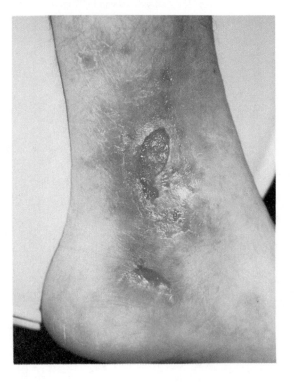

Hair and Nails

Hair is found on all body parts except the palms, soles, lips, nipples, and portions of the external reproductive organs. Most of this hair is fine and downy, but the hair on the head is usually obvious. After puberty, there is noticeable hair in the axillary and pelvic regions of both sexes. In the male, a beard develops and other parts of the body may also become quite hairy.

During development, a bud of epidermal cells pushes down into the dermis and comes to lie at the base of each **hair follicle** (fig. 4.4). These cells continually divide, producing the cells that make up a hair. At first, the cells are nourished by dermal blood vessels, but as the hair grows up and out of the follicle, they get further away from this source of nutrients, become keratinized and die. The part of a hair within the follicle is called the root, and the part that extends beyond the skin is the shaft. Electrolysis kills the root of a hair so that it does not regrow. The life of any particular hair is usually three to five months, and then it regrows. In males, baldness occurs when the hair on the head fails to regrow.

Each hair has one or more sebaceous glands whose ducts empty into the neck of the follicle. A smooth muscle, the **arrector pili,** attaches to the follicle in such a way that contraction of the muscle causes the hair to "stand on end." When one has had a scare or is cold, goose bumps develop due to contraction of these muscles.

Nails grow from special epithelial cells in the region of the half-moon shaped base. These cells become keratinized as they grow out over the nail bed. The pink color of nails is due to the vascularized dermal tissue beneath the nail.

Ordinarily, nails grow only about one millimeter a week.

Both hair and nails are produced by the division of epidermal cells and consist of keratinized cells.

Glands

Sebaceous glands and sweat glands are normally found in skin, but the mammary glands are quite localized.

Sebaceous Glands

Most **sebaceous glands** grow out from a hair follicle. These glands secrete an oily substance called **sebum** and this, along with dead cells, flows into the follicle and then out onto the skin surface. This secretion lubricates the hair and skin, and helps waterproof them.

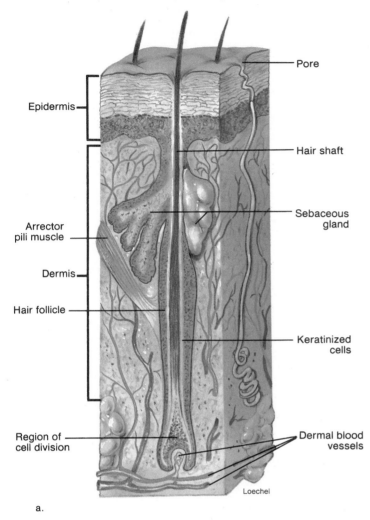

Epidermis

Pore

Hair shaft

Arrector pili muscle

Sebaceous gland

Dermis

Hair follicle

Keratinized cells

Region of cell division

Dermal blood vessels

Loechel

a.

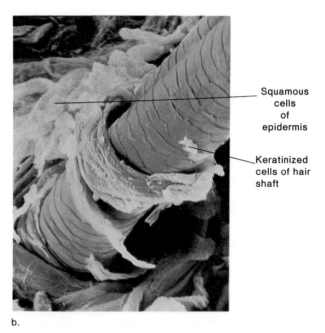

Squamous cells of epidermis

Keratinized cells of hair shaft

b.

Particularly on the nose and cheeks, the sebaceous glands may fail to discharge and the secretions collect forming "whiteheads or blackheads." The color of blackheads is due to the presence of oxidized sebum. If pus-inducing bacteria are also present, a boil or pimple may result.

Sebaceous glands are associated with a hair follicle and produce a secretion that lubricates the hair and skin.

Sweat Glands

Sweat glands are quite numerous and present in all regions of the skin. There can be as many as 90 per cubic centimeter on the leg and 400 per cubic centimeter on the palms and soles, and an even greater number on the fingertips. A sweat gland begins as a coiled tubule that straightens out near its opening.

Some sweat glands (*apocrine glands,* fig. 4.5) open into hair follicles in the anal region, groin, and armpits. The development of these glands begins at puberty and some believe that their secretion acts as a sex attractant. Also, these glands become active when one is under stress.

Other sweat glands (*eccrine glands*) open onto the surface of the skin. They become active when we are hot and help regulate body temperature. The sweat (perspiration) produced by these glands is mostly water, but it does contain salts and some urea, the substance excreted by the kidneys. Therefore, sweat is a form of excretion.

Only in the axillary and pelvic regions are the sweat glands associated with hair follicles. Sweat is mostly water, but also contains salts and urea, and, therefore, is a form of excretion. Sweating helps the body regulate body temperature.

Figure 4.5 Types of skin glands.

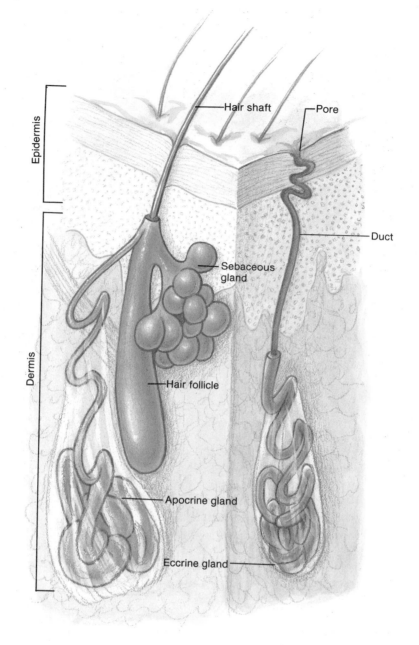

Mammary Glands

The mammary glands are modified sweat glands located within the breasts. A female breast contains fifteen to twenty-five lobules (fig. 16.10). Each lobule has its own milk duct, which begins at the nipple and divides into numerous other ducts until they finally end in blind sacs called alveoli. Cells within the alveoli produce the milk only after the birth of a child.

Mammary glands are modified sweat glands that produce milk after the birth of a child.

Functions of the Skin

Protection

The skin forms a protective covering over the entire body. It protects underlying parts from physical traumas, and the outer dead cells help prevent microbial invasions. The oily secretions from sebaceous glands are acidic and this also retards the growth of pathogenic organisms. The skin is also waterproof. If the cells were alive, they would continuously lose water to the air. Since they are dead and keratinized, fluid (water)

Figure 4.6 An X ray of rickets in a ten-month-old child. Rickets develops from improper diets and also from lack of UV light in the sunlight necessary to synthesize vitamin D.

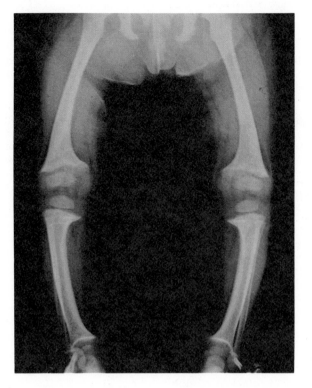

loss is prevented. On the other hand, if we should be immersed in water, the skin prevents the water from entering the body.

Skin protects the body from physical trauma, bacterial invasion, and fluid loss.

Sensory Reception and Communication

Small sense organs are present in the dermis (fig. 4.1). There are different ones for touch, pressure, pain, hot, and cold. The fingertips contain the most touch receptors and these add to our ability to use our fingers for delicate tasks. The receptors also account for the use of the skin as a means of communication between people.

The skin contains sense receptors for touch, pressure, pain, hot, and cold. These help us to be aware of our surroundings. The skin also serves as a means of communication between people.

Synthesis of Vitamin D

When skin cells are exposed to sunlight, the ultraviolet (UV) rays assist them in producing vitamin D. The cells contain a precursor molecule that is converted to vi-

tamin D in the body after it is exposed to UV light. It only takes a small amount of UV radiation to change the precursor molecule to vitamin D, so this shouldn't be used as an excuse to expose the skin unnecessarily. Vitamin D leaves the skin and is converted to a hormone in the liver and kidneys. This hormone circulates about the body where it functions in regulating calcium and phosphorus metabolism. Calcium and phosphorus are very important to the proper development and maintenance of the bones. Most milk today is fortified with vitamin D, and this helps prevent the occurrence of rickets (fig. 4.6).

The skin contains a precursor molecule that is converted to vitamin D following exposure to UV radiation. A hormone derived from vitamin D helps regulate calcium and phosphorus metabolism involved in bone development.

Regulation of Body Temperature

Previously, we emphasized that the energy content of nutrient molecules like glucose is converted to ATP energy within mitochondria (p. 33). When this conversion occurs, heat is released. Only about 40 percent of the energy available in a glucose molecule becomes ATP energy; the rest escapes as heat. Also, when ATP is

Figure 4.7 Temperature control. When the body temperature rises, the blood vessels dilate and the sweat glands become active. When the body temperature lowers, the blood vessels constrict and shivering may occur. In between these extremes the receptor is not stimulated and, therefore, body temperature fluctuates above and below normal.

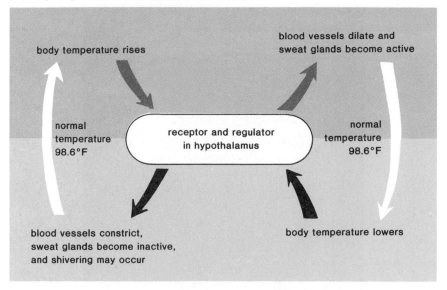

broken down, as occurs when muscles contract; heat is released. In other words, the heat that warms the body comes from ATP buildup and breakdown.

One of the best examples of homeostasis is regulation of normal body temperature (36.2—37.7° C [97—100° F]). The hypothalamus is the portion of the brain concerned with homeostasis and it is involved in regulating body temperature. As you can see from figure 4.7, the skin, too, plays a prominent role in regulating body temperature. The skin has two main avenues by which it assists in regulating body temperatures: (1) the blood vessels in the dermis can constrict to receive less blood or become dilated to receive more blood and (2) the sweat glands can remain inactive or excrete sweat as needed.

If body temperature starts to rise, the blood vessels will dilate so that more blood is brought to the surface of the skin for cooling and the sweat glands will be active. The evaporation of sweat uses up body heat and this helps cool the body. If the weather is humid, evaporation is hindered, but cooling can, then, be assisted by a cool breeze.

If the outer temperature is cool, the sweat glands remain inactive and the blood vessels constrict so that less blood is brought to the surface. Whenever the internal temperature continues to fall below normal, the muscles start to contract, causing shivering, which produces heat. The arrector pili muscles attached to hair follicles are also involved in this reaction, and this is why goose bumps occur when we are cold. If the out-

side temperature is extremely cold and blood flow is severely restricted for an extended period of time, a portion of the skin will die. This is called frostbite.

The skin is involved in temperature regulation. Surface blood vessels dilate and the sweat glands excrete when we are too hot. Surface blood vessels constrict and the sweat glands are inactive when we are cold.

Hyperthermia

Hyperthermia, a body temperature above normal, indicates that the body's regulatory mechanisms have been overcome. If *heat exhaustion* sets in, the individual becomes tired, complains of a headache, and may experience vomiting. Blood pressure may be low and there may have been a loss of salts due to profuse sweating. Bed rest, and increased fluid and salt intake are helpful. *Heat stroke* is recognized by an elevated temperature up to 43° C (110° F). Dizziness, confusion, and delusions may occur. It is important to cool the body off immediately by immersing the person in cool water. Medical care is needed to restore the body's proper fluid and salt balance.

Fever is a special case of hyperthermia. Fever can be brought on by an illness such as an infection. Bacteria release toxic substances. Some of these substances are called *pyrogens* because they cause fever. During the first stages of an infection, the body's thermostat in the hypothalamus is reset at a higher level.

Now the person feels chilly even though the body's temperature is increasing. Once the temperature has risen to the new setting, the body will continue to maintain this temperature (i.e., the fever will continue) until the infection has been brought to an end. At that time, the fever will break as the skin becomes flushed and sweating occurs. The patient is said to have passed the *crisis.*

Hypothermia

Hypothermia, a body temperature below normal, also indicates that the body's regulatory mechanisms have been overcome. It's possible to recognize hypothermia by uncontrollable shivering, incoherent speech, and a lack of coordination. With continued hypothermia, body functions slow down, the person feels sleepy, and death will occur when metabolism stops completely. To correct hypothermia, it is necessary to warm the person immediately. (Placing the person in a warm sleeping bag with a naked individual is a recommended procedure for campers!)

Disorders

The skin is subject to many disorders. Some of these are more annoying than life threatening. For example, *athlete's foot* is caused by a fungus infection usually involving the skin of the toes and soles. *Impetigo* is an infection by bacteria that results in pustules that crust over. *Eczema* and *psoriasis* are due not to an infection, but to overactive cell division, resulting in areas of scaling and itching. When one has *dandruff,* the rate of keratinization is two or three times the normal rate in certain areas of the scalp.

There are three conditions that we are going to discuss in more detail. These are cancer of the skin, wound healing, and burns.

Cancer of the Skin

Raised growths on the skin such as moles and warts are not cancerous. Moles are due to an overgrowth of melanocytes, and warts are due to a viral infection.

The most dangerous of the skin cancers is **malignant melanoma,** a death-causing cancer, which appears as darkly pigmented spots that resemble nonmalignant moles. Though melanoma tends to occur on such sun-exposed areas as the chest of men and the legs of women, its relationship to the sun is unclear.

The two more common types of skin cancer, **basal-cell carcinoma** and **squamous-cell carcinoma,** are definitely related to exposure to sunlight—UV radiation from the sun causes dividing skin cells to become cancerous. During the development of squamous-cell carcinoma, precancerous dark patches known as actinic keratosis precede the rough scaly patches of skin cancer. In basal-cell carcinoma, cancerous germinal cells no longer produce cells that rise to the surface and become keratinized. Instead, these cells invade the dermis and ulcers develop. Both of these types of skin cancer can usually be removed surgically.

In recent years, a great increase of skin cancers due to sunbathing has prompted physicians to strongly recommend that everyone stay out of the sun—or at least, use sunscreen lotions to protect the skin.

Skin cancer has been associated with ultraviolet radiation and consists of three types. Malignant melanoma is the most dangerous. Basal-cell carcinoma and squamous-cell carcinoma can usually be removed surgically.

Wound Healing

Wound healing allows us to observe the regenerative powers of epidermal tissue. The trauma may be extensive enough to open the blood vessels in the dermis. If so, the blood clot that forms is gradually converted to a scab. In the meantime, the germinal layer of the epidermis begins to produce new cells at a faster rate than usual. Eventually, the wound fills in (fig. 4.8). If the wound is deep, there will be scarring. A scar is tissue in which there are many collagen fibers arranged in a way to provide maximum strength. A scar does not contain the accessory organs of the skin.

The skin has regenerative powers and can grow back on its own accord if a wound is not too extensive.

Burns

The epidermal injury known as a burn is usually caused by heat, but can also be caused by radioactive, chemical, or electrical agents. Two factors affect the severity of a burn: the depth of the burn and the extent of the burned area. One way to classify burns is according to the depth of the burned area. In **first-degree burns,** only the epidermis is affected. There is redness and pain, but there are no blisters nor swelling. A classic example of a first-degree burn is moderate sunburn. The pain subsides within forty-eight to seventy-two hours, and the injury heals without further complications or scarring. The damaged skin peels off in about a week.

A **second-degree burn** extends through the entire epidermis and part of the dermis. There is not only redness and pain, but blisters develop in the region of the

Figure 4.8 The process of wound healing. *a.* Normal uninjured skin. *b.* A deep wound ruptures blood vessels in the dermis. *c.* Blood flows out of the vessels and fills the wound. *d.* A blood clot forms and blocks the flow of blood. *e.* A protective scab develops and fibroblasts begin the process of repair. *f.* New blood vessels form and fibroblasts continue to promote tissue regeneration. *g.* The scab sloughs off when the epidermal layer of the skin is back to normal.

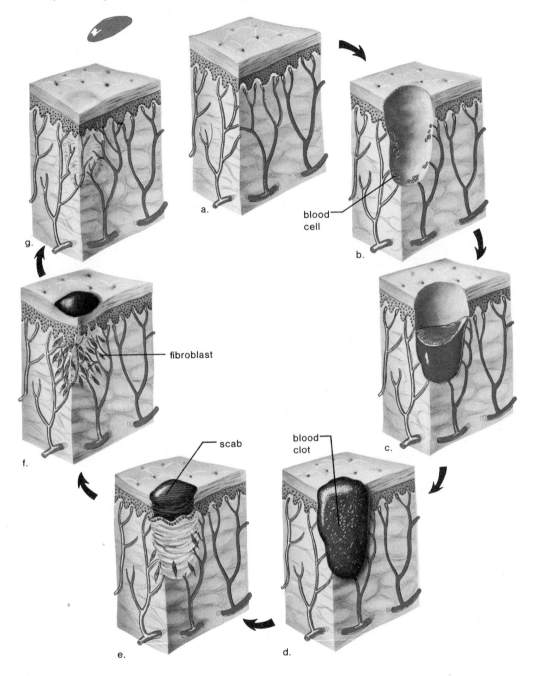

damaged tissue. The deeper the burn, the more prevalent the blisters, which increase in size during the hours after the injury. Unless they become infected, most second-degree burns heal without complications and with little scarring in 10 to 14 days. If the burn extends deep into the dermis, it heals more slowly over a period of 30 to 105 days. The healing epidermis is extremely fragile and scarring is common.

Third-degree burns destroy the entire thickness of the skin. The surface of the wound is leathery and may be brown, tan, black, white, or red. There is no pain because the pain receptors have been destroyed. Blood vessels, sweat glands, sebaceous glands, and hair follicles are all destroyed. There are even **fourth-degree burns** that involve tissues down to the bone. Obviously, the chances of a person surviving are not good unless a very limited area of the body is affected.

Major concerns in the case of severe burns is fluid loss, heat loss, and bacterial infection. Fluid loss is counteracted by intravenous administration of a balanced salt solution. Heat loss is minimized by placing the burn patient in a warm environment. Bacterial infection is treated by the application of an antibacterial dressing.

As soon as possible, the damaged tissue is removed and the skin grafting is begun. Usually the skin needed for grafting is taken from other parts of the patient's body. This is called autografting, as opposed to heterografting, in which the graft is received from another person. Autografting is preferred because there is little chance of rejection of the graft. If the burned area is quite extensive, it may be difficult to acquire enough skin for autografting.

Figure 4.9 Researchers at Shriners Burn Institute in Boston have devised a method of growing large sheets of skin in the laboratory from cells of severely burned patients. In July 1983, they successfully used cultured-skin grafts to treat two young brothers, both of whom had massive burns covering their entire bodies.

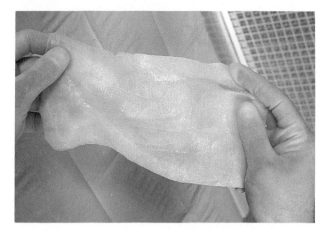

Experimentation is going forward to provide burn patients with artificial skin or extensive skin grown in the laboratory from only a few cells taken from the patient (fig. 4.9).

The severity of a burn is dependent on how deep it is and how extensive it is. First-degree burns affect only the epidermis; second-degree burns affect the entire epidermis and a portion of the dermis; third-degree burns affect the entire epidermis and dermis.

Summary

1. Structure of the Skin. The skin has three layers: the epidermis, the dermis, and the subcutaneous layer.
    A. The epidermis, the outer layer of skin, is a stratified squamous epithelium. New cells continually produced in the innermost layer of the epidermis push outward and become dead, keratinized cells that are sloughed off.
    B. The dermis is composed of dense connective tissue and lies beneath the epidermis. It contains blood vessels and nerve fibers.
    C. The subcutaneous layer of skin is made up of loose connective and adipose tissue that insulates the body.
II. Accessory Structures of the Skin. These structures are categorized as hair, nails, and glands.
    A. Both hair and nails are produced by the division of epidermal cells and consist of keratinized cells.
    B. Sebaceous glands are associated with a hair follicle and produce a secretion that lubricates the hair and skin.
        Sweat glands are associated with hair follicles only in the axillary and pelvic regions. Sweat is mostly water, but also contains salts and urea. Sweating helps the body regulate body temperature.
        Mammary glands are modified sweat glands that produce milk after the birth of a child.
III. Functions of the Skin
    A. Protection. Skin protects the body from physical trauma, bacterial invasion, and fluid loss.
    B. Sensory reception and communication. The skin contains sense receptors for

Summary—*continued*

touch, pressure, pain, hot, and cold. These help us to be aware of our surroundings. The skin also serves as a means of communication between people.

C. Synthesis of vitamin D. The skin contains a precursor molecule that is converted to vitamin D in the presence of UV radiation. Vitamin D is part of a hormone that helps regulate the calcium and phosphorus metabolism involved in bone development.

D. Temperature regulation. Surface blood vessels dilate and the sweat glands excrete when we are too hot. Surface blood vessels constrict and the sweat glands are inactive when we are cold.

IV. Disorders
A. Skin cancer has been associated with ultraviolet radiation and consists of three types. Malignant melanoma is the most dangerous. Basal-cell carcinoma and squamous-cell carcinoma can usually be removed surgically.

B. Wound healing. The skin has regenerative powers and can grow back on its own accord if a wound is not too extensive.

C. Burns. The severity of a burn is dependent on how deep it is and how extensive it is. First-degree burns affect only the epidermis; second-degree burns affect the entire epidermis and a portion of the dermis; third-degree burns affect the entire epidermis and dermis.

Study Questions

1. In general, describe the three layers of the skin.
2. Describe the process by which epidermal tissue continually renews itself.
3. The dermis has what function in relation to the epidermis?
4. What primary role is played by the adipose tissue in the subcutaneous layer of the skin?
5. Describe, in general, the structure of a hair follicle and a nail. How do these structures grow?
6. Describe the structure and function of sebaceous and sweat glands.
7. Describe the structure of a mammary gland.
8. List and describe four functions of the skin.
9. Name the three types of skin cancer and explain why sunlight causes skin cancer.
10. Explain how you determine the severity of a burn. Describe the proper type of treatment for burns.

Objective Questions

I. For questions 1–5, match the layers of skin in the key to the items below.

Key: a. epidermis
b. dermis
c. subcutaneous

1. blood vessels and nerve fibers
2. fat cells
3. basal cells
4. location of sweat glands
5. many collagen and elastic fibers

II. Fill in the blanks.

6. Sebaceous glands are associated with _____ in the dermis and they secrete an oily substance called _____ .
7. Sweat glands are involved in _____ regulation of the body.
8. Skin cells produce vitamin _____ needed for strong bones.
9. Three types of protection afforded by skin are _____ , _____ , and _____ .
10. The severity of a burn is determined by _____ and _____ .

Medical Terminology Reinforcement Exercise

Pronounce, dissect, and analyze the meaning of the following terms:

1. epidermomycosis (ep″ĭ-der″mo-mi-ko′sis)
2. melanogenesis (mel″ah-no-jen′ĕ-sis)
3. acrodermatosis (ak″ro-der″mah-to′sis)
4. pilonidal cyst (pi″lo-ni′dal)
5. mammoplasty (mam′o-plas″te)
6. antipyretic (an″tĭ-pi-ret′ik)
7. dermatome (der′mah-tōm)
8. hypodermic (hy″po-der′mik)
9. trichophagia (trik″o-fa′je-ah)
10. onychocryptosis (on″ĭ-ko-krip-to′sis)
11. hyperhidrosis (hi″per-hi-dro′sis)
12. rhytidectomy (rit″ĭ-dek′to-me)

II

Support and Movement

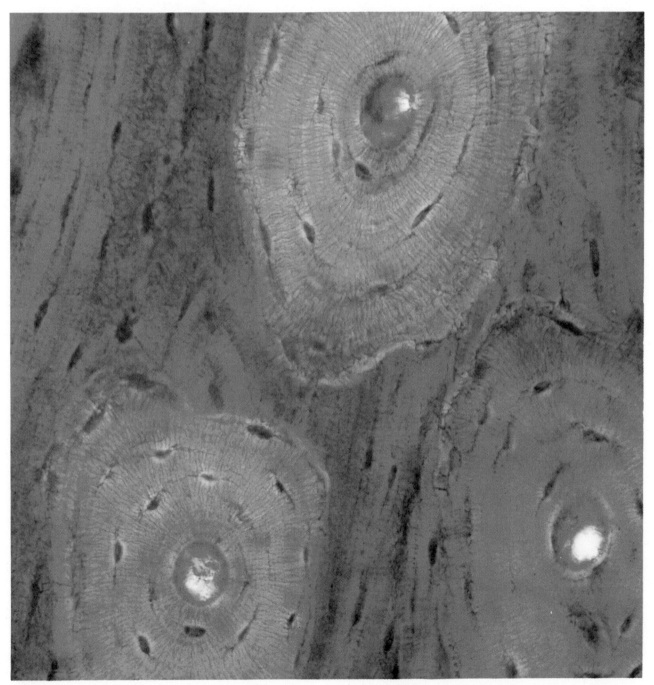

Compact bone of the skeleton contains Haversian systems—concentric circles of osteocyte-containing lacunae around a central canal, which appears white here. The osteocytes secrete the mineralized matrix that gives compact bone its strength.

5

Skeletal System

Chapter Outline

Learning Objectives

After you have studied this chapter, you should be able to:

1. Describe the growth and development of bones.
2. Name at least five functions of the skeleton.
3. Give a classification of bones based on their shapes.
4. Describe the anatomy of a long bone.
5. Distinguish between the axial and appendicular skeletons.
6. Give examples of the surface features of bones.
7. Name the bones and be able to label a diagram of the skull. State the important features of each bone.
8. Name the bones of the vertebral column and be able to label a diagram of the vertebral column. Describe a typical vertebra, the atlas and axis, and the sacrum.
9. Name the bones of the thoracic cage and be able to label a diagram of the thoracic cage. Name the three types of ribs and the three parts of the sternum.
10. Name the bones of the pectoral girdle and be able to label diagrams (including the special features—listed in table 5.1) of these bones.
11. Name the bones of the upper limb (arm) and be able to label a diagram (including the special features—listed in table 5.1) of these bones.
12. Name the bones of the pelvic girdle and be able to label diagrams of these bones. Distinguish between the false and true pelvis. Cite at least five differences between the female and male pelvises.
13. Name the bones of the lower limb (leg) and be able to label diagrams (including the special features—listed in table 5.1) of these bones.
14. Explain how joints are classified and give examples of each type of joint.

Skeleton: Overview

Growth and Development

Most of the bones of the skeleton are cartilaginous during prenatal development. Later, the cartilage is replaced by bone due to the action of bone-forming cells known as **osteoblasts.** At first, there is only a primary ossification center at the middle of a long bone, but later secondary centers form at the ends of the bones. A cartilaginous disk, called the **epiphyseal disk,** remains between the primary ossification center and each secondary center. The length of a bone is dependent on how long the cartilage cells within the disk continue to divide. Eventually, though, the disks disappear, and the bone stops growing as the individual attains adult height.

In the adult, bone is continually being broken down and built up again. After bone-absorbing cells, called **osteoclasts,** break down bone, they remove worn cells and deposit calcium in the blood. Apparently, after a period of about three weeks, the osteoclasts disappear. Then destruction caused by the osteoclasts is repaired by osteoblasts. As they form new bone, they take calcium from the blood. Eventually some of these cells get caught in the matrix they secrete and are converted to **osteocytes,** the cells found within Haversian systems (p. 46).

Because of continual renewal, the thickness of bones can change according to the amount of physical use or due to a change in certain hormone balances (see chapter 9). In most adults, the bones become weaker due to a loss of mineral content. Strange as it may seem, adults seem to require more calcium in the diet than do children in order to promote the work of osteoblasts. Many older women, due to a lack of estrogen, suffer from *osteoporosis,* a condition in which weak and thin bones cause aches and pains, and tend to fracture easily. A tendency toward osteoporosis may also be augmented by lack of exercise and too little calcium in the diet.

Bone is a living tissue, and it is always being rejuvenated.

Functions

The skeleton has the following functions:

The skeleton, notably the large heavy bones of the legs, supports the body against the pull of gravity.

The skeleton protects soft body parts. For example, the skull forms a protective encasement for the brain.

Flat bones (fig. 5.1) such as those of the skull, ribs, and breastbone produce red blood cells in both adults and children.

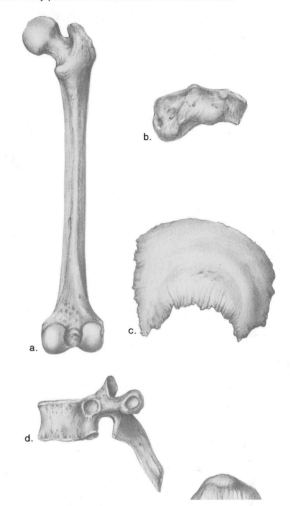

Figure 5.1 Classification of bones. *a.* Long bones are longer than they are wide. *b.* Short bones are cube-shaped; their lengths and widths are about equal. *c.* Flat bones are platelike and have broad surfaces. *d.* Irregular bones have varied shapes with many places for connections with other bones.

Bones are storage areas for inorganic calcium and phosphorus salts.

Bones provide sites for muscle attachment and permit flexible body movement, especially the bones of the legs and arms.

Anatomy of a Long Bone

A long bone, such as one in the arm or leg, can be used to illustrate certain principles of bone anatomy (fig. 5.2). The bone is enclosed in a tough, fibrous connective tissue covering called the **periosteum.** At both ends, there is an expanded portion called an **epiphysis;** the portion between the epiphyses is called the **diaphysis.**

When the bone is split open, as in the figure, the longitudinal section shows that the diaphysis is not solid, but has a **medullary cavity** containing yellow marrow.

Figure 5.2   Anatomy of a long bone.

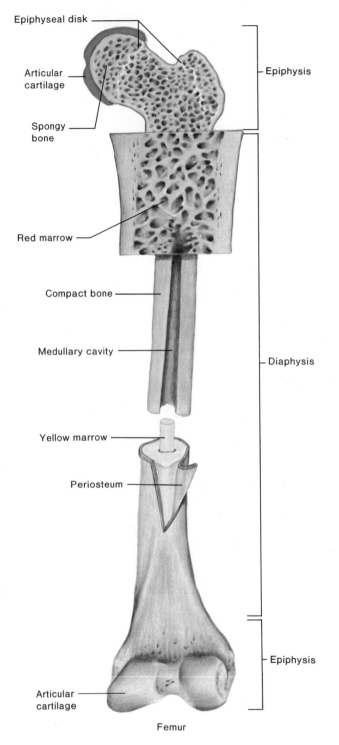

Epiphyseal disk

Articular
cartilage

Spongy
bone

Red marrow

Compact bone

Medullary cavity

Yellow marrow

Periosteum

Articular
cartilage

Epiphysis

Diaphysis

Epiphysis

Femur

The yellow marrow is so named because it contains large amounts of fat. The medullary cavity is bounded at the sides by compact bone.

The epiphyses contain spongy bone. Beyond the spongy bone, there is a thin shell of compact bone and, finally, a layer of cartilage called the **articular cartilage.** The articular cartilage is so named because this is where the bone articulates (meets) another bone.

Compact bone, as discussed previously (p. 46), contains bone cells in tiny chambers called lacunae, that are arranged in concentric circles around Haversian canals, which contain blood vessels and nerves. The lacunae are separated by a matrix that contains protein fibers of collagen and mineral deposits, primarily of calcium and phosphorus salts.

Spongy bone contains numerous bony bars and plates, separated by irregular spaces. Although lighter than compact bone, spongy bone is still designed for strength. Just as braces are used for support in buildings, the solid portions of spongy bone follow lines of stress. The spaces in spongy bone are often filled with **red marrow,** a specialized tissue that produces blood cells.

A long bone has a shaft (diaphysis) and two ends (epiphyses). The shaft contains yellow marrow and is bounded by compact bone. The ends are covered by cartilage and contain spongy bone.

Bones of the Skeleton

The bones of the skeleton are not smooth; they have protuberances called processes and indentations called depressions. Various terms are used to refer to processes and depressions and many of these are listed in table 5.1 for your reference.

The skeleton is divided into the axial skeleton and the appendicular skeleton (fig. 5.3). The **axial skeleton** lies in the midline of the body and contains the bones of the skull, vertebral column, and thoracic cage. The **appendicular skeleton** contains the bones of the pectoral girdle, upper limbs (arms), pelvic girdle, and lower limbs (legs).

Skull

The skull is formed by the cranium and the facial bones. These bones sometimes contain **sinuses** (fig. 5.4), air spaces lined by mucous membranes, which reduce the

Table 5.1 Surface Features of Bone

Processes	
Articulating Surfaces	
Condyle	A large, rounded, articulating knob (the occipital condyle of the occipital bone)
Head	A prominent, rounded, articulating proximal end of a bone (the head of the femur)
Projections for Muscle Attachment	
Crest	A narrow, ridgelike projection (the iliac crest of the coxal bone)
Spine	A sharp, slender process (the spine of the scapula)
Trochanter	A massive process found only on the femur (the greater trochanter of the femur)
Tubercle	A small rounded process (the greater tubercle of the humerus)
Tuberosity	A large roughened process (the radial tuberosity of the radius)
Depressions and Openings	
Fissure	A narrow, slitlike opening (the superior orbital fissure of the sphenoid bone)
Foramen, pl. foramina	A rounded opening through a bone (the foramen magnum of the occipital bone)
Fossa	A flattened or shallow surface (the mandibular fossa of the temporal bone)
Meatus, or canal	A tubelike passageway through a bone (the external auditory meatus of the temporal bone)
Sinus	A cavity or hollow space in a bone (the frontal sinus of the frontal bone)

From Kent M. Van De Graaff and Stuart Ira Fox, *Concepts of Human Anatomy and Physiology,* 2nd ed. Copyright © 1989 Wm. C. Brown Publishers, Dubuque, Iowa. All Rights Reserved. Reprinted by permission.

weight of the skull and give a resonant sound to the voice. The paranasal sinuses empty into the nose and take their names from their location; there are also the maxillary, frontal sphenoidal and ethmoidal sinuses. Two sinuses called the mastoid sinuses drain into the middle ear. Mastoiditis, a condition that can lead to deafness, is an inflammation of these sinuses.

Figure 5.3 Major bones of the skeleton. Note that the axial skeleton is shaded darker than the appendicular skeleton. *a.* Anterior view. *b.* Posterior view.

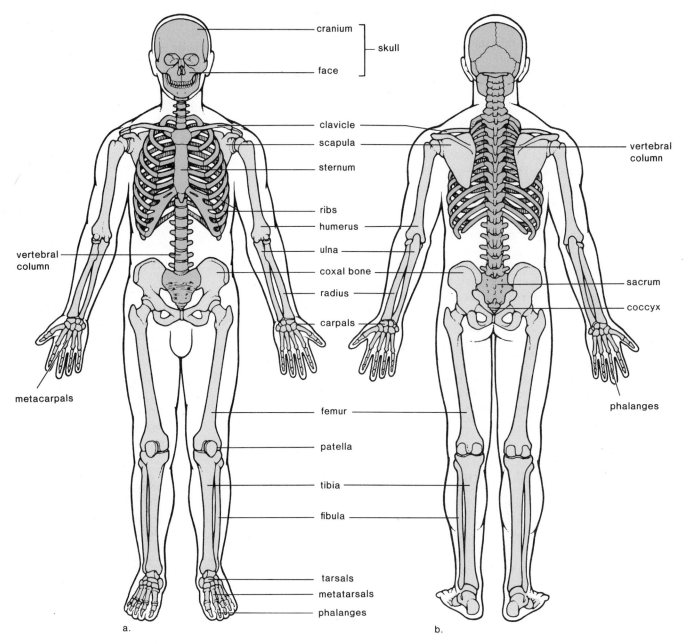

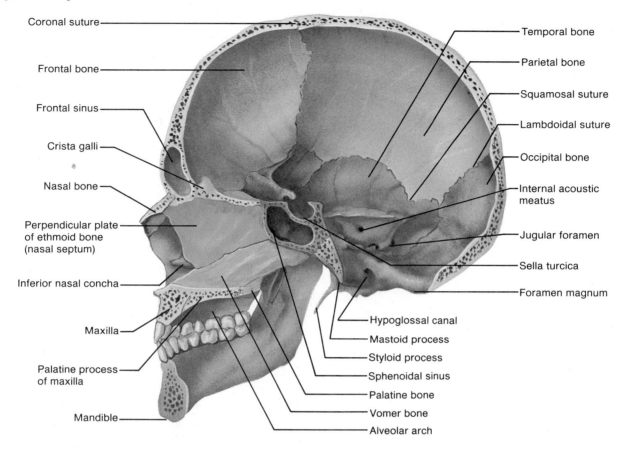

Figure 5.4 Sagittal section of the skull.

Coronal suture

Frontal bone

Frontal sinus

Crista galli

Nasal bone

Perpendicular plate
of ethmoid bone
(nasal septum)

Inferior nasal concha

Maxilla

Palatine process
of maxilla

Mandible

Temporal bone

Parietal bone

Squamosal suture

Lambdoidal suture

Occipital bone

Internal acoustic
meatus

Jugular foramen

Sella turcica

Foramen magnum

Hypoglossal canal

Mastoid process

Styloid process

Sphenoidal sinus

Palatine bone

Vomer bone

Alveolar arch

Cranium

The cranium protects the brain and is composed of eight bones fitted tightly together in adults (figs. 5.5 and 5.6). In newborns, certain bones are not completely formed and instead are joined by membranous regions called **fontanels,** all of which usually close by the age of sixteen months. The largest of these is the anterior or frontal fontanel. The posterior or occipital fontanel is small and triangular in shape. The two lateral pairs of fontanels are the anterolateral or sphenoidal, and posteriolateral or mastoidal. They are small and irregular in shape.

Frontal Bone The frontal bone forms the forehead, a portion of the nose, and the superior portions of the orbits (bony sockets of the eyes).

Parietal Bones The parietal bones are just dorsal to the frontal bone. They form the roof of the cranium and also help form its sides.

Occipital Bone The occipital bone forms the most dorsal part of the skull and base of the cranium. The spinal cord joins the brain by passing through a large opening in the occipital bone called the foramen magnum. The occipital condyles are rounded processes on either side of the foramen magnum that articulate with the first vertebra of the spinal column.

Temporal Bones The temporal bones are just inferior to the parietal bones on the sides of the cranium. They also help form the base of the cranium (fig. 5.6). Each temporal bone has the following:

External auditory meatus, a canal that leads to the middle ear.

Mandibular fossa, which articulates with the mandible.

Mastoid process, which provides a place of attachment for certain neck muscles.

Figure 5.5 Skull anatomy. *a.* Anterior view. *b.* Lateral view.

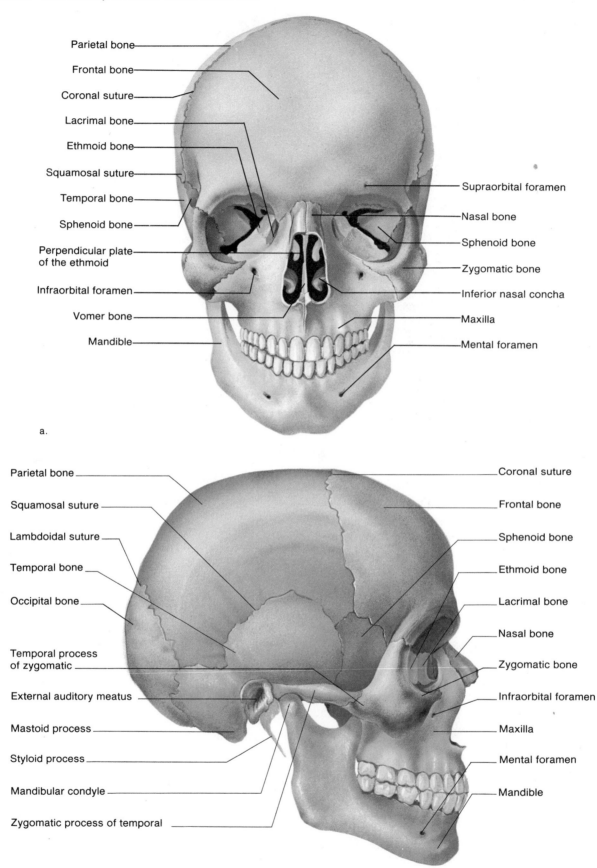

Parietal bone

Frontal bone

Coronal suture

Lacrimal bone

Ethmoid bone

Squamosal suture

Temporal bone

Sphenoid bone

Perpendicular plate
of the ethmoid

Infraorbital foramen

Vomer bone

Mandible

Supraorbital foramen

Nasal bone

Sphenoid bone

Zygomatic bone

Inferior nasal concha

Maxilla

Mental foramen

a.

Parietal bone

Squamosal suture

Lambdoidal suture

Temporal bone

Occipital bone

Temporal process
of zygomatic

External auditory meatus

Mastoid process

Styloid process

Mandibular condyle

Zygomatic process of temporal

Coronal suture

Frontal bone

Sphenoid bone

Ethmoid bone

Lacrimal bone

Nasal bone

Zygomatic bone

Infraorbital foramen

Maxilla

Mental foramen

Mandible

b.

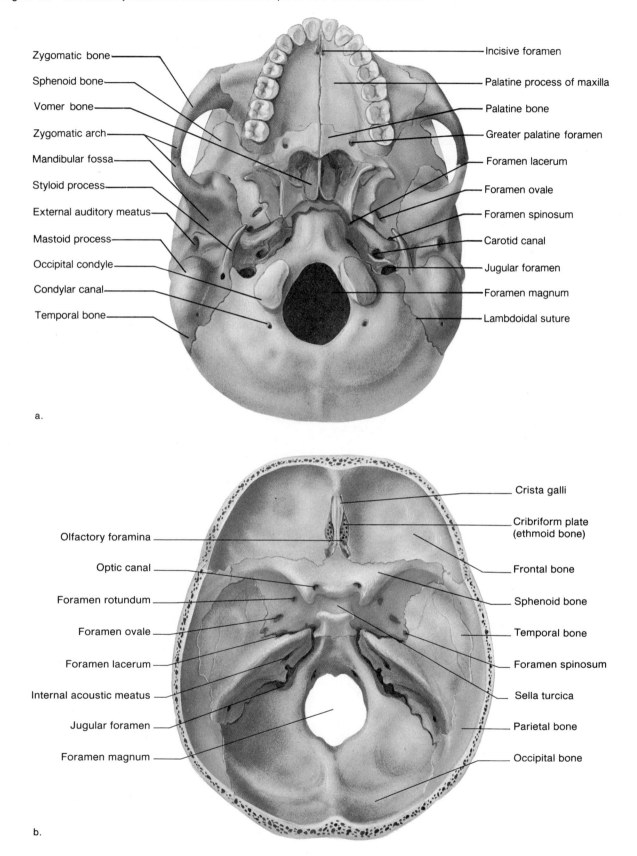

Zygomatic bone

Sphenoid bone

Vomer bone

Zygomatic arch

Mandibular fossa

Styloid process

External auditory meatus

Mastoid process

Occipital condyle

Condylar canal

Temporal bone

Incisive foramen

Palatine process of maxilla

Palatine bone

Greater palatine foramen

Foramen lacerum

Foramen ovale

Foramen spinosum

Carotid canal

Jugular foramen

Foramen magnum

Lambdoidal suture

a.

Olfactory foramina

Optic canal

Foramen rotundum

Foramen ovale

Foramen lacerum

Internal acoustic meatus

Jugular foramen

Foramen magnum

Crista galli

Cribriform plate
(ethmoid bone)

Frontal bone

Sphenoid bone

Temporal bone

Foramen spinosum

Sella turcica

Parietal bone

Occipital bone

b.

Styloid process, which provides a place of attachment for muscles associated with the tongue and larynx.

Zygomatic process, which projects anteriorly and helps form the "cheekbone."

Sphenoid Bone The sphenoid bone helps form the sides and base of the cranium, and the floors and sides of the orbits. Within the cranial cavity (fig. 5.6b), the sphenoid bone has a saddle-shaped midportion called sella turcica where the pituitary gland is found within a depression.

Ethmoid Bone The ethmoid bone forms part of the roof of the nasal cavity (figs. 5.5 and 5.6b). The ethmoid bone has the following:

Crista galli (cock's comb), a triangular process that serves as an attachment for membranes that enclose the brain.

Cribriform plates with tiny holes that serve as passageways for nerve fibers from the olfactory receptors.

Perpendicular plate (fig. 5.4), which projects downward to form the nasal septum.

Superior and middle nasal conchae, which project toward the perpendicular plate. These projections support mucous membranes that line the nasal cavity.

The cranium contains eight bones; the frontal, two parietal, the occipital, two temporal, the sphenoid, and the ethmoid.

Facial Bones

Maxillae The two maxillae form the upper jaw, and each has an alveolar process where the teeth are located. Other processes, called the palatine processes form the anterior portion of the *hard palate,* the roof of the mouth. The maxillae also contribute to the floors of the orbits, and the sides and floor of the nasal cavity.

Palatine Bones The palatine bones make up the posterior portion of the hard palate and the floor of the nasal cavity. A cleft palate results when the palatine bones have failed to fuse.

Zygomatic Bones The zygomatic bones form the sides of the orbits and also help give us our cheekbones. Each bone has a temporal process that joins the zygomatic process of a temporal bone.

Lacrimal Bones The small, thin lacrimal bones are located on the medial walls of the orbits. A small groove lies between the orbit and the nasal cavity, and this serves as a pathway for a tube that carries tears from the eyes to the nose.

Nasal Bones The nasal bones are small rectangular bones that form the bridge of the nose. The ventral portion of the nose is cartilage.

Vomer Bone The vomer bone joins with the perpendicular plate of the ethmoid bone to form the nasal septum (fig. 5.4)

Inferior Nasal Conchae The inferior conchae are thin curved bones that project into the nasal cavity and are attached to the lateral walls of the nasal cavity. Like the other conchae mentioned previously, they support the mucous membranes within the nasal cavity.

Mandible The mandible, or lower jaw, is the only movable portion of the skull. Its horseshoe-shaped body forms the chin. It also contains two upright projections called rami. Each ramus has a mandibular condyle that articulates with a temporal bone and a coronoid process, which serves as a place of attachment for the muscles that allow us to chew. The lower teeth are located on the alveolar arch of the mandible.

The facial bones include the mandible, two maxillary, two palatine, two zygomatic, two lacrimal, two nasal, and the vomer.

Vertebral Column

The **vertebral column** (fig. 5.7) extends from the skull to the pelvis. Normally, the vertebral column has four curvatures that provide more resiliency and strength than a straight column could. These are the cervical, thoracic, lumbar, and pelvic curvatures.

Intervertebral Disks

There are *disks* made of fibrocartilage between the vertebrae that act as a kind of padding. They prevent the vertebrae from grinding against one another and absorb shock caused by movements such as running, jumping, and even walking. Unfortunately, these disks become weakened with age, and can slip or even rupture. This causes pain when the damaged disk presses up against the spinal cord and/or spinal nerves. The body may heal itself, or else the disk can be removed surgically. If the latter occurs, the vertebrae can be fused together, but this will limit the flexibility of the body. The presence of the disks allows motion between the vertebrae so that we can bend forward, backward, and from side to side.

Figure 5.7 Curvatures of the spine. The vertebrae are named for their location in the body. The spinal nerves leave the spine at the intervertebral foramina.

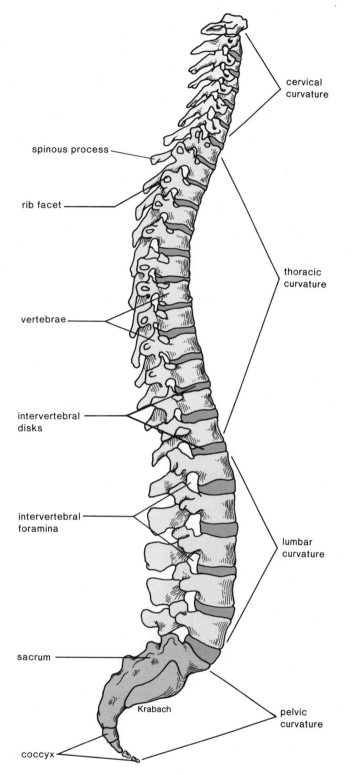

cervical curvature

spinous process

rib facet

thoracic curvature

vertebrae

intervertebral disks

intervertebral foramina

lumbar curvature

sacrum

Krabach

pelvic curvature

coccyx

Figure 5.8 Vertebrae. *a.* Typical vertebra in articulated position. The vertebral canal where the spinal cord is found is formed by adjacent vertebral foramen. *b.* Atlas and axis showing how they articulate with one another. The odontoid process of the axis is the pivot around which the atlas turns as when we shake our head "no."

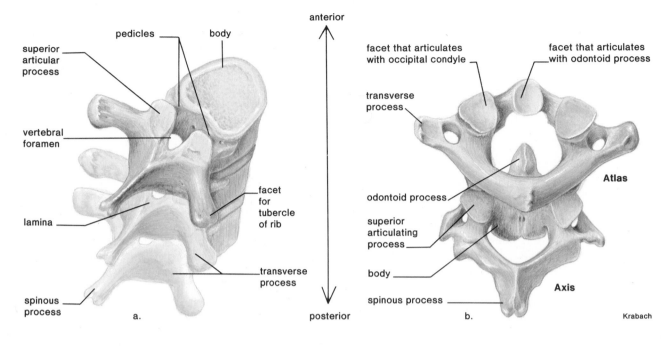

The vertebral column contains the vertebrae separated by intervertebral disks, and has four curvatures.

Typical Vertebra

A typical vertebra is shown in figure 5.8a. The *pedicles* and the *lamina* form a **vertebral arch** that surrounds the **vertebral foramen.** When the vertebrae join, these foramen form a canal through which the nerve cord passes. The **spinous processes** of the vertebrae can be felt as bony projections along the midline of the back.

Atlas and Axis

The first two cervical vertebrae are not typical at all (fig. 5.8b). The **atlas** supports and balances the head. It has two depressions that articulate with the occipital condyles. This allows you to move your head forward and back. The **axis** has an *odontoid process* that projects into the ring of the atlas. When you move your head from side to side, the atlas pivots around the odontoid process.

Sacrum

The **sacrum** is composed of five vertebrae that are fused together. The sacrum articulates with the pelvic girdle and forms the posterior wall of the pelvic cavity (fig.

5.17). The **coccyx,** or tailbone, is the last part of the vertebral column. The coccyx is formed from a fusion of four vertebrae.

The first two vertebrae are the atlas and axis; then there follows the cervical, thoracic, and lumbar vertebrae, and finally the sacrum and the coccyx.

Thoracic Cage

The **thoracic cage** (fig. 5.9) protects the heart and lungs, plays a role in breathing, and supports the bones of the shoulders.

Ribs

Within the thoracic cage, there are twelve pairs of **ribs** that connect directly to the thoracic vertebrae. The *true ribs* join the sternum directly; the *false ribs* join the sternum indirectly via shafts of cartilage, called *costal cartilages.* The lower two pairs of ribs are called *floating ribs* because they do not attach to the sternum.

Sternum

The **sternum** contains three parts: the *manubrium, body,* and *xiphoid process.* The ribs articulate with the manubrium and body of the sternum. In addition, the manubrium articulates with the clavicles.

Figure 5.9 The thoracic cage includes the thoracic vertebrae, the ribs, and the sternum. The three bones that make up the sternum are named.

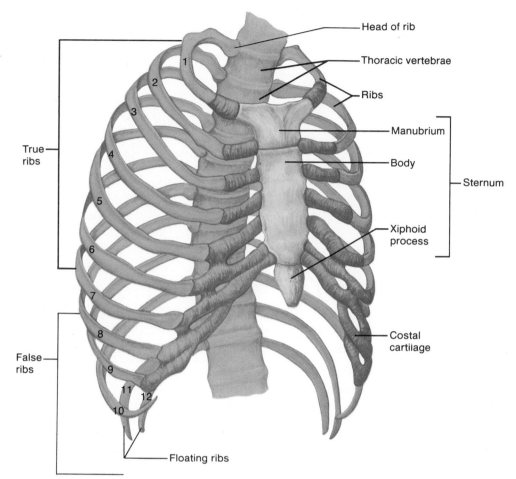

The thoracic cage contains the thoracic vertebrae, ribs, and sternum.

Pectoral Girdle

The pectoral girdle (shoulder girdle) contains four bones: two clavicles and two scapulae (figs. 5.10 and 5.11). The pectoral girdle supports the arms and serves as a place of attachment for muscles that move the arms. The bones of this girdle are not held tightly together and instead are weakly attached and held in place by ligaments and muscles. This allows great flexibility, but means that the shoulder girdle is easily dislocated.

Clavicles

The **clavicles** (collarbones) are slender and S-shaped. Each clavicle articulates with the manubrium of the sternum medially. This is the only place of attachment of the pectoral girdle to the axial skeleton.

Each clavicle also articulates with a scapula. The clavicles serve as braces for the scapulae and help stabilize the shoulder; but they are structurally weak and if undue force is applied to the shoulder, they will fracture (see "More about Fractures").

Scapulae

The **scapulae** (shoulder blades) are broad bones that somewhat resemble triangles (fig. 5.12). Notice that the flexibility of the pectoral girdle is exemplified by the fact that the scapulae are not joined one to the other.

Each scapula has a spine that leads laterally to a head having the following:

Acromion process, which articulates with a clavicle and provides a place of attachment for arm and chest muscles.

Coracoid process that serves as a place of attachment for arm and chest muscles.

Glenoid cavity that articulates with the head of the upper arm bone (humerus). The flexibility of

Figure 5.10 The bones of the left arm and the left portion of the pectoral girdle. The humerus becomes the "funny bone" of the elbow.

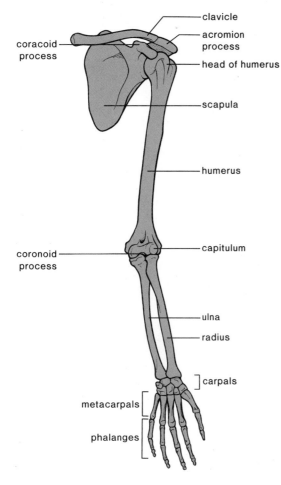

Figure 5.11 Articulation of the right humerus with the right portion of the pectoral girdle.

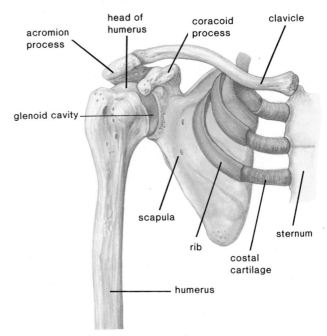

the pectoral girdle is also exemplified by the fact that the glenoid cavity is smaller than the head of the humerus.

The pectoral (shoulder) girdle contains two clavicles and two scapulae.

Upper Limb (Arm)

The upper limb (fig. 5.10) includes the bones of the upper arm (humerus), the forearm (radius and ulna), and hand (carpals, metacarpals, and phalanges).

Humerus

The **humerus** (fig. 5.13) is the bone of the upper arm. It is a long bone with these features at the proximal end:

Head, which fits into the glenoid cavity of the scapula.

Figure 5.12 Scapula. *a.* Posterior surface. *b.* Lateral view.

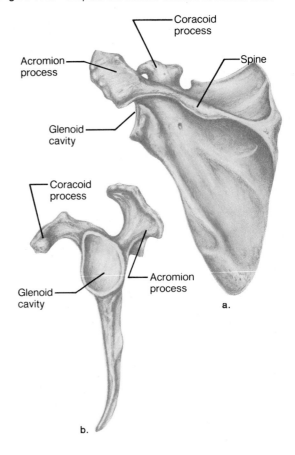

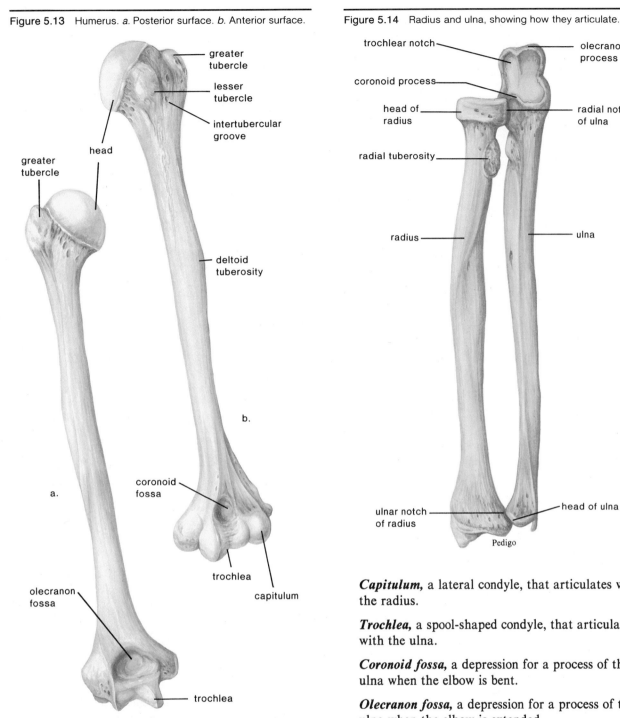

Figure 5.13 Humerus. *a.* Posterior surface. *b.* Anterior surface.

greater
tubercle

lesser
tubercle

intertubercular
groove

head

greater
tubercle

head

deltoid
tuberosity

b.

coronoid
fossa

a.

olecranon
fossa

trochlea

capitulum

trochlea

Figure 5.14 Radius and ulna, showing how they articulate.

trochlear notch

olecranon
process

coronoid process

head of
radius

radial notch
of ulna

radial tuberosity

radius

ulna

ulnar notch
of radius

head of ulna

Pedigo

Greater and **lesser tubercle** that provide attachments for muscles that move arm and shoulder.

Intertubercular groove holds the tendon from the biceps brachii, a muscle of the upper arm.

Deltoid tuberosity, which provides an attachment for the deltoid, a muscle that covers the shoulder joint.

The humerus has these features at the distal end:

Capitulum, a lateral condyle, that articulates with the radius.

Trochlea, a spool-shaped condyle, that articulates with the ulna.

Coronoid fossa, a depression for a process of the ulna when the elbow is bent.

Olecranon fossa, a depression for a process of the ulna when the elbow is extended.

Radius

The **radius** and **ulna** (figs. 5.10 and 5.14) are the bones of the forearm (lower arm). The radius is on the thumb side when the palm faces forward, but crosses over the ulna when the hand is turned so that the palm faces backward. Proximally, the radius has these features:

Head, which articulates with the capitulum of the humerus and fits into the radial notch of the ulna.

Figure 5.15 Posterior view of the left hand.

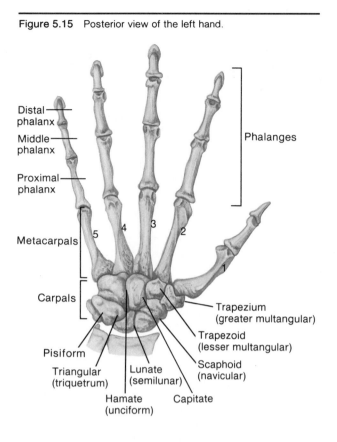

Figure 5.16 The bones of the left leg and the left portion of the pelvic girdle.

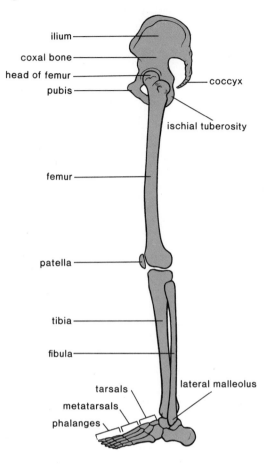

Radial tuberosity, which serves as a place of attachment for a tendon from the biceps brachii.

Ulna

The ulna is the longer bone of the forearm. Proximally, the ulna has these features:

Coronoid process, which articulates with the coronoid fossa of the humerus when the elbow is bent.

Olecranon process, which articulates with the olecranon fossa of the humerus when the elbow is extended.

Trochlear notch, which articulates with the trochlea of the humerus.

Hand

Each hand (figs. 5.10 and 5.15) has a wrist, a palm, and five fingers with the following features:

Wrist contains eight small **carpal bones** (tightly bound by ligaments in two rows of four each.

Palm has five **metacarpal bones** that form the knuckles when you make a fist.

Fingers contain the **phalanges.** The thumb has only two phalanges but the other fingers have three each.

The upper limb contains the humerus, radius, and ulna; and the bones of the hand: carpals, metacarpals, and phalanges.

Pelvic Girdle

The **pelvic girdle,** or pelvis, contains two coxal bones (hipbones) plus the sacrum and the coccyx (figs. 5.16 and 5.17).

The strong bones of the pelvic girdle are firmly attached to one another and bear the weight of the body. The pelvis also serves as the place of attachment for the legs and protects the urinary bladder, the internal reproductive organs, and a portion of the large intestine.

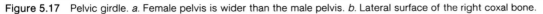

Figure 5.17 Pelvic girdle. *a.* Female pelvis is wider than the male pelvis. *b.* Lateral surface of the right coxal bone.

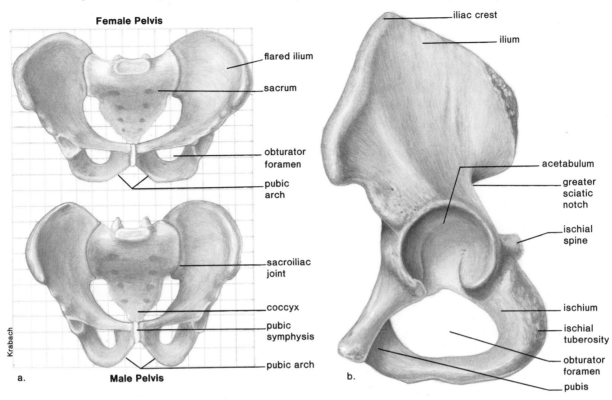

Female Pelvis

- flared ilium
- sacrum
- obturator foramen
- pubic arch
- sacroiliac joint
- coccyx
- pubic symphysis
- pubic arch

a.   **Male Pelvis**

Krabach

- iliac crest
- ilium
- acetabulum
- greater sciatic notch
- ischial spine
- ischium
- ischial tuberosity
- obturator foramen
- pubis

b.

Coxal Bones

Each **coxal bone** has three parts: an ilium, ischium, and a pubis (fig. 5.17b). Where the three bones meet, there is a depression called the **acetabulum,** which receives the rounded head of the femur.

Ilium   The ilium is the largest part of a coxal bone and it flares outward to give the hip prominence. The margin of the ilium is called the **iliac crest.** Each ilium connects posteriorly with the sacrum at a **sacroiliac joint.**

Ischium   The ischium is the most inferior part of a coxal bone and has a posterior region, the **ischial tuberosity,** on which we sit. Near the junction of the ilium and ischium is the **ischial spine,** which projects into the pelvic cavity. The distance between the ischial spines tells the size of the pelvic cavity—important information for childbearing purposes. The **greater sciatic notch** is where blood vessels and the large sciatic nerve pass posteriorly into the lower leg.

Pubis   The pubis is the anterior part of a coxal bone. The two pubic bones join together at the *pubic symphysis.* Posterior to the place where the pubis and the

ischium join together, there is a large opening, the *obturator foramen* through which blood vessels and nerves pass anteriorly into the lower leg.

The pelvic girdle contains two coxal bones plus the sacrum and the coccyx.

False and True Pelvis

The so-called false pelvis is bounded laterally by the flared parts of the iliac bones. This space is much larger than that of the so-called true pelvis. The true pelvis is inferior to the false pelvis, and is the ring formed by the sacrum, lower ilium, ischium, and pubic bones. The true pelvis is said to have an upper inlet and lower outlet. The dimensions of these outlets are important because they must be large enough to allow a baby to pass through during the birth process.

Sex Differences

Several differences usually exist between female and male pelvises (fig. 5.17a) including the following:

Female iliac bones are more flared than those of the male; therefore, the female has broader hips.

Female pelvis is wider between the ischial spines and ischial tuberosities.

Female inlet and outlet of the true pelvis is wider.

Female pelvic cavity is more shallow, while the male pelvic cavity is more funnel shaped.

Female bones are lighter and thinner.

Female pubic arch (angle at the pubic symphysis) is wider.

Lower Limb (Leg)

The lower limb includes the bones of the thigh (femur), the lower leg (tibia and fibula) and those of the foot (tarsals, metatarsals, and phalanges).

Femur

The **femur** (fig. 5.18), or thighbone, is the longest and strongest bone in the body. Proximally, the femur has the following:

Head that fits into the acetabulum of the coxal bone.

Greater and *lesser trochanter* that provide a place of attachment for the muscles of the legs and buttocks.

Linea aspera, a crest that serves as a place of attachment for several muscles.

Distally, the femur articulates with the **patella** (kneecap) and has **lateral** and **medial condyles** that articulate with the tibia.

Tibia

The **tibia** and **fibula** (fig. 5.19) are the bones of the lower leg. The tibia, or shinbone, is medial to the fibula and has the following:

Medial and *lateral condyles,* which articulate with the femur.

Tibial tuberosity where the patellar (kneecap) ligaments attach.

Anterior crest, commonly called the shin.

Medial malleolus, the bulge of the inner ankle.

Fibula

The fibula is lateral to the tibia and is more slender. It has an upper head that articulates with the tibia just below the **lateral condyle** and a lower **lateral malleolus** that forms the outer part of the ankle.

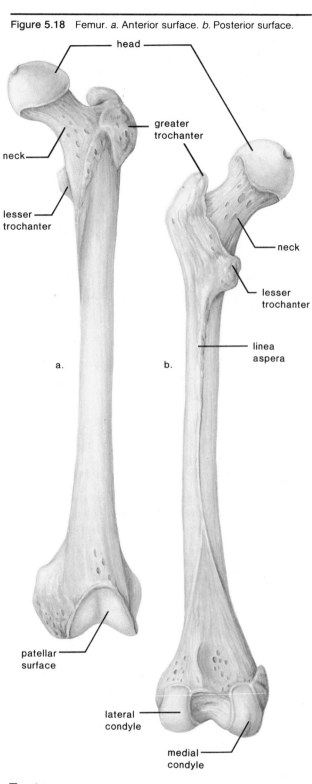

Figure 5.18 Femur. *a.* Anterior surface. *b.* Posterior surface.

head

greater
trochanter

neck

lesser
trochanter

neck

lesser
trochanter

linea
aspera

a.

b.

patellar
surface

lateral
condyle

medial
condyle

Foot

Each foot (fig. 5.20) has bones that are found in the ankle, instep, and five toes.

The ankle has seven **tarsal bones;** together they are called the tarsus. Only one of the seven, the **talus,** can move freely where it joins the tibia and fibula. The largest of the ankle bones is the **calcaneus,** or heel bone, which, along with the talus, supports the weight of the body.

Figure 5.19   Tibia and fibula, showing how they articulate.

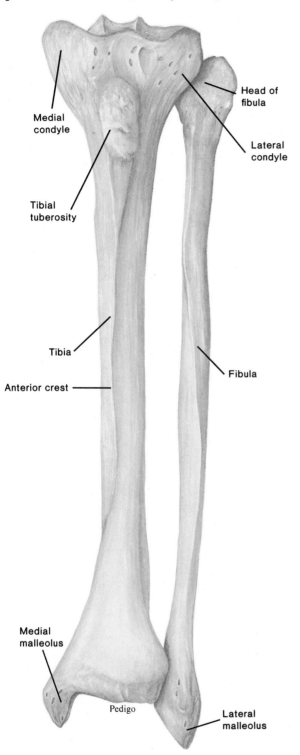

Medial condyle

Head of fibula

Lateral condyle

Tibial tuberosity

Tibia

Anterior crest

Fibula

Medial malleolus

Pedigo

Lateral malleolus

The instep has five elongated **metatarsal bones.** The distal end of the metatarsals form the ball of the foot. Along with the tarsals, these bones form the arches of the foot (longitudinal and transverse) that give spring to our step. If the ligaments and tendons holding these bones together weaken, the individual has fallen arches or flat feet.

Figure 5.20   Anterior view of the left foot.

Calcaneus

Talus

Navicular

Cuboid

Lateral cuneiform

Intermediate cuneiform

Medial cuneiform

Proximal phalanx

Middle phalanx

Distal phalanx

Tarsals

Metatarsals

Phalanges

1 2 3 4 5

Toes contain the **phalanges.** The big toe has only two phalanges, but the other toes have three each.

The lower limb contains the femur, tibia, and fibula; and the bones of the foot: tarsals, metatarsals, and phalanges.

Joints

Bones articulate at the **joints,** which are often classified according to the amount of movement they allow.

Classification

Some bones, such as those that make up the cranium, are sutured together by a thin layer of fibrous connective tissue and are **immovable.** Review figures 5.5 and 5.6, and note that the **sutures** have been named:

Sagittal suture occurs between the parietal bones.

Coronal suture occurs between parietal bones and frontal bone.

Lambdoidal suture occurs between the parietal bones and the occipital bone.

Squamosal suture occurs between each parietal bone and each temporal bone.

Fractures
A Clinical Application

Although a **fracture** may involve injury to cartilaginous structures, it is usually defined as a break in a bone. A fracture can be classified according to its cause and the nature of the break sustained. For example, a break due to injury is a *traumatic* fracture, while one resulting from disease is a *spontaneous* or *pathologic* fracture.

If a broken bone is exposed to the outside by an opening in the skin, the injury is termed a *compound fracture.* Such a fracture is accompanied by the added danger of infection, since microorganisms almost surely enter through the broken skin. On the other hand, if the break is protected by uninjured skin, it is called a *simple fracture.* Figure a shows several types of fractures.

Repair of a Fracture

Whenever a bone is broken, blood vessels within the bone and its periosteum are ruptured, and the periosteum is likely to be torn. Blood escaping from the broken vessels spreads through the damaged area and soon forms a blood clot, or *hematoma.* As vessels in surrounding tissues dilate, these tissues become swollen and inflamed.

Within days or weeks the hematoma is invaded by developing blood vessels and large numbers of osteoblasts, originating from the periosteum. The osteoblasts multiply rapidly in the regions close to the new blood vessels, building spongy bone nearby. Granulation tissue develops, and in regions further from a blood supply, fibroblasts produce masses of fibrocartilage.

Figure a Various types of traumatic fractures.

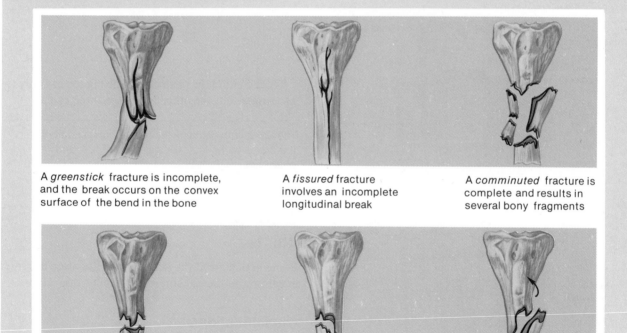

A *greenstick* fracture is incomplete, and the break occurs on the convex surface of the bend in the bone

A *fissured* fracture involves an incomplete longitudinal break

A *comminuted* fracture is complete and results in several bony fragments

A *transverse* fracture is complete, and the break occurs at a right angle to the axis of the bone

An *oblique* fracture occurs at an angle other than a right angle to the axis of the bone

A *spiral* fracture is caused by twisting a bone excessively

Meanwhile, phagocytic cells begin to remove the blood clot as well as any dead or damaged cells in the affected area. Osteoclasts also appear and resorb bone fragments, thus aiding in "cleaning up" debris.

In time, a large amount of fibrocartilage fills the gap between the ends of the broken bone, and this mass is termed a cartilaginous *callus*. The callus is later replaced by bone tissue in much the same way as the hyaline cartilage of a developing endochondral bone is replaced. That is, the cartilaginous callus is broken down, the area is invaded by blood vessels and osteoblasts, and the space is filled with a bony callus.

Usually more bone is produced at the site of a healing fracture than is needed to replace the damaged tissues. However, osteoclasts are able to remove the excess, and the final result of the repair process is a bone shaped very much like the original one. Figure b shows the steps in the healing of a fracture.

The rate at which a fracture is repaired depends on several factors. For instance, if the ends of the broken bone are close together, healing is more rapid than if they are far apart. This is the reason for setting fractured bones and for using casts or metal pins to keep the broken ends together. Also, some bones naturally heal more rapidly than others. The long bones of the arms, for example, may heal in half the time required by the leg bones. Furthermore, as age increases, so does the time required for healing.

Figure b   Major steps in the repair of a fracture.

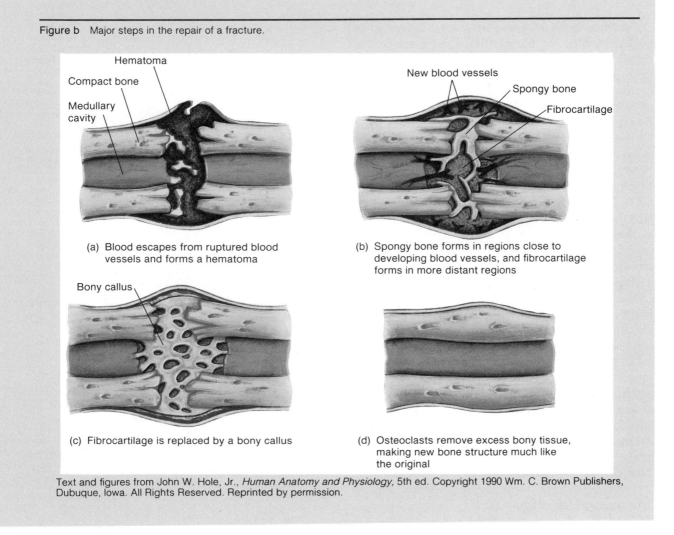

(a) Blood escapes from ruptured blood vessels and forms a hematoma

(b) Spongy bone forms in regions close to developing blood vessels, and fibrocartilage forms in more distant regions

(c) Fibrocartilage is replaced by a bony callus

(d) Osteoclasts remove excess bony tissue, making new bone structure much like the original

Figure 5.21 Generalized freely movable joint.

- Joint cavity filled with synovial fluid
- Joint capsule
- Articular cartilage
- Synovial membrane
- Spongy bone

Figure 5.22 Knee joint. *a.* Longitudinal section. *b.* Posterior view showing some of the ligaments that stabilize the joint.

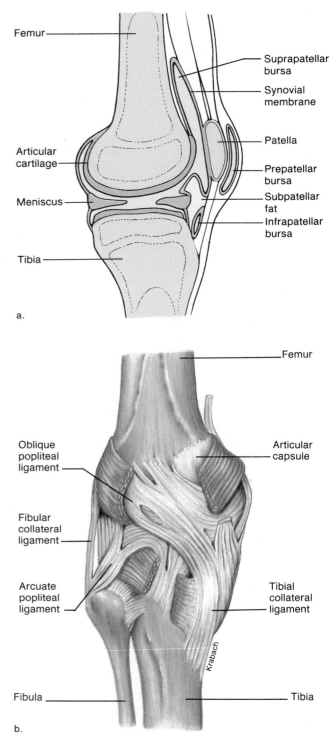

- Femur
- Suprapatellar bursa
- Synovial membrane
- Articular cartilage
- Patella
- Prepatellar bursa
- Subpatellar fat
- Infrapatellar bursa
- Meniscus
- Tibia

a.

- Femur
- Oblique popliteal ligament
- Articular capsule
- Fibular collateral ligament
- Arcuate popliteal ligament
- Tibial collateral ligament
- Fibula
- Tibia

Krabach

b.

Other joints are **slightly movable,** and are connected by hyaline cartilage or fibrocartilage. For example, the vertebrae are separated by disks (fig. 5.7) that increase their flexibility. Also, the ribs are joined to the sternum by the costal cartilage (fig. 5.9) and the pubic symphysis occurs between the pubic bones (fig. 5.17). Owing to hormonal changes, this joint becomes more flexible during late pregnancy, which allows the pelvis to expand during childbirth.

Most joints are **freely movable synovial joints,** in which the two bones are separated by a cavity (fig. 5.21). **Ligaments** composed of fibrous connective tissue bind the two bones to one another, holding them in place as they form a capsule. In a "double-jointed" individual, the ligaments are unusually loose. The joint capsule is lined by a **synovial membrane,** which produces **synovial fluid,** a lubricant for the joint.

The knee is an example of a synovial joint (fig. 5.22). In the knee, as in other freely movable joints, the bones are capped by cartilage, but in the knee, there are also crescent-shaped pieces of cartilage between the bones, called **menisci.** These give added stability, helping to support the weight placed on the knee joint. Unfortunately, athletes often suffer injury of the menisci, known as torn cartilage. The knee joint also contains thirteen fluid-filled sacs called **bursae,** which ease friction between tendons and ligaments, and between tendons and bones. Inflammation of the bursae is called bursitis. Tennis elbow is a form of bursitis.

There are different types of movable joints as listed here and depicted in figure 5.23.

Figure 5.23 Types and examples of freely movable joints.

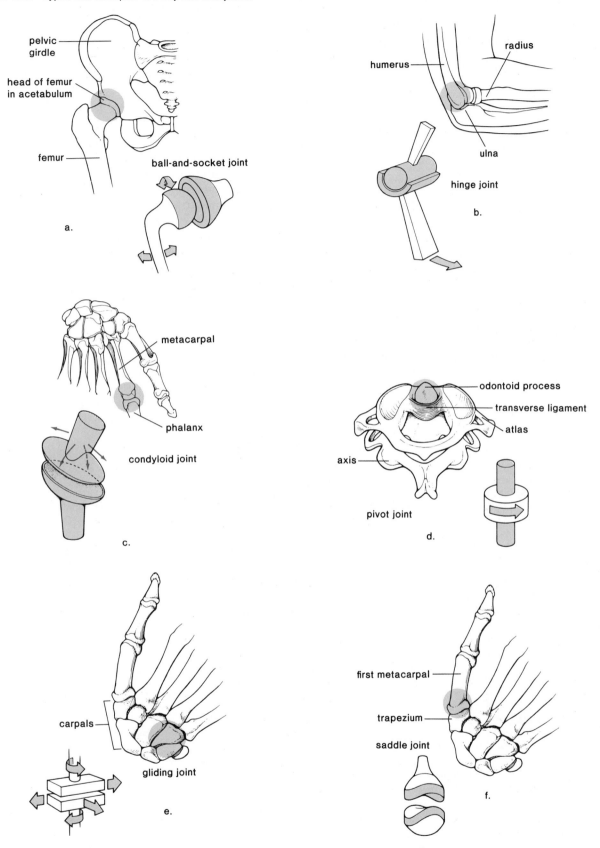

Ball-and-socket joint Ball-shaped head of one bone fits into cup-shaped socket of another. Movement in all planes and rotation is possible. For example, the shoulder and hip joint.

Hinge joint Convex surface of one bone articulates with concave surface of another. Up and down motion in one plane is possible. For example, the elbow and knee joint.

Condyloid joint Oval-shaped condyle of one bone fits into the elliptical cavity of another. Movement in different planes is possible, but no rotation. For example, the joints between the metacarpals and phalanges.

Pivot joint Small cylindrical projection of one bone pivots within a ring formed of bone and ligament of another. Rotation only is possible. For example, joint between proximal ends of radius and ulna, and joint between atlas and axis.

Gliding joint Flat or slightly curved surfaces of bones are articulating. Sliding or twisting in various planes is possible. For example, joints between the bones of the wrist and between bones of ankle.

Saddle joint Each bone is saddle-shaped and fits into the complementary regions of the other. A variety of movements is possible. For example, joint between the carpal and metacarpal bones of the thumb.

The terms that are used to describe the various movements of body parts at joints are depicted in figure 6.7.

Synovial joints are subject to various disorders. A sudden movement that twists or wrenches a joint can cause a stretching or tearing of a ligament, which is called a *sprain. Arthritis* is a more serious disorder. In rheumatoid arthritis, the synovial membrane becomes inflamed and grows thicker. Degenerative changes take place that make the joint almost immovable and painful to use. There is evidence that these effects are brought on by an autoimmune reaction. In old-age arthritis, or osteoarthritis, the cartilage at the ends of the bones disintegrates so that the two bones become rough and irregular. This type of arthritis is apt to affect the joints that have received the greatest use over the years.

Joints are classified according to the degree of movement. Some joints are immovable, some are slightly movable, and some are freely movable.

Summary

I. Skeleton: Overview
   A. The skeleton not only permits flexible movement, it also supports and protects the body, produces red blood cells, and serves as a storehouse for certain inorganic salts.
   B. A long bone has a shaft, or diaphysis, and two ends, or epiphyses, covered by articular cartilage. The diaphysis contains a medullary cavity and compact bone. The epiphyses contain spongy bone where red blood cells are produced.
   C. Bone is a living tissue, and it is always being rejuvenated. Osteoblasts produce bone; osteoclasts break down bone.
II. Bones of the Skeleton. The axial skeleton lies in the midline of the body and consists of the skull, vertebral column, and the thoracic cage. The appendicular skeleton consists of the pectoral girdle, bones of the upper limbs, and pelvic girdle and bones of the lower limbs.
   A. The skull contains the cranium and facial bones. The cranium includes the frontal, parietals, occipital, temporals, sphenoid, and ethmoid bone. The facial bones are the maxillae, palatine, zygomatic, lacrimal, nasal, vomer, inferior nasal conchae bones, and the mandible.
   B. The vertebral column contains the cervical, thoracic, and lumbar vertebrae separated by intervertebral disks. The typical vertebra has a body, vertebral arch about the vertebral foramen, and a spinous process. The first two vertebra are the atlas and axis.
   C. The thoracic cage contains the thoracic vertebrae, ribs and sternum.
   D. The pectoral (shoulder) girdle contains two clavicles and two scapulae.
   E. The upper limb contains the humerus, radius, and ulna; and the bones of the hand: carpals, metacarpals, and phalanges.
   F. The pelvic girdle contains two coxal bones plus the sacrum and the coccyx. The female pelvis, in general, is wider and more shallow than the male pelvis.
   G. The lower limb contains the femur, tibia, and fibula; and the bones of the foot: tarsals, metatarsals, and phalanges.
III. Joints. Joints are regions of articulations between bones.
   There are immovable joints, slightly movable joints, and freely movable synovial joints. The different kinds of freely movable joints are ball-and-socket, hinge, condyloid, pivot, gliding, and saddle.

Study Questions

1. What are five functions of the skeleton?
2. What are four major categories of bones based on their shapes?
3. What are the parts of a long bone? What are some differences between compact bone and of spongy bone?
4. How does bone grow in young people, and how is it rejuvenated in all age groups?
5. What is the difference between the axial and the appendicular skeleton?
6. What are the bones of the cranium and the face? What are the special features of the temporal bones, sphenoid bone, and ethmoid bone?
7. What are the parts of the vertebral column, and what are its curvatures? Distinguish between the atlas, axis, sacrum, and coccyx.
8. What are the bones of the thoracic cage, and what are several functions of the thoracic cage?
9. What are the bones of the pectoral girdle? Give examples to demonstrate the flexibility of the pectoral girdle. What are the special features of each scapula?
10. What are the bones of the upper limb, and what are the special features of these bones?
11. What are the bones of the pelvic girdle and what are their functions? Give examples to demonstrate the strength and stability of the pelvic girdle.
12. What is the false and true pelvis, and what are several differences between the male and female pelvises?
13. What are the bones of the lower limb? Describe the special features of these bones.
14. How are joints classified? Give examples of each type of joint.

Objective Questions

I. For questions 1–6, match the items in the key to the correct bone.

Key: a. forehead
b. chin
c. cheekbone
d. elbow
e. shoulder blade
f. hip
g. ankle

1. temporal and zygomatic
2. tibia and fibula
3. frontal bone
4. humerus
5. coxal bone
6. scapula

II. For questions 7–13, match the items in the key to the correct bone.

Key: a. external auditory meatus
b. cribiform plates
c. xiphoid process
d. glenoid cavity
e. olecranon process
f. acetabulum
g. greater and lesser trochanter

7. humerus
8. sternum
9. femur
10. temporal
11. coxal bone
12. ethmoid
13. ulna

III. Fill in the blanks.

14. Long bones are _____ than they are wide.
15. The epiphysis of a long bone contains _____ bone where red blood cells are produced.
16. The _____ are the air-filled spaces in the cranium.
17. The sacrum is a part of the _____ and the sternum is a part of the _____ .
18. The knee is a freely movable joint of the _____ type.
19. The pectoral girdle is specialized for _____ , while the pelvic girdle is specialized for _____ .
20. The term phalanges is used for both the _____ and the _____ .

Medical Terminology Reinforcement Exercise

Pronounce, dissect, and analyze the following terms:

1. chondromalacia (kon''dro-mah-la'she-ah)
2. osteomyelitis (os''te-o-mi''e-li'tis)
3. craniosynostosis (kra''ne-o-sin''os-to'sis)
4. myelography (mi''ĕ-log'rah-fe)
5. acrocyanosis (ak''ro-si''ah-no'sis)
6. syndactylism (sin-dak'tĭ-lizm)
7. orthopedist (or''tho-pe'dist)
8. prognathism (prog'nah-thizm)
9. micropodia (mi''kro-po'de-ah)
10. arthroscopic surgery (ar''thro-skop'ik)

6

Muscular System

Chapter Outline

Skeletal Muscle Structure
  Whole Skeletal Muscle
  Muscle Fiber
  Oxygen Debt
  Contraction of a Whole Muscle
  Muscle Movements and Actions
Skeletal Muscles
  Naming Muscles
  Muscles of the Head
  Muscles of the Neck and Trunk
  Muscles of the Abdominal Wall
  Muscles of the Upper Limb (Arm)
  Muscles of the Lower Limb (Leg)

Learning Objectives

After you have studied this chapter, you should be able to:

1. Describe the anatomy of a whole muscle and muscle fiber.
2. Describe the sliding filament theory of muscle fiber contraction.
3. Explain how a nerve fiber innervates a muscle fiber to contract.
4. Describe how aspects of whole muscle contraction are dependent on myofibril contraction.
5. List the types of movements that occur at joints as muscles contract.
6. Discuss the manner in which smooth coordination is achieved as muscles contract.
7. Name the superficial muscles of the head, neck, trunk, upper limb (arm), and lower limb (leg), indicating their origins and insertions, and give their functions.

Skeletal Muscle Structure

This chapter is concerned with skeletal muscles—those muscles that make up the bulk of the human body. Skeletal muscles are attached to the skeleton and their contraction causes the movement of bones. Nerve impulses originating in the brain and spinal cord innervate skeletal muscles by way of nerves and, through a series of steps, bring about contraction. Blood vessels serve the muscles, bringing to the mitochondria of the muscle cells, the oxygen and nutrients needed to produce a supply of ATP for muscle contraction.

Skeletal muscles make up the bulk of the body and their contraction accounts for the movement of bones. Nerves innervate skeletal muscles and blood vessels bring them oxygen and nutrients.

Whole Skeletal Muscle

Muscles are covered by several layers of fibrous connective tissue called the **fascia,** which extends beyond the muscle to become its tendon. The fascia layer closest to the muscle is called the **epimysium.** The **perimysium** divides up the muscle tissue into units called **fascicles.** Within the fascicles, there are individual muscle cells called **muscle fibers.** Each muscle fiber is covered by another layer of connective tissue called the endomysium (fig. 6.1).

Muscle Fiber

Each muscle fiber is a cell and, therefore, it contains the usual cellular components; but special terminology has been assigned to some of the cellular components, as indicated in table 6.1. In muscle fibers, the *sarcolemma,* or cell membrane, forms tubules that penetrate or dip down into the cell so that they come into contact with expanded portions of modified ER, which is called the *sarcoplasmic reticulum* in muscle cells. These tubules comprise the *T* (for transverse) *system* (fig. 6.2). The expanded portions of the sarcoplasmic reticulum are *calcium storage sacs;* here the element calcium, Ca^{++}, is stored. The sarcoplasmic reticulum encases hundreds and sometimes even thousands of **myofibrils,** which are cylindrical in shape and run the length of a fiber. They have light and dark bands called striations, and that is why skeletal muscle has a striated appearance. Each myofibril is divided into units called sarcomeres, which shorten when they contract.

Myofibrils are the contractile portions of muscle fibers.

Sarcomere Anatomy

A sarcomere (fig. 6.3a and b) contains two types of filaments, thin filaments and thick filaments. A thin filament is a twisted double strand of the protein *actin.* A thick filament is a strand of the larger protein *myosin.*

A sarcomere extends between two dark lines called Z lines. Electron micrographs show that actin filaments alone are attached to the Z lines, accounting for the lightest regions of a myofibril (I bands). The actin filaments extend between myosin filaments, and the darkest region of each sarcomere contains both actin and myosin filaments (A band). In the H zone, there are only myosin filaments.

Sarcomere Contraction

When a sarcomere contracts (fig. 6.3c and d) it shortens because the actin filaments slide past the myosin filaments and approach one another. This causes the H zone to disappear. This is called the **sliding filament theory.**

The sliding occurs because myosin has cross bridges that attach to and pull the actin filaments toward the center of the sarcomere. For attachment to occur, calcium (Ca^{++}) and ATP must be present. Following attachment, ATP is broken down as detachment occurs. The cross bridges attach and detach some fifty to one hundred times as the thin filaments are pulled to the center of a sarcomere. If, by chance, ATP molecules are not available, detachment cannot occur. This explains rigor mortis, a temporary stiffening of the muscles immediately after death.

The sliding filament theory states that actin filaments slide past myosin filaments because myosin has cross bridges that pull the actin filaments inward.

Muscles need a steady supply of ATP for muscle contraction. Ultimately, this ATP is made in mitochondria, but there is also a storage compound, creatine phosphate, which quickly gives more ATP. Creatine phosphate does not participate directly in muscle contraction. Instead, it is used to regenerate ATP by the following reaction:

$$creatine \sim P + ADP \rightarrow ATP + creatine$$

You may have heard of creatinine, the break down product of creatine which is excreted by the kidneys. One test for kidney function is a serum creatinine test. Creatinine is chosen for this test because it is normally released into the bloodstream at a constant rate. If the normal amount of creatinine in the blood should rise, it indicates that the kidneys are malfunctioning.

Figure 6.1 Muscle anatomy. *a.* Fascia (connective tissue) covers the surface of the muscle. *b.* Epimysium is the layer of connective tissue closest to the muscle, and perimysium divides up the muscle tissue into units called fascicles. *c.* Endomysium surrounds each muscle fiber. *d.* Enlargement of muscle fiber and cutaway section reveals the myofibrils and their filaments.

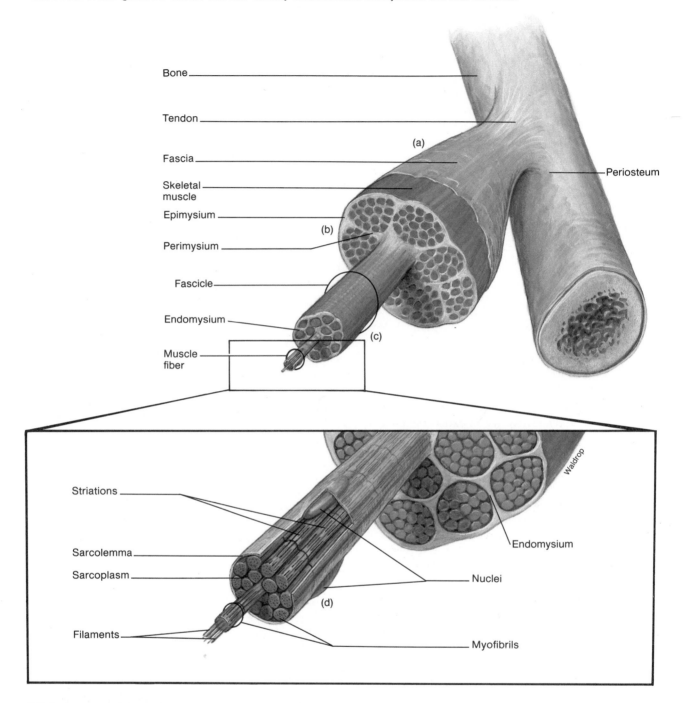

Table 6.1	Muscle Cell
Component	**Term**
Cell membrane	Sarcolemma
Cytoplasm	Sarcoplasm
Endoplasmic reticulum	Sarcoplasmic reticulum

Muscle contraction requires a steady supply of ATP. Creatine phosphate is used to generate ATP rapidly.

Oxygen Debt

When all of the creatine phosphate has been depleted and there is no oxygen available for aerobic respiration, anaerobic respiration, which does not require oxygen, can supply some ATP. Anaerobic respiration, which is apt to occur during strenuous exercise, stops after a while because of *lactic acid buildup*. Lactic acid, an end product of anaerobic respiration, interferes with muscle contraction, and causes temporary muscular cramping and aching.

Figure 6.2 Anatomy of a muscle fiber as revealed by the electron microscope. A muscle fiber contains numerous myofibrils, each of which is enclosed by sarcoplasmic reticulum. The sarcolemma forms tubules that dip down and come in contact with the sarcoplasmic reticulum.

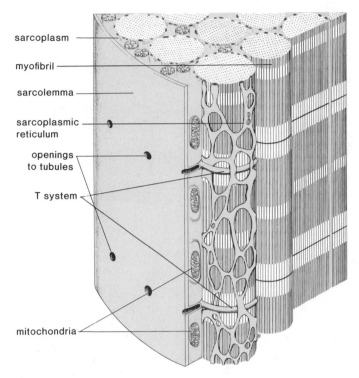

sarcoplasm

myofibril

sarcolemma

sarcoplasmic reticulum

openings to tubules

T system

mitochondria

Figure 6.3 Sliding filament theory. *a.* Sarcomere relaxed. *b.* Enlargement of filaments. *c.* Sarcomere contracted. *d.* Enlargement of filaments.

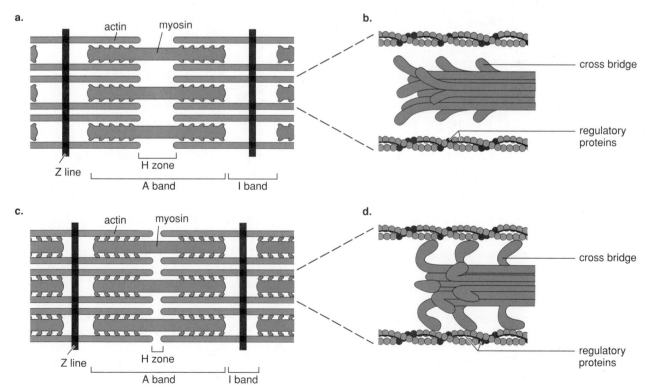

a.

actin   myosin

Z line   H zone

A band   I band

b.

cross bridge

regulatory proteins

c.

actin   myosin

Z line   H zone

A band   I band

d.

cross bridge

regulatory proteins

Figure 6.4 Anatomy of a neuromuscular junction. *a.* A neuromuscular junction occurs where a synaptic ending of a nerve fiber (an axon branch) comes in close proximity to a muscle fiber. *b.* The synaptic ending contains synaptic vesicles filled with ACh. When these vesicles fuse with the presynaptic membrane, ACh diffuses across the synaptic cleft to initiate a muscle action potential. Then the muscle contracts.

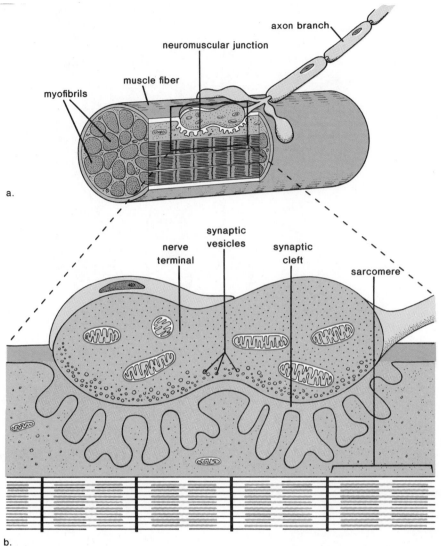

We all have had the experience of having to continue deep breathing following strenuous exercise. This continued need for extra oxygen is required to complete the metabolism of lactic acid. It represents an **oxygen debt** that the body must pay to rid itself of lactic acid.

Lactic acid buildup due to anaerobic respiration causes the muscles to ache and brings on oxygen debt.

Muscle Fiber Contraction and Relaxation

Nerves innervate muscles, and nerve impulses cause muscles to contract. The region where a branch of a nerve fiber approaches a muscle fiber is called a **neuromuscular (myoneural) junction** (fig. 6.4). The space between the nerve fiber and muscle fiber is called a synaptic cleft.

There are small vesicles at the end of the nerve fiber called synaptic vesicles that contain ACh (acetylcholine), a neurotransmitter substance. When nerve impulses travel down a nerve fiber, synaptic vesicles fuse with the presynaptic membrane, releasing ACh, which diffuses across the synaptic cleft and attaches to the sarcolemma. This starts a muscle action potential (p. 124) that spreads over the sarcolemma and down the T system to where calcium ions (Ca^{++}) are stored in the calcium storage sacs (fig. 6.5). Calcium is now released and sarcomere (myofibril) contraction occurs.

Table 6.2 lists the events that precede sarcomere contraction.

Innervation of a muscle fiber occurs at a neuromuscular (myoneural) junction. Nervous stimulation results in sarcomere contraction.

Figure 6.5 The role of calcium in muscle contraction. *a.* When the sarcomere is resting, calcium ions are stored in expanded ends of sarcoplasmic reticulum, called calcium storage sacs. *b.* The muscle action potential, illustrated here by the change in polarity (+ and −) of the sarcolemma and T system, prompts the release of calcium (pink), which stimulates muscle contraction.

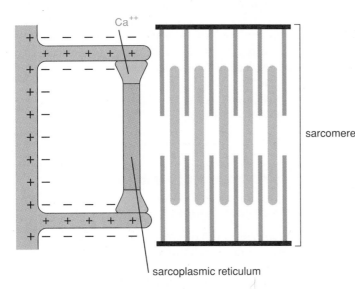

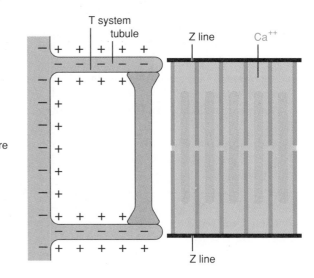

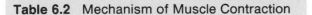

Table 6.2 Mechanism of Muscle Contraction

1. Nerve impulses move down motor nerve fibers.
2. Synaptic vesicles release ACh.
3. ACh moves across synaptic cleft.
4. Muscle action potential begins at sarcolemma and passes down T system.
5. Calcium storage sacs release calcium.
6. Sarcomeres contract.

Contraction of a Whole Muscle

Thus far, we have been speaking of sarcomere contraction within a myofibril (fig. 6.3), and now we wish to get back to discussing whole muscles again.

All-or-None Law

When a muscle fiber is innervated, all the sarcomeres contract. In other words, a muscle fiber behaves in an all-or-none manner. Contrary to this, the strength of the contraction of a whole muscle can increase according to how many muscle fibers are contracted. In other words, a whole muscle does not obey the *all-or-none law* because the total amount of contraction depends on how many muscle fibers are contracted at that time.

Muscle Twitch, Summation, and Tetanus

A single stimulus would cause a muscle to contract and then relaxation would follow. This is called a *muscle twitch*. Figure 6.6a shows that a muscle twitch can be

Figure 6.6 Physiology of muscle contraction. *a.* Simple muscle twitch is composed of three periods: latent, contraction, and relaxation. *b.* Summation and tetanic contraction. When a muscle is not allowed to relax completely between stimuli, the contractions increase in size and then the muscle remains contracted until it fatigues.

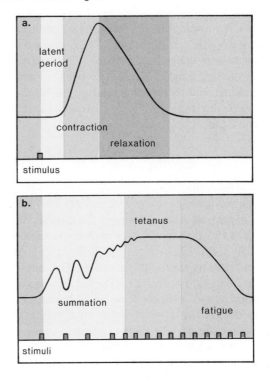

divided into the latent period, or the period of time between stimulus and initiation of contraction; the period of contraction; and the period of relaxation. Normally, in the body, a muscle receives many impulses in rapid succession. Because of this, tension summates until maximal sustained tetanic contraction is achieved (fig. 6.6b). If the muscle is not allowed to rest, fatigue will set in. Fatigue is apparent when a muscle relaxes, even though stimulation is continued. This rarely happens in the body because certain fibers of a muscle are usually contracting while others are relaxing. Still, at any one time, there are always some fibers that are experiencing **tetanus.** This maintains the tone of a muscle and allows muscles to work in a prolonged smooth manner.

When a muscle fiber fatigues, it indicates that the fiber has run out of ATP. The muscles of long distance runners have been known to fatigue because most of the fibers have run out of ATP. At that point, the runner simply collapses.

The muscles of the body are normally always partially contracted; the muscle fibers experience tetanus in turn.

Muscle Tone

As mentioned, whole skeletal muscles have **tone,** a condition in which there are always some fibers contracted. Muscle tone is particularly important in maintaining posture. If the muscles of the neck, trunk, and legs suddenly become relaxed, the body collapses.

The maintenance of the right amount of tone requires the use of special sense receptors called **muscle spindles.** A muscle spindle consists of a bundle of modified muscle fibers with sensory nerve fibers wrapped around a short, specialized region somewhere near the middle of their length. A spindle contracts along with muscle fibers, but thereafter it sends stimuli to the CNS that enable it to regulate muscle contraction so that tone is maintained.

Effect of Contraction on Size of Muscle

Forceful muscular activity over a prolonged period of time causes muscles to increase in size. This increase, called *hypertrophy,* occurs only if the muscle contracts to at least 75 percent of its maximum tension. However, only a few minutes of forceful exercise a day are required for hypertrophy to occur. The number of muscle fibers do not increase. Instead the size of each fiber increases. The fibers show a gain in metabolic potential as well as in the number of myofibrils. This means that the muscle can work longer before it gets tired. Some athletes take steroids, either testosterone or related chemicals, to promote muscle growth. This practice has many side effects, as discussed on page 175.

When muscles are not used or are used for only very weak contractions, they decrease in size, or atrophy. Atrophy can occur when a limb is placed in a cast or when the nerve serving a muscle is damaged. If nerve stimulation is not restored, the muscle fibers will gradually be replaced by fat and fibrous tissue. Unfortunately, atrophy causes the fibers to shorten progressively, leaving body parts in contorted positions.

All muscles have tone, but forceful contraction can cause them to increase in size; lack of contraction weakens them.

Muscle Movements and Actions

Types of Movements

Intact skeletal muscles are attached to bones by **tendons** that span joints. When a muscle contracts, a bone moves in relation to another bone. Figure 6.7 shows some of the following movements.

Angular movements increase or decrease the joint angle between the bones of a joint.

Flexion decreases the joint angle. Flexion of the elbow moves the forearm toward the upper arm; flexion of the knee moves the lower leg toward the upper leg. Special terms are applied to the feet. *Dorsiflexion* occurs when you stand on your heels; *plantar flexion* occurs when you stand on your toes.

Extension increases the joint angle. Extension of the flexed elbow straightens the arm so that there is now a 180° angle at the elbow. Hyperextension occurs when a portion of the body parts are extended beyond 180°. It is possible to hyperextend the head and the trunk of the body.

Abduction is the movement of a body part laterally away from the midline. Abduction of the arm or leg moves it to the side, away from the body.

Adduction is the opposite of abduction. For example, the arms have been adducted when they are next to the trunk and the legs have been adducted when they are next to one another.

Circular movements occur at ball-and-socket joints.

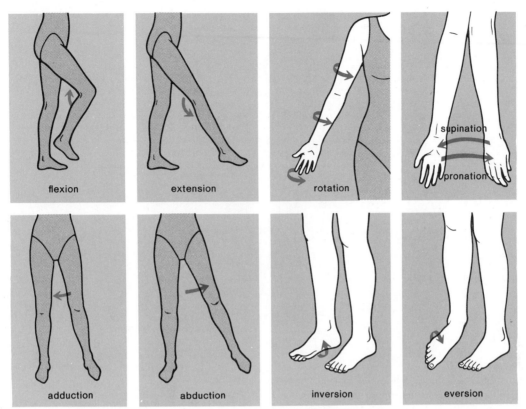

Rotation is the movement of a part around its own axis as when your head turns to answer no or when the arm is twisted one way and then the other. *Supination* is a special term meaning that the lower arm has been rotated so that the palm is upward, and *pronation,* is the opposite—the lower arm has been rotated so that the palm is downward.

Circumduction is the movement of a body part in a wide circle as when you move the arm in a wide circular swing. If you observe the motion carefully, you can see that because the proximal end of the arm is stationary, the shape outlined by the arm is actually a cone.

Other movements are described below.

Inversion and eversion are terms that apply only to the feet. Inversion is turning the foot so that the sole is inward and eversion is turning the foot so that the sole is outward.

Elevation and depression is a lifting up and down of a body part as when you shrug your shoulders.

Movements at joints are broadly classified as angular and circular. A few other types of movements are also possible.

Action of Movement

When the central portion of the muscle, called the belly, contracts, one bone remains fairly stationary and the other one moves. The origin of the muscle is on the stationary bone, and the insertion of the muscle is on the bone that moves.

Often a body part is moved by a group of muscles working together. Even so, one muscle does most of the work and it is called the prime mover. The assisting muscles are called the synergists. When a muscle contracts, it usually shortens. In other words, muscles can only pull; they cannot push. However, body parts often

Figure 6.8 Attachment of skeletal muscles as exemplified by the biceps and triceps. The origin of a muscle remains stationary, while the insertion moves. These muscles are antagonistic. When the biceps contract, the forearm is raised, and when the triceps contracts, the forearm is lowered.

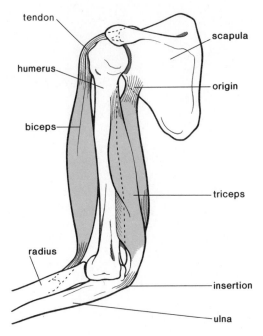

move in opposite directions as when the arm is flexed or extended. Therefore, muscles have **antagonists,** or work in antagonistic pairs as do the biceps and triceps when they move the lower arm up and down, respectively (fig. 6.8).

When muscles cooperate to achieve movement, some act as prime movers, some as synergists, and some are antagonists.

Isotonic Versus Isometric Contraction

Ordinarily, when muscles contract, they shorten and a movement occurs. This is called **isotonic contraction.** On occasion, muscles contract, but they do not shorten and no movement occurs. This is called an **isometric contraction.** For example, if we move a barbell weight up, the biceps contracts isotonically; however, when we pull on a stationary bar, the biceps contracts isometrically and no movement occurs.

Skeletal Muscles

Naming Muscles

The names of the various skeletal muscles (fig. 6.9 and 6.10) may indicate the muscle's:

a. Size. For example, the gluteus maximus is the large muscle that makes up the buttocks.
b. Shape. For example, the deltoid is shaped like a delta or triangle.
c. Direction of fibers. For example, the rectus abdominus is a longitudinal muscle of the abdomen (rectus means straight).
d. Location. For example, the frontalis overlies the frontal bone.
e. Number of attachments. For example, the biceps brachii has two attachments or origins.
f. Action. For example, the extensor digitorum extends the fingers.

Figure 6.9 Anterior view of superficial skeletal muscles.

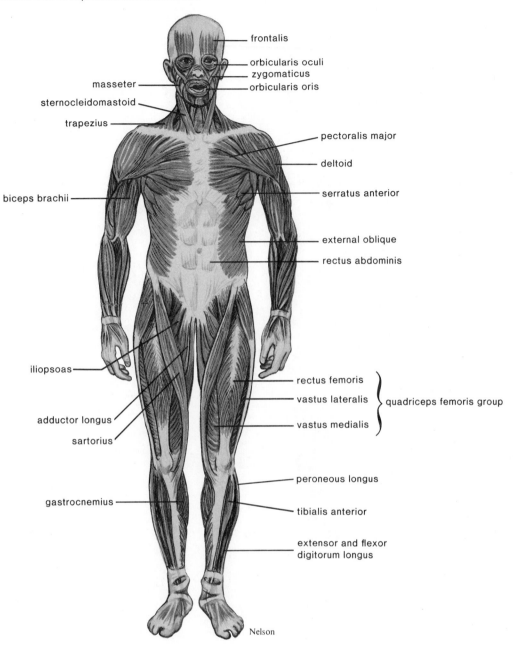

frontalis
orbicularis oculi
zygomaticus
masseter
orbicularis oris
sternocleidomastoid
trapezius
pectoralis major
deltoid
biceps brachii
serratus anterior
external oblique
rectus abdominis
iliopsoas
rectus femoris
vastus lateralis } quadriceps femoris group
adductor longus
vastus medialis
sartorius
peroneous longus
gastrocnemius
tibialis anterior
extensor and flexor
digitorum longus

Nelson

Figure 6.10 Posterior view of superficial skeletal muscles.

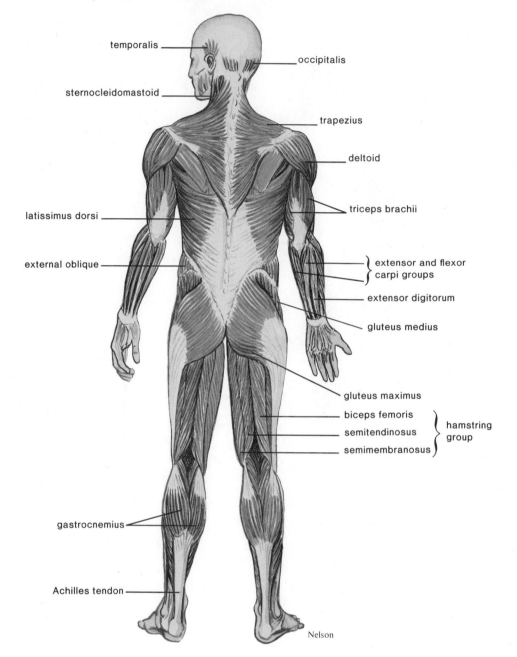

temporalis

occipitalis

sternocleidomastoid

trapezius

deltoid

triceps brachii

latissimus dorsi

external oblique

extensor and flexor carpi groups

extensor digitorum

gluteus medius

gluteus maximus

biceps femoris

semitendinosus

semimembranosus

hamstring group

gastrocnemius

Achilles tendon

Nelson

The names of muscles often include information about the size, shape, direction of fibers, attachments, location, and action of a muscle.

Muscles of the Head

The muscles of the head (fig. 6.11) are divided into the muscles of facial expression and the chewing muscles.

Facial Expression

Frontalis lies over the frontal bone and raises the eyebrows and wrinkles the brow.

Orbicularis oculi is a ringlike band of muscle that encircles (forms an orbit about) the eye. It causes the eye to close or blink and is responsible for what we call "crow's feet" at the corners of the eyes.

Orbicularis oris encircles the mouth and is used to pucker the lips as in forming a kiss.

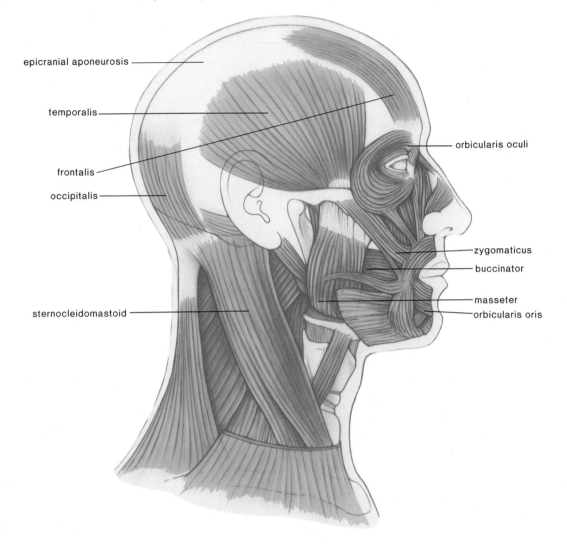

epicranial aponeurosis

temporalis

frontalis

occipitalis

sternocleidomastoid

orbicularis oculi

zygomaticus

buccinator

masseter

orbicularis oris

Buccinator is located in the cheek area, and when it contracts, the cheek is compressed as when you whistle or blow out air. (Sometimes this muscle is called the trumpeter's muscle.) Most importantly, this muscle helps hold our food in contact with the teeth when we are chewing.

Zygomaticus extends from the zygomatic arch (cheekbone) to the corners of the mouth. It raises the corners of the mouth when we smile.

Chewing Muscles

Masseter extends from the zygomatic arch to the mandible. It's a muscle of mastication (chewing) because it raises the mandible.

Temporalis is a fan-shaped muscle that overlies the temporal bone. It acts as a synergist to the masseter.

The muscles of the head are divided into those for facial expression and those for chewing.

Muscles of the Neck and Trunk

The posterior muscles are shown in figure 6.12 and the anterior muscles are shown in figure 6.13.

Muscles that Move the Head, Pectoral Girdle, and Upper Arm

Sternocleidomastoid muscles occur in the sides of the neck and extend from the sternum to the mastoid process—they are named for their attachments. When both of these muscles contract, the neck is flexed and the head is bent toward the chest. When only one contracts, the head turns to the opposite side.

Figure 6.12 Posterior muscles of the shoulder. (The trapezius is removed on the right.)

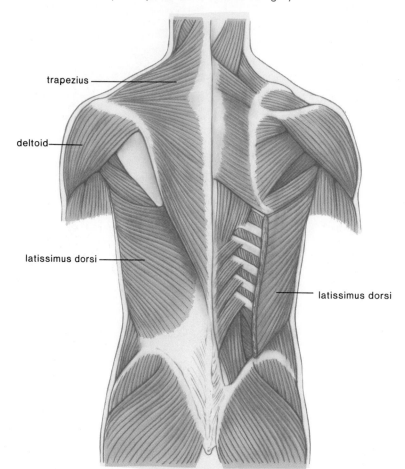

trapezius

deltoid

latissimus dorsi

latissimus dorsi

Trapezius muscles are triangular, but together they take on a diamond shape. Perhaps to some they appear to resemble a trapezoid. Each trapezius muscle runs from the base of the skull down to the end of the thoracic vertebrae and also inserts on a scapula laterally. The trapezius muscles move the scapulae and the head. They help move the scapulae as we shrug the shoulders or pull them back. They also extend the head as we pull it back.

The sternocleidomastoid and trapezius muscles are antagonistic in that one flexes the neck and the other extends the neck.

Deltoid is a large fleshy triangular muscle (deltoid in Greek means triangular) that covers the shoulder and causes the bulge of the upper arm. It runs from both the clavicle and the scapula to the humerus. This muscle abducts the arm (raises the arm laterally) to the horizontal position.

Pectoralis major is a large (major) anterior muscle of the upper chest. It does originate from the clavicles of the pectoral girdle, but also from

the sternum and ribs. It inserts on the humerus. If you press your hands together forcefully, you can feel these muscles contract isometrically. The pectoralis major flexes the arm (raises it anteriorly) and adducts the arm, pulling it across the chest.

Latissimus dorsi (located both laterally and dorsally) is a wide triangular muscle of the lower back. This muscle originates from the lower spine and sweeps upward to insert on the humerus. The latissimus dorsi extends and adducts the arm (brings it down from a raised position). This is a very important muscle for swimming, rowing, and climbing a rope.

Serratus anterior is located below the axilla (armpit) on the side of the chest. It runs between the upper ribs and the scapula. It pulls the scapula downward and forward as when we push something. It also helps to raise the arm above the horizontal level.

The deltoid, pectoralis major, latissimus dorsi, and serratus anterior muscles all function to move the humerus and, therefore, the arm in relation to the trunk.

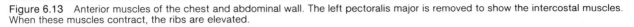

Figure 6.13   Anterior muscles of the chest and abdominal wall. The left pectoralis major is removed to show the intercostal muscles. When these muscles contract, the ribs are elevated.

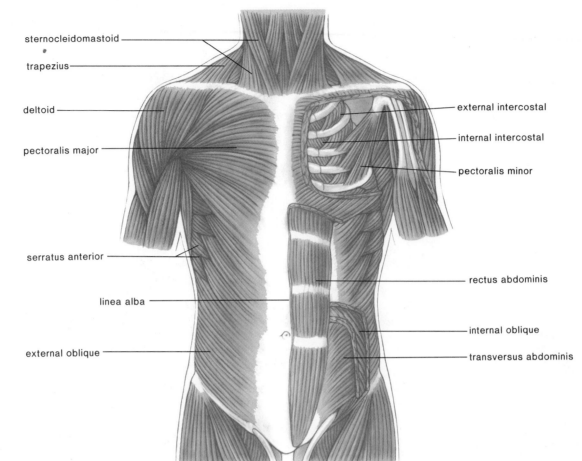

Muscles of the Abdominal Wall

External and *internal obliques,* and the *transversus abdominis* are located in the abdominal wall. The external obliques and the internal obliques run at a slant and at right angles to one another between the lower ribs to the pelvic girdle (fig. 6.13). Below the obliques, the transversus abdominis runs horizontally across the abdomen. Just like plywood, the abdominal wall is strengthened by having muscle fibers that run in different directions. All of these muscles tense and support the abdominal wall.

Rectus abdominis has a straplike appearance but takes its name from the fact that it runs straight (rectus means straight) up from the pubic bones to the ribs and sternum. It is the outermost muscle that compresses the contents of the abdominal cavity, but it also helps flex the vertebral column.

The obliques (external and internal), the transversus abdominis, and the rectus abdominis all function to provide a sturdy abdominal wall.

Muscles of the Upper Limb (Arm)

These muscles are illustrated in fig. 6.14.

Muscles that Move the Forearm

Biceps brachii is a muscle of the anterior forearm that is familiar to everyone because it bulges when the upper arm is flexed. It also supinates the hand when you turn a doorknob or twist the cap of a jar. The name of the muscle refers to its two heads that attach to the scapula where it originates. It inserts on the radius.

Triceps brachii is the only muscle of the posterior upper arm. It has three heads that attach to the scapula and humerus, and it inserts on the ulna. The triceps extends the forearm, and is sometimes called the boxer's muscle because it straightens the elbow when hitting something. It is also used when you push something.

The biceps and triceps are antagonistic muscles because one flexes the forearm and the other extends the forearm.

Figure 6.14 Cross section of the upper arm.

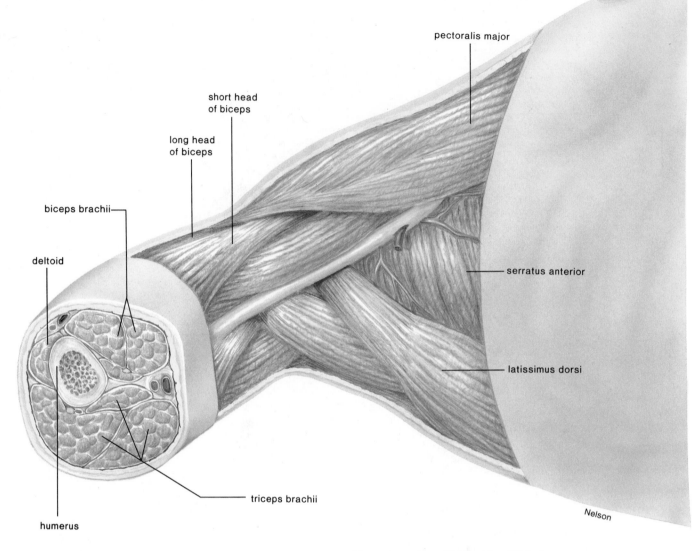

pectoralis major

short head
of biceps

long head
of biceps

biceps brachii

deltoid

serratus anterior

latissimus dorsi

triceps brachii

Nelson

humerus

Muscles that Move the Hand

Flexor carpi and *extensor carpi muscles* (fig. 6.10) originate on the bones of the forearm and insert on the bones of the hand. They move the wrist and hand.

Flexor digitorum and *extensor digitorum muscles* also originate on the bones of the forearm and insert on the bones of the hand. They move the fingers.

The muscles that move the hand and fingers span the wrist.

Muscles of the Lower Limb (Leg)

These muscles tend to be large and heavy because they are used to move the entire weight of the body and to resist the force of gravity. Therefore they are important for locomotion of all types and for balance.

Muscles that Move the Thigh

These muscles are shown in figures 6.15 and 6.16.

Iliopsoas originates from the ilium and the bodies of the lumbar vertebrae and inserts on the femur anteriorly. This muscle flexes the thigh (raises the leg[1] anteriorly) and is important to the process of walking. It also helps the trunk from falling backward when you are standing erect.

Gluteus maximus is the largest muscle in the body and covers a large part of the buttock (gluteus means buttocks in Greek). It originates at the ilium and sacrum, and inserts on the femur. It acts to straighten the leg at the hip (and in that way, to extend the thigh) when one is walking, climbing stairs, or jumping from a crouched position.

1. The term leg is used in this section to refer to the entire lower limb.

Figure 6.15   Anterior muscles of the right thigh.

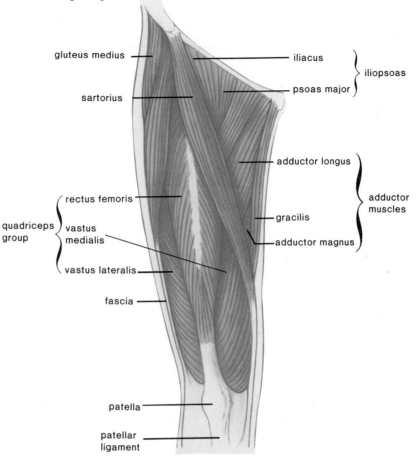

The iliopsoas and the gluteus maximus are antagonistic muscles because one flexes the thigh and the other extends the thigh.

Gluteus medius lies partly behind the gluteus maximus. It runs between the ilium and the femur, and functions to abduct the thigh (raise the leg sideways to a horizontal position).

Adductor muscles are located on the medial part of the thigh. They originate from the pubic bone and ischium, and insert on the femur. These muscles adduct the thigh (bring down the leg from a horizontal position) and press the thighs together.

The gluteus medius and adductor muscles are antagonistic because one acts to abduct the thigh and the others act to adduct the thigh.

Muscles that Move the Lower Leg

These muscles are shown in figures 6.15 and 6.16.

Quadriceps femoris group is composed of several muscles that are found in the front and sides of the thigh. One member of the group (rectus femoris) originates from the ilium, and the others (vastus muscles) originate from the femur. This group of muscles is the primary extensor of the lower leg as when you kick a ball.

Hamstring group is composed of several muscles that are found at the back of the thigh. All have origins on the ischium and insert on the tibia. Their strong tendons can be felt behind the knee; these same tendons are present in pigs and were used by butchers as strings to hang up hams for smoking. This group of muscles helps in flexing the lower leg and extending thigh.

The quadriceps femoris group of muscles and the hamstring group are antagonistic because one extends the lower leg and the other flexes it.

Sartorius is a long straplike muscle that begins at the iliac spine and then passes inward across the front of the thigh to descend over the medial side of the knee. It is used when one sits cross-legged as tailors were accustomed to do in another era. Therefore, it is sometimes called the tailor's muscle and, in fact, sartor means tailor in Latin.

Figure 6.16 Posterior muscles of the right thigh.

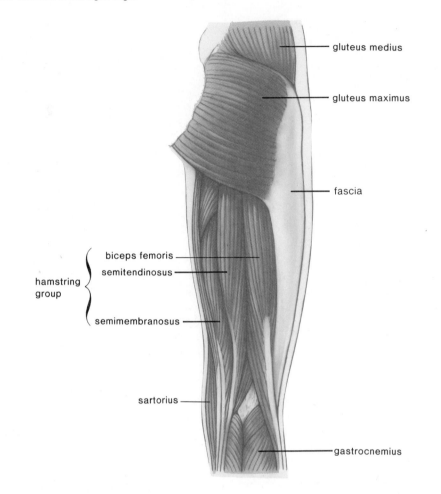

- gluteus medius
- gluteus maximus
- fascia

hamstring group {
- biceps femoris
- semitendinosus
- semimembranosus
}

- sartorius
- gastrocnemius

Muscles that Move the Ankle and Foot

These muscles are shown in figures 6.17 and 6.18.

Gastrocnemius is located on the back of the leg where it forms a large part of the calf. It arises from the femur; distally, the muscle joins the strong Achilles tendon, which attaches behind the calcaneus (heel bone). The gastrocnemius is a powerful plantar flexor of the foot that aids in pushing the body forward when you walk or run. It is sometimes called the toe dancer's muscle because it allows you to stand on tiptoe.

Tibialis anterior is a long spindle-shaped muscle located on the front of the lower leg. It arises from the surface of the tibia, and attaches to the bones of the ankle and foot. Contraction of this muscle causes dorsiflexion and inversion of the foot.

The gastrocnemius and the tibialis anterior are antagonistic muscles because one brings about plantar flexion and the other, dorsiflexion of the foot.

Peroneus muscles are found on the lateral side of the leg where they connect the fibula to the metatarsal bones of the foot. These muscles evert the foot and also help bring about plantar flexion.

Flexor and extensor digitorum longus muscles are found on the lateral and posterior portion of the leg. They arise mostly from the tibia and insert on the toes. They flex and extend the toes, and assist in other movements of the feet.

The peroneus muscles, and the flexor and extensor digitorum muscles move the feet and toes.

Table 6.3 lists all of the muscles discussed in this chapter and, in particular, gives more complete information regarding their origin and insertions.

Figure 6.17 Anterior muscles of the right lower leg.

Figure 6.18 Lateral muscles of the right lower leg.

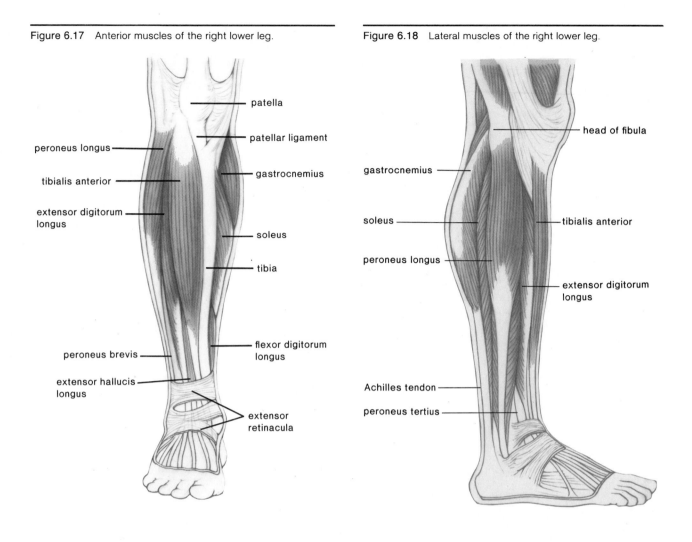

Table 6.3 Skeletal Muscles

Name	Origin/Insertion	Function
Muscles of the Head		
Frontalis[a]	Epicranial aponeurosis/skin and muscles around eye	Raises eyebrows
Orbicularis oculi[a]	Maxillary and frontal bones/skin around eye	Closes eyes
Orbicularis oris[a]	Muscles near the mouth/skin around mouth	Closes and protrudes lips
Buccinator	Outer surfaces of maxilla and mandible/orbicularis oris	Compresses cheeks inward
Zygomaticus[a]	Zygomatic bone/orbicularis oris	Raises corner of mouth
Masseter[a]	Zygomatic arch/mandible	Closes jaw
Temporalis[b]	Temporal bone/coronoid process of mandible	Closes jaw

[*continued*]

Table 6.3—*continued*

Name	Origin/Insertion	Function
Muscles of the Neck and Trunk		
Sternocleidomastoid[b]	Sternum/mastoid process of temporal bone	Flexes neck and pulls head to one side
Deltoid[a,b]	Acromion process, spine of scapula, and the clavicle/deltoid tuberosity of humerus	Abducts arm[c]
Pectoralis major[a]	Clavicle, sternum, upper ribs/intertubercular groove of humerus	Flexes and adducts arm
Latissimus dorsi[b]	Lower spine, iliac crest/intertubercular groove of humerus	Extends and adducts arm
Serratus anterior[a]	Upper ribs/scapula	Pulls scapula downward and forward
Muscles of the Abdominal Wall		
External oblique[a]	Lower ribs/iliac crest	Tenses abdominal wall
Internal oblique	Crest of ilium/lower ribs, crest of pubis	Tenses abdominal wall
Transversus abdominis	Lower ribs, lumbar vertebrae, iliac crest/crest of pubis	Tenses abdominal wall
Rectus abdominis[a]	Crest of pubis, symphysis pubis/xiphoid process of sternum, costal cartilages	Tenses abdominal wall and flexes vertebral column
Muscles of the Upper Limb		
Biceps brachii[a]	Scapula/tuberosity of the radius	Flexes elbow and supinates hand
Triceps brachii[b]	Scapula, proximal humerus/olecranon process of ulna	Extends elbow
Extensor carpi and flexor carpi[b]	Humerus/carpal and metacarpal	Move wrist and hand
Extensor digitorum and flexor digitorum[b]	Humerus, radius, ulna/phalanges	Move fingers
Muscles of the Lower Limb		
Iliopsoas[a]	Lumbar vertebrae, ilium/lesser trochanter of femur	Flexes thigh
Gluteus maximus[b]	Posterior ilium, sacrum/posterior femur	Extends leg[d]
Gluteus medius	Ilium/greater trochanter of femur	Abducts thigh
Adductor muscles[a]	Pubis, ischium/femur and tibia	Adducts thigh
Quadriceps femoris group[a]	Ilium, femur/patella tendon that continues as a ligament to tibial tuberosity	Extends knee
Hamstring group[b]	Ischial tuberosity/lateral and medial tibia	Flexes knee and extends thigh
Sartorius[a]	Ilium/medial tibia	Flexes, abducts, and rotates leg[d]
Gastrocnemius[b]	Condyles of femur/calcaneus by way of Achilles tendon	Plantar flexion of foot and flexes knee
Tibialis anterior[a]	Condyles of tibia/tarsal and metatarsal bones	Dorsiflexion and inversion of foot
Peroneus muscles[a]	Fibula/tarsal and metatarsal bones	Plantar flexion and eversion of foot
Flexor and extensor digitorum longus[a]	Tibia, fibula/phalanges	Move toes

[a] These muscles are shown in figure 6.9, anterior superficial muscles.

[b] These muscles are shown in figure 6.10, posterior superficial muscles.

[c] Refers to the entire upper limb.

[d] Refers to the entire lower limb.

Summary

I. Skeletal Muscle Structure
   A. Muscle anatomy. Fascia (connective tissue) covers the surface of the muscle; epimysium is the layer of connective tissue closest to the muscle and perimysium divides up the muscle tissue into units called fascicles; endomysium surrounds each muscle fiber.
   B. Muscle fibers. Myofibrils are the contractile portions of muscle fibers. The sliding filament theory states that actin filaments slide past myosin filaments because myosin has cross bridges that pull the actin filaments inward.
      Muscle contraction requires a ready supply of ATP. Creatine phosphate is used to generate ATP rapidly. If oxygen is in limited supply, anaerobic respiration produces ATP and results in oxygen debt.
      When a nerve fiber innervates a muscle fiber at a neuromuscular (myoneural) junction, a muscle action potential occurs that causes calcium to be released from calcium storage sacs, and thereafter contraction occurs.
   C. Contraction of a Whole Muscle. Muscle fibers obey the all-or-none law, but whole muscles do not obey this law.
      Muscle twitch, summation, and tetanus are related to the frequency with which a muscle is stimulated.
      Intact muscles. Movements at joints are broadly classified as angular and circular. A few other types of movements are also possible.
      When muscles cooperate to achieve movement, some act as prime movers, some as synergists, and some are antagonists.
      All muscles have tone, but forceful contraction can cause them to increase in size; lack of contraction weakens them.

II. Skeletal Muscles
   A. Naming muscles. The names of muscles often include information about the size, shape, attachments, location, and action of a muscle.
   B. Muscles of the head. The muscles of the head are divided into those for facial expression and those for chewing.
   C. Muscles of the neck and trunk.
      The sternocleidomastoid and trapezius muscles are antagonistic in that one flexes the neck and the other extends the neck.
      The deltoid, pectoralis major, latissimus dorsi, and serratus anterior muscles all function to move the humerus and, therefore, the arm in relation to the trunk.
   D. Muscles of the abdominal wall. The obliques (external and internal), the transverse abdominis, and the rectus abdominis, all function to provide a sturdy abdominal wall.
   E. Muscles of the upper limb.
      The biceps and triceps are antagonistic muscles because one flexes the forearm and the other extends the forearm.
      The muscles that move the hand and fingers span the wrist.
   F. Muscles of the lower limb.
      The iliopsoas and the gluteus maximus are antagonistic muscles because one flexes the thigh and the other extends the thigh.
      The gluteus medius and adductor muscles are antagonistic because one acts to abduct the thigh and the others act to adduct the thighs.
      The quadriceps femoris group of muscles and the hamstring group are antagonistic because one group extends the lower leg and the other flexes it.
      The gastrocnemius and the tibialis anterior are antagonistic muscles because one brings about plantar flexion and the other, dorsiflexion of the foot.
      The peroneus muscles, and the flexor and extensor digitorum muscles move the feet and toes.

Study Questions

1. Name the layers of fascia that cover and divide up muscle tissue.
2. Discuss the microscopic anatomy of a muscle fiber and the structure of a sarcomere. What is the sliding filament theory?
3. What is oxygen debt, and how is it repaid?
4. What causes a muscle action potential? How does the muscle action potential bring about sarcomere and muscle fiber contraction?
5. The study of whole muscle physiology often includes observing both threshold and maximal stimulus, muscle twitch, summation, and tetanic contraction. Describe the significance of each of these.
6. How are joint movements classified? Give examples of each type.
7. Describe how muscles are attached to bones. Define prime mover, synergist, and antagonist.
8. What is tone? Compare isotonic versus isometric contraction.
9. How do muscles get their names? Give an example for each method of naming a muscle.
10. Which of the head muscles is used for facial expression; for chewing?
11. Which of the muscles of the neck and trunk flex and extend the head? Name the muscles that move the humerus and tell their actions.
12. What are the muscles of the abdominal wall?
13. Which of the muscles of the upper limb move the forearm, and what are their actions? Name the muscles that move the hand and fingers.
14. Which of the muscles of the lower limb move the thigh, and what are their actions? Which of the muscles of the lower limb move the lower leg, and what are their actions? Which of the muscles of the lower limb move the feet?

Objective Questions

I. Fill in the blanks.

1. Actin and myosin filaments are found within cell inclusions called _____ , which are divided into units called _____ .

2. The molecule _____ serves as an immediate source of high-energy phosphate for ATP production in muscle cells.

3. The region between an axon ending and muscle cell sarcolemma is called a _____ junction.

4. The angular movement, _____ , decreases the joint angle, and _____ increases the joint angle.

5. When muscles contract, the _____ does most of the work, but the _____ help.

6. Whole muscles have _____ , a condition in which there are always some fibers contracted.

7. The _____ is a muscle in the upper arm that has two origins.

8. The _____ act as the origin of the latissimus dorsi, and the _____ acts as the insertion during most activities.

II. Name the muscle indicated by the following combinations of origin and insertion.

Origin	Insertion	Muscle
9. temporal bone	coronoid process of mandible	_____
10. scapula, clavicle	humerus	_____
11. scapula, proximal humerus	olecranon process of ulna	_____
12. posterior ilium, sacrum	posterior femur	_____

III. For questions 13–20, match the muscle in the key to the actions below.

Key: a. orbicularis oculi
 b. zygomaticus
 c. deltoid
 d. serratus anterior
 e. rectus abdominis
 f. iliopsoas
 g. gluteus maximus
 h. gastrocnemius

13. stand on tiptoe
14. tense abdominal wall
15. abducts arm
16. flexes thigh
17. raises corner of mouth
18. closes eyes
19. extends leg
20. pulls scapula downward and forward

Medical Terminology Reinforcement Exercise

Pronounce, dissect, and analyze the meaning of the following terms:

1. hyperkinesis (hi''per-ki-ne'sis)
2. dystrophy (dis'tro-fe)
3. electromyogram (e-lek''tro-mi'-o-gram)
4. meninsectomy (men''i-sek'to-me)
5. tenorrhaphy (ten-or'ah-fe)
6. myatrophy (mi-at'ro-fe)
7. leiomyoma (li''o-mi-o'mah)
8. kinesiotherapy (ki-ni''se-o-ther'ah-pe)
9. myocardiopathy (mi''o-kar''de-op'ah-the)
10. myasthenia (mi''as-the'ne ah)

Integration and Coordination

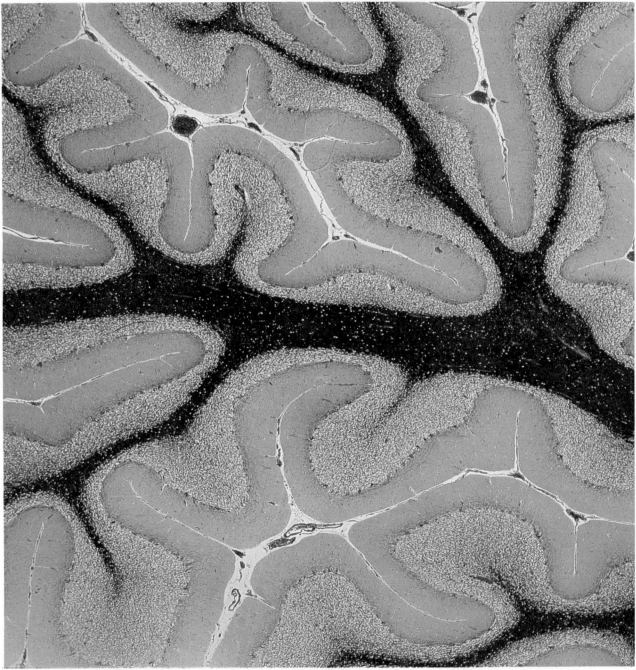

Tracts of white matter pervade the convoluted gray matter of cerebellum, that portion of the brain responsible for skeletal muscle coordination.

7

Nervous System

Chapter Outline

Learning Objectives

After you have studied this chapter, you should be able to:

1. State and explain the divisions of the nervous system.
2. Describe the structure of three types of neurons and the components of the nerve impulse.
3. Explain how the nerve impulse is transmitted across a synapse, and name two common neurotransmitters.
4. Describe a reflex arc that consists of three neurons.
5. Describe the three layers of meninges and state their function.
6. Describe in detail the structure of the spinal cord.
7. Name the major parts of the brain and state functions for each part.
8. Describe the structure of the cerebral cortex and the major functions of its lobes.
9. Describe the limbic system, and its relationship to learning and memory.
10. Explain the relationship between neurotransmitters in the brain, and neurological illnesses and drugs.
11. Name and describe the differences between the three different types of nerves.
12. Name the twelve cranial nerves and give a function for each.
13. Describe the classification of spinal nerves and, in general, their relationship to plexuses.
14. Distinguish between the sympathetic and parasympathetic systems in four ways, and give examples of their respective effects on specific organs.

Nervous System

The nervous system and the endocrine system are responsible for regulation and coordination of body parts. The nervous system permits a quick response to external and internal stimuli while the endocrine system is slower to act, but the effect is longer lasting. This chapter concerns the nervous system; the endocrine system is discussed in chapter 9.

Divisions of the Nervous System

The nervous system has two major divisions: the central and peripheral nervous system (fig. 7.1). The **central nervous system (CNS)** includes the brain and spinal cord, which lie in the midline of the body where the brain is protected by the skull, and the spinal cord is protected by the vertebrae. The **peripheral nervous system (PNS)**, which is further divided into the somatic division and the autonomic division, includes all the cranial and spinal nerves. These nerves project out from the central nervous system and this is why it is called the peripheral nervous system. Figure 7.2 illustrates what is meant by the central nervous system and the peripheral nervous system. The division is arbitrary; the two systems work together and are connected to one another.

The CNS lies in the midline of the body and the PNS is located peripherally to the CNS.

Nervous Tissue

Nerve tissue contains nerve cells which are called neurons. **Neurons** are specialized to carry nerve impulses. **Glial cells** are also present in nerve tissue. These cells support, protect, and nourish the neurons. One type of glial cell in the PNS, called a *Schwann cell,* wraps itself around long nerve fibers, forming an insulating sheath known as the **myelin sheath** (fig. 7.3). Outside the myelin sheath is the **neurilemmal sheath** containing the cytoplasm, nucleus, and the exposed cell membrane of the Schwann cell. It is these protective coverings that give nerves a white glistening appearance. Notice in figure 7.4 that there are gaps in between the Schwann cells; these are called *nodes of Ranvier,* after the person who first observed them.

The neurilemma is important to nerve regeneration. The nerve fiber and myelin sheath distal to a cut deteriorate and are removed by macrophages, but the outer layer of neurilemma remains and forms a hollow tube. The healthy proximal end of the nerve fiber develops sprouts, and one of these may grow into and be guided by this tube.

Figure 7.1 Overall organization of the nervous system. The central nervous system is at the top of the diagram and the peripheral nervous system is below. These portions of the nervous system take their names from their locations in the body.

Figure 7.2 Location of central nervous system (*brain* and *cord*) and peripheral nervous system (*nerves*). The CNS (central nervous system) lies in the center of the body and the PNS (peripheral nervous system) lies to either side.

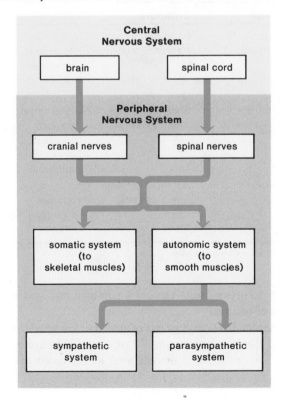

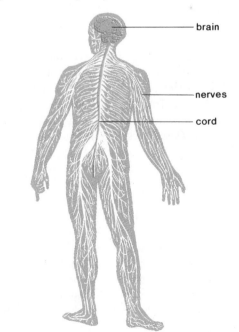

Figure 7.3 The myelin sheath forms when Schwann cells wrap themselves about a nerve fiber in the manner shown. The neurilemmal sheath contains the cytoplasm, nucleus, and outer cell membrane of the Schwann cell.

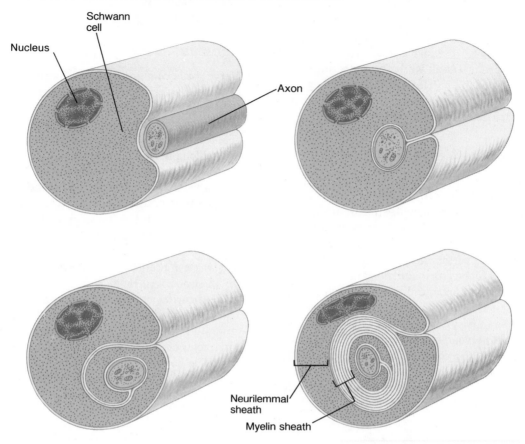

Neurons

Structure

All neurons (fig. 7.4) have three parts: a dendrite(s), a cell body, and an axon. The **cell body** contains the nucleus and other organelles. A **dendrite** conducts nerve impulses toward the cell body, and an **axon** conducts nerve impulses away from the cell body. There are three types of neurons: sensory, motor, and interneuron. A sensory neuron takes a message from a sense organ to the CNS, and has a long dendrite and a short axon. A **motor neuron,** however, takes a message away from the CNS to a muscle fiber or gland, and has short dendrites and a long axon. Because motor neurons cause muscle fibers and glands to react, they are said to innervate these structures.

Sometimes a sensory neuron is referred to as an *afferent neuron,* and the motor neuron is called an *efferent neuron.* These words, which are derived from Latin, mean running to and running away from, respectively. Obviously, they refer to the relationship of these neurons to the CNS.

An **interneuron** is always found completely within the CNS and conveys messages between parts of the system. An interneuron has short dendrites but the axon

can be short or long. Table 7.1 summarizes the three types of neurons.

The dendrites and axons of neurons are sometimes called **fibers,** or processes. Most long fibers are protected by myelin and neurilemma as discussed previously.

Table 7.1	Neurons	
Neuron	**Structure**	**Function**
Sensory (afferent)	Long dendrites, short axon	Carries nerve impulses (message) from periphery to CNS[a]
Motor (efferent)	Short dendrites, long axon	Carries nerve impulses (message) from CNS to periphery
Interneuron	Short dendrites, long or short axon	Carries nerve impulses (message) within CNS

[a]CNS = central nervous system.

Figure 7.4 All neurons have three parts: dendrites, a cell body, and an axon. *a.* Motor neuron. *b.* Sensory neuron.

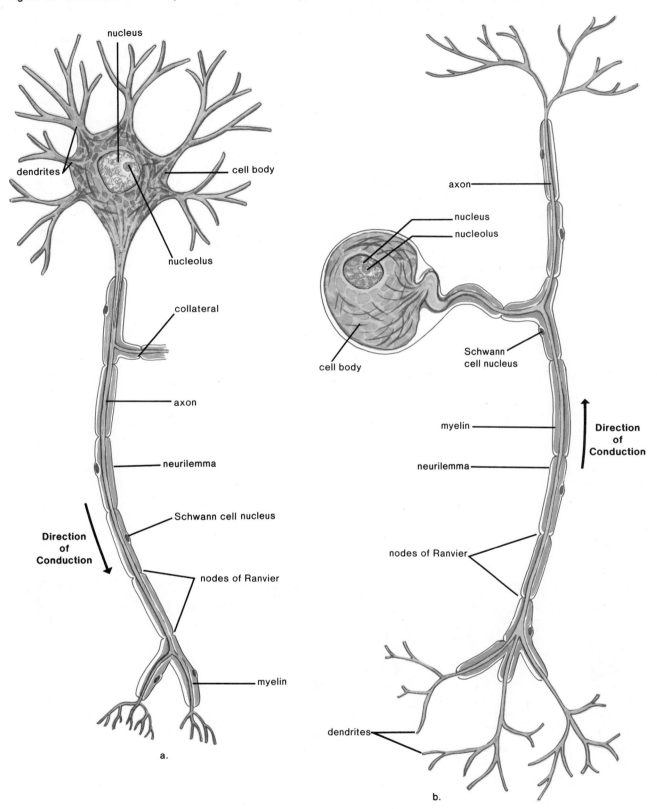

a.

b.

Although all neurons have the same three parts, there are three different types of neurons. Each one is specialized for one job in relation to the CNS.

Nerve Impulse

When an axon is not conducting a nerve impulse, the outside is positive compared to the inside which is negative (fig. 7.5). This difference in polarity (charge) is due to the accumulation of positive charges outside the axon. First, there is a carrier in the membrane (see p. 36) called the sodium–potassium pump, which pumps sodium (Na^+) out of the axon and potassium (K^+) into the axon. Thereafter, each of these ions diffuses down its concentration gradient; however, the membrane is more permeable to K^+ than to Na^+, so the K^+ ions "leak" across the membrane at a faster rate than Na^+. Another factor that causes the inside of the axon to be negative compared to the outside is the presence of large, negatively charged protein ions inside an axon.

The difference in charge across an axon that is not conducting impulses is called the *resting potential*. Another name for a nerve impulse is the *action potential*.

Action Potential

When the nerve fiber is conducting a nerve impulse, there is a change in polarity that flows along the axomembrane. As the nerve impulse passes by, the inside

Figure 7.5   The resting and action potential. *a.* Resting potential. The outside of the membrane is positive and the inside is negative due to *b.* the presence of large negative organic ions inside the axoplasm. Note also the unequal distribution of Na^+ and K^+ across the membrane. *c.* Action potential. The action potential is a change in polarity that may be explained by *d.* first the movement of Na^+ to the inside and second, by the movement of K^+ to the outside of the axon.

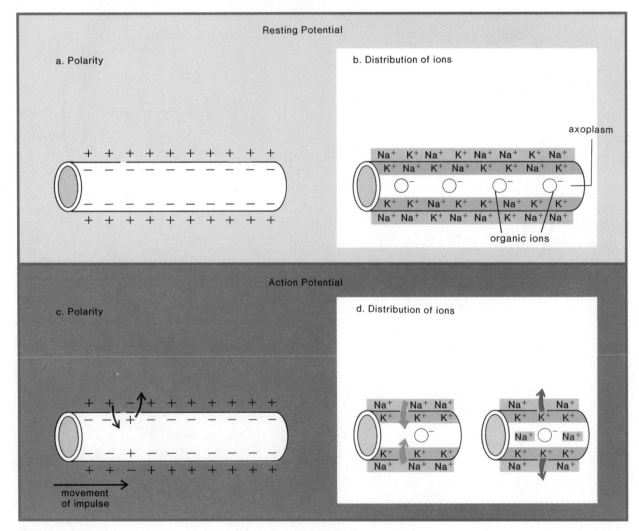

of an axon first becomes positive compared to the outside (this is called *depolarization*) and then the inside becomes negative again (this is called *repolarization*). During depolarization, Na$^+$ ions move to the inside of the axon and during repolarization, K$^+$ ions move to the outside. There are special channels in the axomembrane that open to allow each ion to temporarily move rapidly across the membrane (fig. 7.6). After the nerve impulse has passed, the normal polarity has been restored, but the distribution of the ions is opposite to before: Na$^+$ is on the inside and K$^+$ is on the outside. The sodium–potassium pump, however, restores the original distribution and the nerve fiber is ready once again to conduct a nerve impulse.

In myelinated fibers, each nerve impulse skips from one node of Ranvier to the next, and this increases the speed with which a fiber can conduct nerve impulses.

All neurons transmit the same type of nerve impulse—a change in polarity that flows along the membrane of a nerve fiber (table 7.2).

Transmission across a Synapse

Individual neurons do not actually touch each other. Instead, there is a small gap between the end of a nerve fiber and the dendrite or cell body of another. This gap is called a synaptic cleft. Transmission of nerve impulses across a **synapse** is carried out by chemicals called **neurotransmitter substances.** A neurotransmitter is stored within synaptic vesicles at the end of axons (fig. 7.7). When nerve impulses reach the end of an axon, calcium ions flow into the axoplasm. These ions activate enzymes that cause the synaptic vesicles to release the neurotransmitter substance into the synaptic cleft. The neurotransmitter diffuses across the cleft and binds to receptors in the membrane of the receiving dendrite (or cell body). Now action potentials (nerve impulses) begin in the next neuron.

Table 7.2 Nerve Impulse Events

1. Resting potential is present.
   Outside is positive and sodium (Na$^+$) ions are outside. Inside is negative and potassium (K$^+$) ions are inside.
2. Na$^+$ channels open and sodium ions move to inside.
3. K$^+$ channels open and potassium ions move to outside.
4. Channels close and the sodium-potassium pump starts to work to restore conditions of resting potential.
5. Nerve impulse moves to next node of Ranvier.

Acetylcholine (ACh) is a well-known stimulatory neurotransmitter that is active in all parts of the nervous system. Once acetylcholine has been received, it is broken down by an enzyme called **acetylcholinesterase (AChE).** This prevents overstimulation of the receiving neuron.

One-Way Propagation

Transmission of a nerve impulse is always from the axon to the dendrite (or cell body) because only the ends of axons have synaptic vesicles. Also, it may be noted that neurons obey the all-or-none law, meaning that a neuron either fires (conducts a nerve impulse) or does not fire at all. A nerve does not obey the all-or-none law because a nerve contains many fibers (fig. 7.18), any number of which may be carrying nerve impulses. Therefore, a nerve may have degrees of performance.

Neurotransmitters and Neurological Disorders

It has been discovered that several neurological illnesses, such as *Parkinson disease* and *Huntington disease* are due to an imbalance in neurotransmitters

Figure 7.6   The nerve cell membrane is believed to have Na$^+$ and K$^+$ channels that act like gates. When a channel is closed, ions cannot pass through; when a channel is open, ions can pass through.

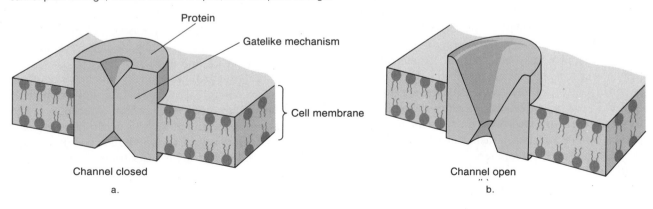

Protein

Gatelike mechanism

Cell membrane

Channel closed

a.

Channel open

b.

Figure 7.7 Transmission of nerve impulse across the synapse. Inflow of calcium leads to enzymatic change in axomembrane permeability so that acetylcholine (ACh) is released and diffuses across the synaptic cleft. After ACh binds to receptors (R), nerve impulses begin and ACh is broken down by AChE (acetylcholinesterase).

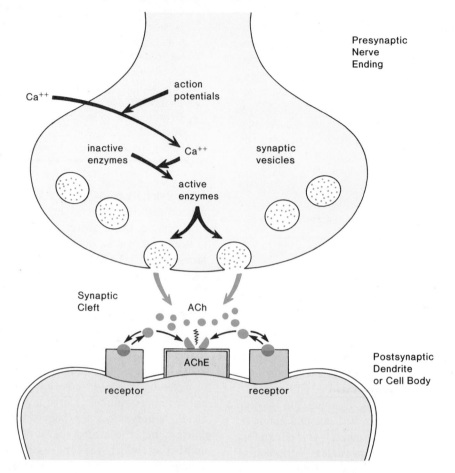

within the brain. Parkinson disease is a condition characterized by a wide-eyed, unblinking expression, an involuntary tremor of the fingers and thumbs, muscular rigidity, and a shuffling gait. All of these symptoms are due to a deficiency of dopamine, a particular neurotransmitter found in the brain. Huntington disease is characterized by a progressive deterioration of the individual's nervous system that eventually leads to constant thrashing and writhing movement, and finally, insanity and death. The problem is believed to be due to a malfunction of GABA, another neurotransmitter of the brain. Most recently, it has been discovered that *Alzheimer disease,* a severe form of senility with marked memory loss found in 5 to 10 percent of all people over age sixty-five is due to the death of basal ganglia (p. 137) neurons that use ACh as a transmitter.

Treatment of individuals with brain disorders has formerly been directed toward trying to restore the proper balance of neurotransmitter substances. More recently, however, researchers are exploring the possibility of transplanting tissue that produces the missing neurotransmitter.

Transmission of a nerve impulse across a synapse is dependent on the release of a transmitter substance into a synaptic cleft.

Reflex Arc

Reflexes are automatic, involuntary responses to changes occurring inside or outside the body. For example, we do not have to consciously think about our heart rate, breathing rate, body temperature, or even digesting our food because these are controlled by reflexes. Reflexes are also involved in swallowing, sneezing, vomiting, and urination.

Other reflexes that we are very much aware of are reactions to painful stimuli (fig. 7.8). For example, if a person touches a very hot object, a receptor in the skin generates nerve impulses that move along the dendrite of a sensory neuron toward the cell body and CNS. From the cell body, the impulses travel along the axon of the sensory neuron and enter the cord; there they

Figure 7.8 Diagram of a reflex arc, the functional unit of the nervous system. Trace the path of a reflex by following the black arrows. Name the three types of neurons that are required for a simple reflex, such as the rapid response to touching a hot object with the hand.

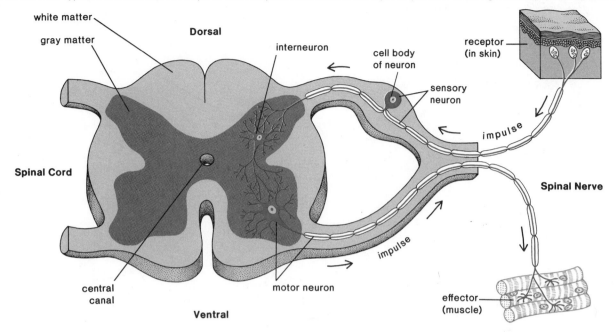

Table 7.3 Path of a Reflex Arc

Part	Description	Function
Receptor	The receptor end of a dendrite or a specialized receptor cell in a sensory organ	Sensitive to a specific type of internal or external change
Sensory neuron	Dendrite, cell body, and axon of a sensory neuron	Transmits nerve impulse from the receptor into the brain or spinal cord
Interneuron	Dendrite, cell body, and axon of a neuron within the brain or spinal cord	Serves as processing center; conducts nerve impulse from the sensory neuron to a motor neuron
Motor neuron	Dendrite, cell body, and axon of a motor neuron	Transmits nerve impulse from the brain or spinal cord out to an effector
Effector	A muscle or gland outside the nervous system	Responds to stimulation by the motor neuron and produces the reflex or behavioral action

From John W. Hole, Jr., *Human Anatomy and Physiology*, 5th ed. Copyright © 1990 Wm. C. Brown Publishers, Dubuque, Iowa. All Rights Reserved. Reprinted by permission.

may pass to many interneurons, one of which connects with a motor neuron. The nerve impulses travel along the axon to muscle fibers, which then contract so that the hand is withdrawn from the hot object. (See table 7.3 for a listing of these events.) Various other reactions usually accompany a reflex response; the person may look in the direction of the object, jump back, and utter appropriate exclamations. This whole series of responses is explained by the fact that the sensory neuron stimulates several interneurons, which take impulses to all parts of the central nervous system, including those responsible for consciousness.

Clinical Application

Reflexes are often used to determine if the nervous system is reacting properly. For example, two reflexes are described here:

Knee-jerk reflex (patellar reflex). This reflex is initiated by striking the patellar ligament just below the patella. The response is a contraction of the quadriceps femoris muscles that cause the lower leg to extend.

Figure 7.9 Central nervous system. *a.* The brain and spinal cord are protected by bone. *b.* The spinal cord is protected by vertebrae, and the spinal nerves are not apparent until they project from between the vertebrae. *c.* Anatomy of the spinal cord and spinal nerve. A spinal nerve arises after the dorsal and ventral roots join.

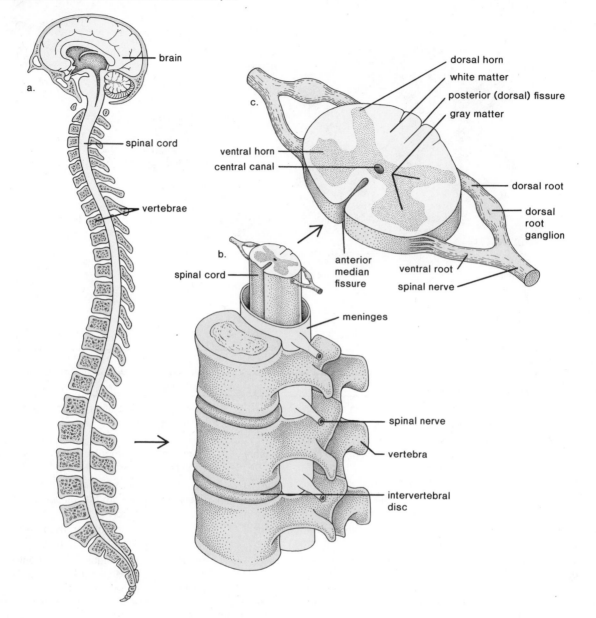

Ankle-jerk reflex. This reflex is initiated by tapping the Achilles tendon just above its insertion on the calcaneus. The response is plantar flexion due to contraction of the gastrocnemius and soleus muscles.

Some reflexes are important for avoiding injury, but these two are important for normal physiological functions. For example, the knee-jerk reflex helps us stand erect. If when we are standing still, the knee begins to bend a bit, the quadriceps femoris is stretched, and thereafter the leg straightens.

Reflexes, automatic reactions to internal and external stimuli, are dependent upon the reflex arc. Some reflexes are important for avoiding injury, and others are necessary for normal physiological functions.

Central Nervous System

The CNS consists of the spinal cord and brain. As figure 7.9 illustrates, the CNS is protected by bone: the brain is enclosed within the skull and the spinal cord is surrounded by vertebrae. Also, both the brain and spinal cord are wrapped in three protective membranes known as **meninges.** The outer meningeal layer, called the **dura mater,** is tough, white fibrous connective tissue that lies next to the interior of the skull and next to the interior of the vertebral canal. It contains blood sinuses (called dural sinuses) that collect venous blood before it is returned to the circulatory system. Bleeding due to a head injury is called an *epidural hematoma* when there is blood between the dura mater and bone, and is called *subdural hematoma* when there is blood beneath the dura mater. The middle meningeal layer is the **arachnoid membrane,** a weblike covering with thin strands that attach to the pia mater, the third meningeal layer. Beneath the arachnoid membrane is the subarachnoid space that is filled with cerebrospinal fluid. The **pia mater** is very thin and follows the contours of the brain and spinal cord closely (fig. 7.10).

Cerebrospinal Fluid

Cerebrospinal fluid is clear fluid that forms a protective cushion around and within the CNS. It also supplies the CNS with some nutrients that have filtered from the blood and collects wastes that are returned to the blood.

Cerebrospinal fluid is contained within the subarachnoid space, and within the **central canal** of the spinal cord and the **ventricles** of the brain. The latter are four spaces within the brain that connect with one another. The roof of each ventricle has a choroid plexus, specialized capillaries that produce cerebrospinal fluid—most cerebrospinal fluid seems to be made in the first two ventricles called the lateral ventricles because they occur on the sides of the brain. The cerebrospinal fluid leaves the fourth ventricle and passes into the subarachnoid space surrounding the brain and cord. From there, it enters the dural sinuses by way of arachnoid villi (fig. 7.11).

The CNS is protected by the meninges and the cerebrospinal fluid.

Normally, the pressure of cerebrospinal fluid within the brain stays constant because it steadily drains into the circulatory system by way of the dural sinuses. However, if there is a blockage of some sort in an infant whose cranial sutures have not closed, the brain can enlarge due to the accumulation of fluid. This is called *hydrocephalus* or "water on the brain." If fluid collects in an adult, the brain cannot enlarge and, instead, it is pushed up against the skull and injury can occur.

Spinal Cord

The spinal cord extends from the base of the brain through the foramen magnum of the skull into the vertebral canal formed by the vertebrae. It has two main functions: (1) it is the center for many reflex actions (fig. 7.8), and (2) it provides a means of communication between the brain and the peripheral nerves that leave the cord (fig. 7.9). Ascending tracts are long fibers of neurons that run together in columns up to the brain and descending tracts are long fibers of neurons that run from the brain to the cord.

The spinal cord terminates at a point called the *conus medullaris* at the level of the first lumbar vertebra (fig. 7.12). Thereafter, there is a fibrous strand composed mostly of pia mater called the *filum terminale* that extends to the coccyx. The nerves that radiate from the conus medullaris within the vertebral canal are called the *cauda equina* because they resemble a horse's tail.

There are two prominent enlargements of the cord. The cervical enlargement is located between the third cervical and second thoracic vertebrae. Nerves from this region serve the arms. The lumbar enlargement is located between the ninth and twelfth thoracic vertebrae. Nerves from this region serve the legs. Spinal nerves are discussed more fully on page 140.

Gray Matter

Spinal reflexes take place within the gray matter of the cord (figs. 7.8 and 7.9). It is gray because it contains short fibers and cells bodies that are unmyelinated. In cross section, the gray matter looks like a butterfly or some say the letter H, therefore, the gray matter is divided into the *dorsal* (anterior) *horns* and the *ventral* (posterior) *horns.* The sensory neurons enter the spinal cord through the dorsal roots, the motor neurons exit the spinal cord through the ventral roots. The dorsal and ventral roots join to form the spinal nerves.

White Matter

The *ascending* and *descending tracts* are found in columns of white matter found dorsally, ventrally, and laterally in the cord (figs. 7.8 and 7.9). Descending tracts are either pyramidal or extrapyramidal tracts. The **pyramidal tracts** are carrying impulses from the cerebral cortex to the motor neurons of the cord and since they cross over at the level of the medulla, the left side of the brain generally controls the right side of the body and vice versa. The **extrapyramidal** tracts originate in the midbrain and brain stem regions. The **spinothalamic tracts** are ascending tracts which carry impulses

Figure 7.10 Anatomy of the human brain. The cerebrum, the highest and largest part of the human brain, is responsible for consciousness. The medulla, the last part of the brain before the spinal cord, controls various internal organs. The enlargement below shows the anatomy of the meninges.

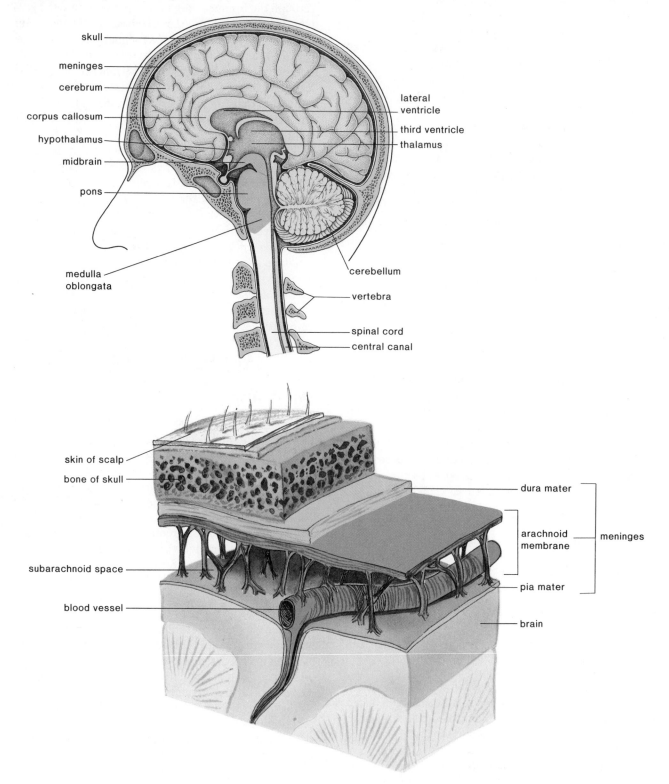

Figure 7.11 Cerebrospinal fluid is produced by choroid plexuses in the roof of the ventricles. The fluid circulates through the ventricles of the brain and central canal of the spinal cord, enters the subarachnoid space, and is reabsorbed into the blood of the dural sinuses through arachnoid villi.

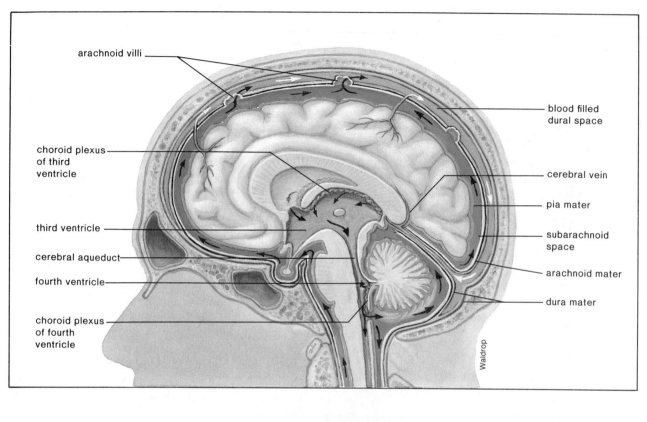

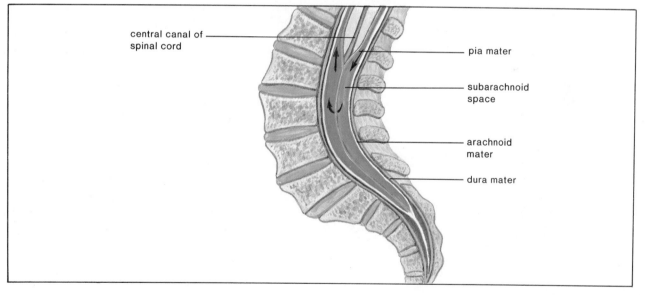

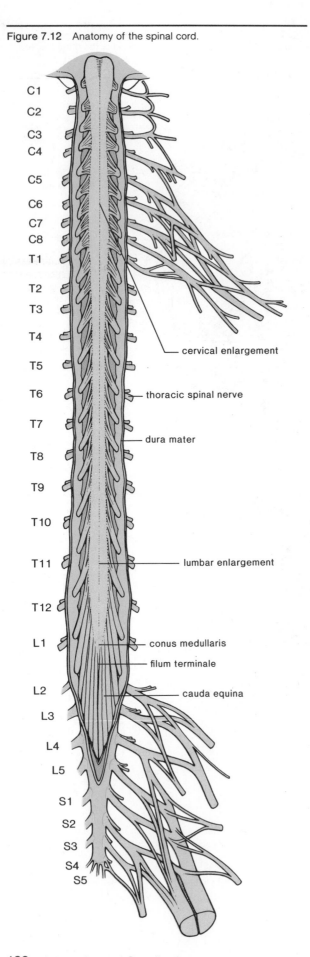

Figure 7.12 Anatomy of the spinal cord.

C1
C2
C3
C4
C5
C6
C7
C8
T1
T2
T3
T4
T5
T6
T7
T8
T9
T10
T11
T12
L1
L2
L3
L4
L5
S1
S2
S3
S4
S5

cervical enlargement

thoracic spinal nerve

dura mater

lumbar enlargement

conus medullaris

filum terminale

cauda equina

from the sensory neurons of the spinal cord to the thalamus and then to the cerebral cortex. These tracts also cross over.

The spinal cord extends from the base of the brain into the vertebral canal formed by the vertebrae. It is a center for reflex action and allows communication between the brain and nerves leaving the cord.

Clinical Considerations

Infections *Meningitis* is an inflammation of the meninges often caused by a viral or bacterial infection. If the inflammation is confined to the spinal cord, it is called spinal meningitis. If the infection should spread to the brain itself, it is called *encephalitis*.

Poliomyelitis is caused by a virus that destroys nerve cell bodies within the ventral horn of the spinal cord, especially those within the cervical and lumbar enlargements. At first, the individual suffers fever, severe headache, stiffness, and pain. Later, the muscles atrophy due to lack of nervous innervation. Death can occur if the respiratory muscles are involved.

Multiple sclerosis is caused by a deterioration of the myelin sheath about the long fibers of neurons found in the CNS. The lesions give way to many scarlike patches and this accounts for the name of the disorder. Nerve conduction cannot occur normally, and there are many symptoms such as motor difficulties, numbness, double vision, and tremors.

Diagnosis of Disorders A routine physical examination usually includes the testing of certain sensory functions and reflexes as mentioned previously. If a disorder of some sort is suspected, further testing can be done. A *lumbar puncture* is performed by inserting a fine hollow needle between the third and fourth lumbar vertebrae, and withdrawing a small amount of cerebrospinal fluid from the subarachnoid space. Testing of this fluid for its chemical composition and pH can be performed. It is also possible to introduce an anesthetic into the subarachnoid space. This is commonly called a spinal or saddle block. A caudal block is when the anesthetic is placed into the epidural space. The latter is common during childbirth. There is no sensation below the block with this type of anesthesia.

X rays have long been used to detect tumors and other disorders of the brain. Today, the *CAT* (computerized axial tomography) scanner projects a cross section of a patient's brain onto a screen. The technique can reveal the size and shape of the ventricles, and any abnormalities present in the brain and spinal cord. The information is stored in a computer and later, it is possible to reconstruct a three-dimensional picture of some portion of the CNS. An even newer technique, called *MRI* (magnetic resonance imaging), allows even better definition of certain neural structures. MRI doesn't make use of X rays; instead, the technique makes use of radiation in the range of radio frequencies. The MRI system is so sensitive that cerebrospinal fluid collection in the brain may be detected.

Brain

The largest and most prominent portion of the human brain (fig. 7.10) is the cerebrum. Consciousness resides only in the cerebrum; the rest of the brain functions below the level of consciousness.

The Unconscious Brain

The **medulla oblongata** lies closest to the spinal cord and contains centers for regulating heartbeat, breathing, and vasoconstriction (blood pressure). It also contains the reflex centers for vomiting, coughing, sneezing, hiccoughing, and swallowing.

The **hypothalamus** is concerned with homeostasis, or the constancy of the internal environment, and contains centers for regulating hunger, sleep, thirst, body temperature, water balance, and blood pressure. The hypothalamus controls the pituitary gland and thereby serves as a link between the nervous and endocrine systems.

The medulla oblongata and the hypothalamus both control the internal organs.

The **midbrain** and **pons** contain tracts that connect the cerebrum with other parts of the brain. In addition, the pons functions with the medulla to regulate breathing rate and the midbrain has reflex centers concerned with head movements in response to visual and auditory stimuli.

The **thalamus** is the last portion of the brain for sensory input before the cerebrum. It serves as a central relay station for sensory impulses traveling upward from other parts of the cord and brain to the cerebrum. It receives all sensory impulses (except those associated with the sense of smell) and channels them to appropriate regions of the cortex for interpretation.

The thalamus has connections to various parts of the brain by way of the diffuse thalamic projection system. This system is the upper part of the ARAS (ascending reticular activating system), a complex network of cell bodies and fibers that extends from the medulla to the thalamus (fig. 7.13). The **ARAS** sorts out incoming stimuli, passing on only those that require immediate attention. For this reason, the thalamus is sometimes called the gatekeeper to the cerebrum because it alerts the cerebrum to only certain sensory input. In this way, it may allow you to concentrate on your homework while the television is on.

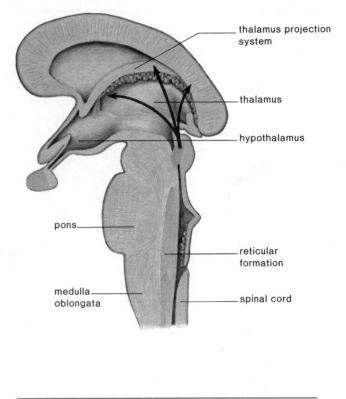

Figure 7.13 The ascending reticular activating system includes the reticular formation and the thalamus projection system. The sensory information received by the reticular formation is sorted out by the thalamus before nerve impulses are sent to other parts of the brain.

The thalamus receives sensory impulses from all parts of the body and channels them to the cerebrum.

The **cerebellum,** a bilobed structure that resembles a butterfly, is the second largest portion of the brain. It functions in muscle coordination, integrating impulses received from higher centers to ensure that all of the skeletal muscles work together to produce smooth and graceful motions. The cerebellum is also responsible for maintaining normal muscle tone and transmitting impulses that maintain posture. It receives information from the inner ear, indicating the position of the body and, thereafter, sends impulses to those muscles whose contraction maintains or restores balance.

The cerebellum controls balance and complex muscular movements.

The Cerebrum

The **cerebrum,** which is the only area of the brain responsible for consciousness, is the largest portion of the brain in humans. The cerebrum is divided into halves known as the right and left **cerebral hemispheres.** The outer layer of the cerebrum, called the cortex, is gray in color and contains cell bodies and short fibers. The

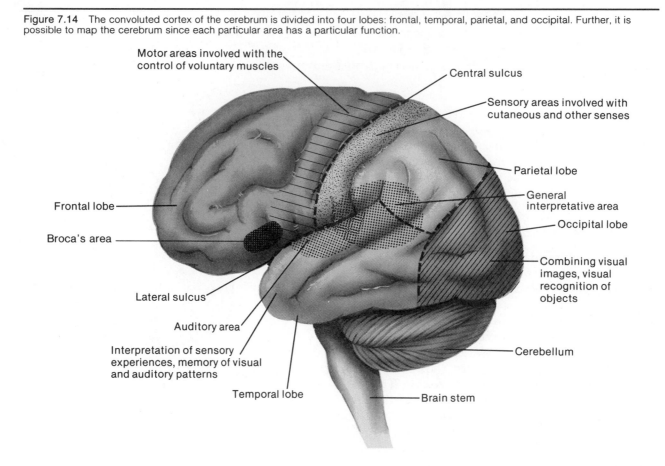

Figure 7.14 The convoluted cortex of the cerebrum is divided into four lobes: frontal, temporal, parietal, and occipital. Further, it is possible to map the cerebrum since each particular area has a particular function.

cortex has convolutions known as *gyri,* which are separated by shallow grooves called *sulci.* There are also deep grooves called fissures. The cerebrum is divided by the deep longitudinal fissure, but the corpus callosum is a bridge of myelinated fibers that joins the two hemispheres.

Each cerebral hemisphere is divided into four lobes: **frontal, parietal, temporal,** and **occipital** (fig. 7.14), which are named for the bones that cover them. Each of the lobes has particular functions (table 7.4). Association areas are believed to be areas for the intellect, artistic and creative ability, and learning. Sensory areas receive nerve impulses from the sense organs and produce what we call sensations. The particular sensation produced is the prerogative of the area of the brain that is stimulated, since the nerve impulse itself always has the same nature (as described previously). Motor areas of the cerebrum initiate nerve impulses that control muscle fibers.

The primary *sensory area* of the cerebral cortex lies in the parietal lobe just dorsal to the central sulcus and the primary *motor area* lies just ventral to the central sulcus. These two areas, in particular, demonstrate how it is possible to map the cerebrum (fig. 7.15). It is possible to associate particular parts of the cerebrum with particular parts of the body. The pyramidial tracts, mentioned previously (see p. 129), run between the motor area of the cerebrum and the cord; the spinothalamic tracts run between the cord and the sensory area.

A momentary lack of oxygen during birth can damage the motor areas of the cerebral cortex so that the individual develops the symptoms of cerebral palsy, a condition characterized by a spastic weakness of the arms and legs.

There is a motor area for speech, called *Broca's area,* at the base of the precentral gyrus (fig. 7.14) usually only in the left cerebral hemisphere. Damage to this

Table 7.4 Functions of the Cerebral Lobes

Lobe	Functions
Frontal lobes	Motor areas control movements of voluntary skeletal muscles. Association areas carry on higher intellectual processes such as those required for concentration, planning, complex problem solving, and judging the consequences of behavior.
Parietal lobes	Sensory areas are responsible for the sensations of temperature, touch, pressure, and pain from the skin. Association areas function in the understanding of speech and in using words to express thoughts and feelings.
Temporal lobes	Sensory areas are responsible for hearing and smelling. Association areas are used in the interpretation of sensory experiences and in the memory of visual scenes, music, and other complex sensory patterns.
Occipital lobes	Sensory areas are responsible for vision. Association areas function in combining visual images with other sensory experiences.

From John W. Hole, Jr., *Human Anatomy and Physiology*, 4th ed. © 1987 Wm. C. Brown Publishers, Dubuque, Iowa. All Rights Reserved. Reprinted by permission.

area can interfere with a person's ability to understand words (written or spoken) and to communicate with others. This explains why a stroke is often accompanied by speech problems. Strokes (p. 206) occur when the blood supply to the brain is temporarily halted due to a blood clot in an artery of the brain.

Consciousness is the province of the cerebrum, the most highly developed portion of the brain, which is responsible for higher mental processes, including the interpretation of sensory input and the initiation of voluntary muscular movements.

There has been a great deal of testing to determine whether the right and left halves of the cerebrum serve different functions. These studies have tended to suggest that the left half of the brain is the verbal (word) half and the right half of the brain is the visual (spatial relation) and artistic half. However, other results indicate that such a strict dichotomy does not always exist between the two halves. In any case, the two cerebral hemispheres normally share information because they are connected by the corpus callosum.

Severing the corpus callosum can control severe epileptic seizures, but results in a person with two brains, each with its own memories and thoughts. Today, use of the laser permits more precise treatment without these side effects. *Epilepsy* is caused by a disturbance of the normal communication between the ARAS and the cortex. In a grand mal seizure, the cerebrum becomes extremely excited. Due to a reverberation of signals within the ARAS and cerebrum, the individual loses consciousness, even while convulsions are occurring. Finally the neurons become fatigued and the signals cease. Following an attack, the brain is so fatigued that the person must sleep for a while.

EEG The electrical activity of the brain can be recorded in the form of an **electroencephalogram (EEG).** Electrodes are taped to different parts of the scalp, and an instrument called the electroencephalograph records the so-called brain waves (fig. 7.16).

When the subject is awake, two types of waves are usual: *alpha waves,* with a frequency of about six to thirteen per second and a potential of about 45 microvolts, predominate when the eyes are closed; and *beta waves,* with higher frequencies, but lower voltage, appear when the eyes are open.

During an eight-hour sleep, there are usually five times when the brain waves become slower and larger than alpha waves. During each of these times, there are irregular flurries as the eyes move back and forth rapidly. When subjects are awakened during this **REM** (rapid eye movement) **sleep,** they always report that they were dreaming. The significance of REM sleep is still being debated, but some studies indicate that REM sleep is needed for memory to occur.

The EEG is a good diagnostic tool; for example, an irregular pattern can signify epilepsy or a brain tumor. A flat EEG signifies lack of electrical activity of the brain, or brain death; and thus it may be used to determine the precise time of death.

Figure 7.15 This diagram shows the portions of the body that are controlled by *a.* the primary motor area, and that send sensory information to *b.* the primary sensory area. Notice that the size of the body part reflects the amount of cerebral cortex devoted to that part of the body.

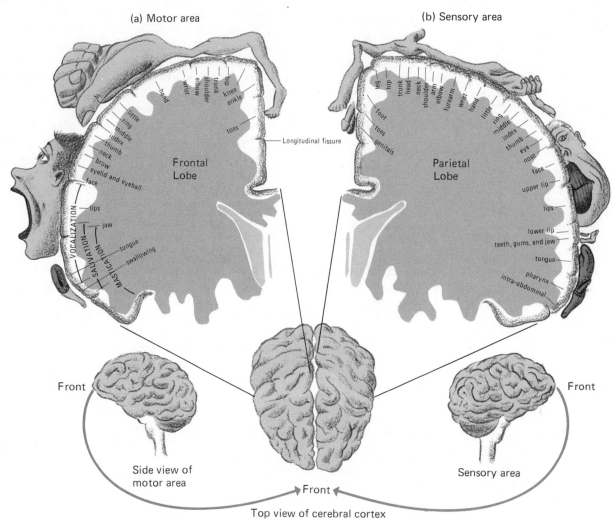

(a) Motor area

(b) Sensory area

Figure 7.16 Encephalograms are recordings of the electrical activity of the brain. The alpha waves, which appear when the subject is awake with eyes closed, are the most common. Second most common are the beta waves recorded when the subject is awake with eyes open. Sleep has various stages, as indicated.

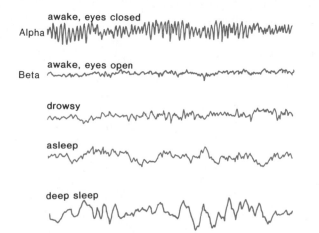

Figure 7.17 The limbic system (color) includes portions of the frontal lobes, temporal lobes, the thalamus, and the hypothalamus. Among other functions, the limbic system is thought to produce endorphins, internal opiates that can cause a feeling of euphoria and reduce the threshold for pain.

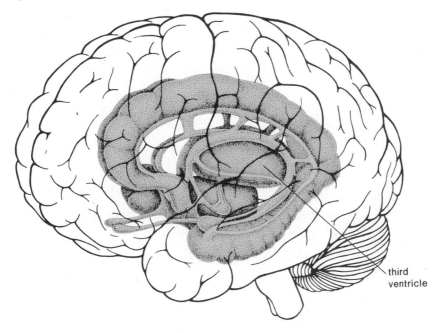

third ventricle

An EEG is a record of the brain's electrical activity and can be used as a diagnostic tool.

Limbic System

The **limbic system** (fig. 7.17) involves portions of both the unconscious and conscious brain. It sits above the third ventricle just beneath the cortex and contains neural pathways that connect portions of the frontal lobes, temporal lobes, thalamus, and hypothalamus. Several masses of gray matter that lie deep within each hemisphere of the cerebrum, termed the *basal ganglia*, are also a part of the limbic system.

Stimulation of different areas of the limbic system causes the subject to experience rage, pain, pleasure, or sorrow. By causing pleasant or unpleasant feelings about experiences, the limbic system apparently guides the individual into behavior that is likely to increase the chance of survival.

Learning and Memory Learning requires memory, but just what permits memory to occur is not definitely known. At the cellular level, research has recently shown that learning is accompanied by an increase in the number of synapses, while forgetting parallels a decrease in the number of synapses. In other words, the nerve-circuit patterns are constantly changing as learning, remembering, and forgetting occur. Within the individual, neuron learning involves a change in gene regulation, nerve protein synthesis, and an increased ability to secrete transmitter substances.

At the organ level, investigators have concluded that the limbic system is absolutely essential to both short-term and long-term memory. An example of short-term memory in humans is the ability to recall a telephone number long enough to dial the number; an example of long-term memory, is the ability to recall the events of the day. It's believed that, at first, impulses move within the limbic circuit, but eventually, the basal ganglia transmit impulses to the sensory areas where memories are stored. The involvement of the limbic system certainly explains why emotionally charged events result in our most vivid memories. The fact that the limbic system communicates with the sensory areas for touch, smell, vision, and so on accounts for the ability of any particular sensory stimulus to awaken a complex memory.

The limbic system is particularly involved in the emotions, and in memory and learning.

Peripheral Nervous System

The **peripheral nervous system** (**PNS**) contains the somatic and the autonomic nervous systems. The somatic nervous system is primarily concerned with reactions to outside stimuli, while the **autonomic nervous system**

Pain

A physiological picture is emerging to explain the phenomenon of pain. When a portion of the body is traumatized, potent chemicals are released including bradykinin and prostaglandins. These are believed to initiate nerve impulses that travel perhaps by special small-diameter fibers to the cord, where they stimulate SG (substantia gelatinosa) cells. Ordinarily, the activity of the SG cells is kept at bay by messages received from large-diameter fibers that also synapse with these cells. The perception of pain depends on which of these fibers (small-diameter or large-diameter) is most active.

SG cells release the neurotransmitter that is responsible for sending on the message that injury has occurred and pain is present. This transmitter substance has been labeled substance P (for pain). On occasion, people have undergone serious bodily injury and yet report that they felt no pain. The potent pain-reliever heroin has provided an answer to this circumstance. It turns out that neurons (and presumably SG cells also) have receptors for heroin because the body produces its own opiates, termed the endorphins and enkephalins. When endorphins and enkephalins occupy the receptors, the message that results in the sensation of pain is not transmitted. In other words, opiates prevent the release of substance P.

Endorphin and enkephalin research gives us an explanation for drug addiction. Narcotic intake causes the CNS to stop producing its own natural opiates. Therefore, the addict must take more and more of the drug in order to get the same effect. Furthermore, when the intake of heroin is stopped, the release of neurotransmitters are no longer opposed, causing withdrawal symptoms. As the body gradually begins to produce its own supply of opiates again, the individual returns to a normal state.

This explanation for the sensation of pain has proven to be fruitful. Aspirin has long been found to alleviate pain, but now it is known that it interferes with prostaglandin production in the tissues. Tylenol and other popular analgesics work in much the same way. Steroids also prevent prostaglandin production, but at a later juncture. The action has been misused—athletes who "break the pain barrier" by cortisone injections have damaged their muscles without feeling a thing.

It is now also possible to explain TENS (transcutaneous electrical nerve stimulation). Stimulating electrodes are placed above the painful area and a mild current is applied. This produces a sensation of tingling or warmth that may relieve pain by increasing large-diameter fiber activity. Acupuncture, the technique of placing fine needles into the skin, most likely works similarly.

Some physicians believe that pain should be stopped at the second link in the pain chain—heroin should be used for medicinal purposes. Physicians testify, however, that there are synthetic opiates just as effective as heroin.

No doubt pain has its psychological aspects and not surprisingly antidepressant drugs have been found to alleviate pain. Still, we must remember that these drugs most likely block the transmission of certain impulses and in this way promote the transmission of others—perhaps the very ones that promote endorphin and/or enkephalin release. Those who engage in athletic activities report an elevation of mood, and this has now been linked to the stimulation of fibers that bring about the release of the natural opiates.

(ANS) is primarily concerned with the proper functioning of the internal organs so that homeostasis is maintained.

Somatic System

The **somatic nervous system** includes all of those nerves that serve the musculoskeletal system and the exterior sense organs, including those of the skin. Exterior sense organs are *receptors* that receive environmental stimuli and then initiate nerve impulses. Muscle fibers are *effectors* that bring about a reaction to the stimulus. Muscle effectors were studied in chapter 6, and receptors are studied in chapter 8.

Nerves

A nerve contains bundles of nerve fibers surrounded by connective tissue (fig. 7.18). A nerve fiber is either a long axon or a long dendrite. The cell bodies of neurons are found only in the brain, spinal cord, and ganglia. **Ganglia** are collections of cell bodies within the PNS.

There are three types of nerves (table 7.5). **Sensory nerves** contain only the long dendrites of sensory neurons, and **motor nerves** contain only the long axons of motor neurons. **Mixed nerves,** however, contain both the long dendrites of sensory neurons and the long axons of motor neurons. Each nerve fiber within a nerve is surrounded by a white myelin sheath (fig. 7.3), and therefore nerves have a white, shiny, glistening appearance.

The gate control theory of pain. When small diameter fibers are active, the gate is open and a person feels pain.

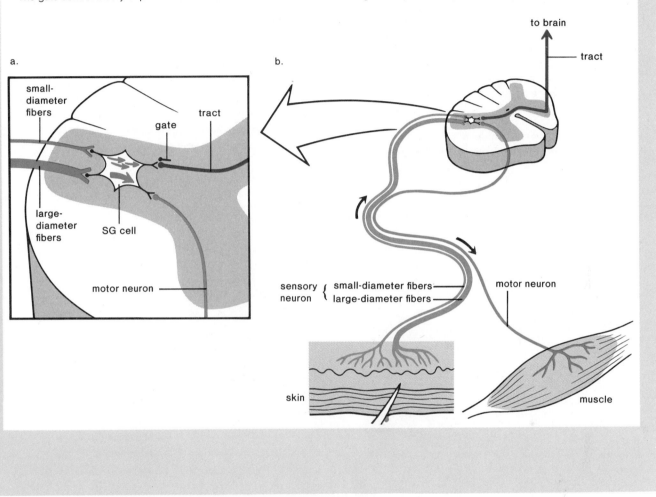

Figure 7.18   Diagram of a cross section of a nerve, with one axon extended to show that each fiber is enclosed by a myelin sheath. Because nerves contain so many fibers, it has been difficult to successfully rejoin them after they are severed in an accident. Investigators have now found that if they hold well-cut pieces together, and then surround them by a solution that resembles cytoplasm, the nerve will repair itself and be functional.

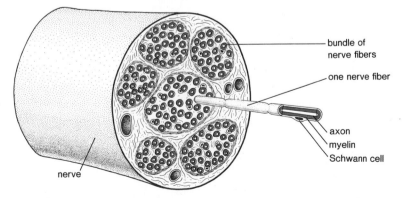

Table 7.5 Nerves

Type of Nerve	Consists of	Function
Sensory nerves	Long dendrites only of sensory neurons	Carry message from receptors to CNS
Motor nerves	Long axons only of motor neurons	Carry message from CNS to effectors
Mixed nerves	Both long dendrites of sensory neurons and long axons of motor neurons	Carry message in dendrites to CNS and away from CNS in axons

Note: Compare this table to table 7.1.

Cranial Nerves Humans have twelve pairs of **cranial nerves** attached to the brain (fig. 7.19 and table 7.6). Some of these are sensory, some are motor, and others are mixed. Notice that although the brain is a part of the CNS, the cranial nerves are a part of the PNS. All cranial nerves, except the vagus, are concerned with the head, neck, and facial regions of the body, but the vagus nerve has many branches to serve the internal organs.

Spinal Nerves Each **spinal nerve** emerges from the cord (fig. 7.9c) by two short branches, called the dorsal and ventral roots. The dorsal root can be identified by the presence of an enlargement called the **dorsal root ganglion.** This ganglion contains the cell bodies of the sensory neurons whose dendrites conduct impulses toward the cord. The ventral root of each spinal nerve

Figure 7.19 There are twelve pairs of cranial nerves, and their functions are listed in table 7.6.

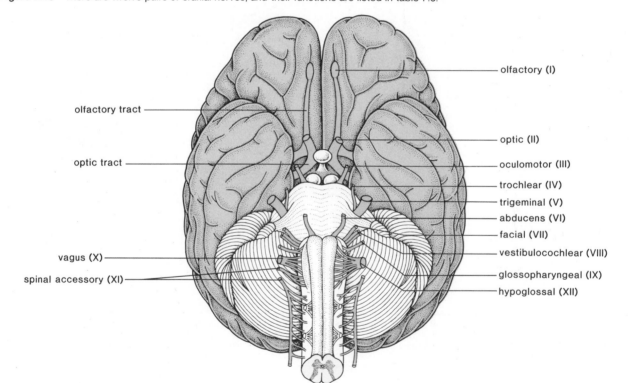

Table 7.6 Cranial Nerves

Nerve	Type	Transmit Nerve Impulses to (Motor) or from (Sensory)
Olfactory (I)	Sense of smell	Olfactory receptors
Optic (II)	Sense of sight	Retina
Oculomotor (III)	Motor	Eye muscles (including eyelids and lens); pupil (parasympathetic system)[a]
Trochlear (IV)	Motor	Eye muscles
Trigeminal (V)	Sensory	Teeth, eyes, skin, and tongue
	Motor	Jaw muscles (chewing)
Abducens (VI)	Motor	Eye muscles
Facial (VII)	Sensory	Taste buds of anterior tongue
	Motor	Facial muscles (facial expression) and glands (tear and salivary)
Vestibulocochlear (VIII)	Sense of balance and hearing	Inner ear
Glossopharyngeal (IX)	Sensory	Pharynx
	Motor	Pharyngeal muscles (swallowing)
Vagus (X)	Sensory	Internal organs
	Motor	Internal organs (parasympathetic system)
Spinal accessory (XI)	Motor	Neck and back muscles
Hypoglossal (XII)	Motor	Tongue muscles

[a]See page 143

contains the axons of motor neurons that conduct impulses away from the cord. These two roots join just before the spinal nerve leaves the vertebral column. Therefore, all spinal nerves are mixed nerves that contain many sensory dendrites and motor axons.

There are thirty-one pairs of spinal nerves (fig. 7.20). Instead of continuing directly to a peripheral body part, spinal nerves (except thoracic nerves) send branches to complex networks called plexuses. Within a plexus, various nerve fibers are rejoined to form nerves that go to nearby body parts. The different types of plexuses are these:

Cervical plexuses serve the muscles and skin of the neck and the diaphragm.

Brachial plexuses serve muscles and skin of the arms and hands.

Lumbosacral plexuses serve muscles and skin of the lower abdomen, genitalia, and the legs and feet.

The thoracic spinal nerves do not send branches to a plexus. The thoracic spinal nerves send branches directly to the ribs, and the upper abdominal wall muscles. They also receive sensory impulses from the skin of the thorax and abdomen.

Cranial nerves take impulses to and/or from the brain. Spinal nerves take impulses to and from the spinal cord.

Autonomic System

The ANS is made up of motor neurons that innervate smooth muscle, cardiac muscle, and glands automatically and usually without the need for conscious intervention. There are two divisions of the autonomic nervous system: the sympathetic and parasympathetic systems. Both of these (1) function automatically and usually subconsciously in an involuntary manner; (2) innervate all internal organs; and (3) utilize two motor neurons and one ganglion for each impulse. The

Figure 7.20 The spinal nerves are divided into the cervical (C1–C8), thoracic (T1–T12), lumbar (L1–L5), sacral (S1–S5), and the coccygeal nerve (C1). Branches from these nerves combine to form complex networks called plexuses. Some of the nerves from these plexuses are labeled. Spinal nerves in the thoracic region give rise to intercostal nerves.

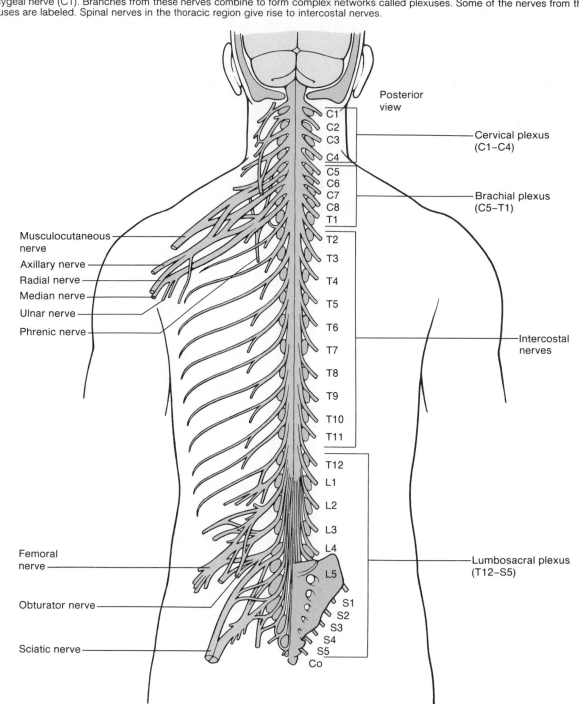

Posterior view

C1
C2
C3
C4

Cervical plexus (C1–C4)

C5
C6
C7
C8
T1

Brachial plexus (C5–T1)

Musculocutaneous nerve

Axillary nerve

Radial nerve

Median nerve

Ulnar nerve

Phrenic nerve

T2
T3
T4
T5
T6
T7
T8
T9
T10
T11

Intercostal nerves

T12
L1
L2
L3
L4

Femoral nerve

L5

Obturator nerve

S1
S2
S3
S4
S5
Co

Sciatic nerve

Lumbosacral plexus (T12–S5)

Figure 7.21 Location of ganglia in the sympathetic and parasympathetic nervous systems. *a.* In the sympathetic nervous system, each ganglion lies close to the spinal cord (CNS), and therefore the preganglionic fiber is short and the postganglionic fiber is long. *b.* In the parasympathetic nervous system, each ganglion lies close to the organ being innervated, and therefore the preganglionic fiber is long and the postganglionic fiber is short.

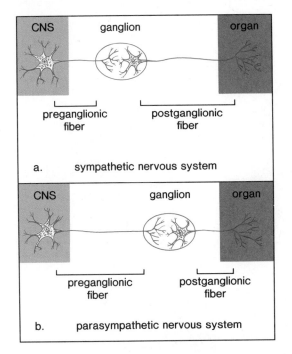

Table 7.7	Sympathetic Versus Parasympathetic System
Sympathetic	**Parasympathetic**
Fight or flight	Normal activity
Norepinephrine is neurotransmitter	Acetylcholine is neurotransmitter
Postganglionic fiber is longer than preganglionic	Preganglionic fiber is longer than postganglionic
Preganglionic fiber arises from thoracolumbar portion of the ANS	Preganglionic fiber arises from craniosacral portion of ANS

first neuron has a cell body within the central nervous system and a **preganglionic axon.** The second neuron has a cell body within the ganglion and a **postganglionic axon.**

The autonomic nervous system controls the functioning of internal organs without need of conscious control.

Sympathetic System

The preganglionic fibers of the **sympathetic nervous system** arise from the middle, or thoracolumbar levels of the cord and almost immediately terminate in ganglia that lie near the cord. Thus, in this system the preganglionic is short, but the postganglionic fiber that makes contact with the organs is long (fig. 7.21a).

The sympathetic nervous system is especially important during emergency situations and is associated with "fight or flight." For example, it inhibits the diges-

tive tract, but dilates the pupil, accelerates the heartbeat, and increases the breathing rate. It is not surprising, then, that the neurotransmitter released by the postganglionic axon is norepinephrine, a chemical close in structure to epinephrine, a well-known stimulant.

The sympathetic nervous system brings about those responses we associate with "fight or flight."

Parasympathetic System

Cranial nerves, including the vagus nerve, and fibers that arise from the sacral levels of the cord form the **parasympathetic nervous system.** Therefore this system is often called the craniosacral portion of the ANS. In the parasympathetic nervous system, the preganglionic fiber is long and the postganglionic fiber is short because the ganglia lie near or within the organ (fig. 7.21b). The parasympathetic system promotes all of those internal responses we associate with a relaxed state; for example, it causes the pupil of the eye to contract, promotes the digestion of food, and retards the heartbeat. The parasympathetic system is the "housekeeper system." The neurotransmitter utilized by the parasympathetic system is acetylcholine.

Figure 7.22 contrasts the sympathetic and parasympathetic systems, and table 7.7 lists all the differences we have noted between these two systems.

The parasympathetic nervous system brings about those responses we associate with normally restful activities.

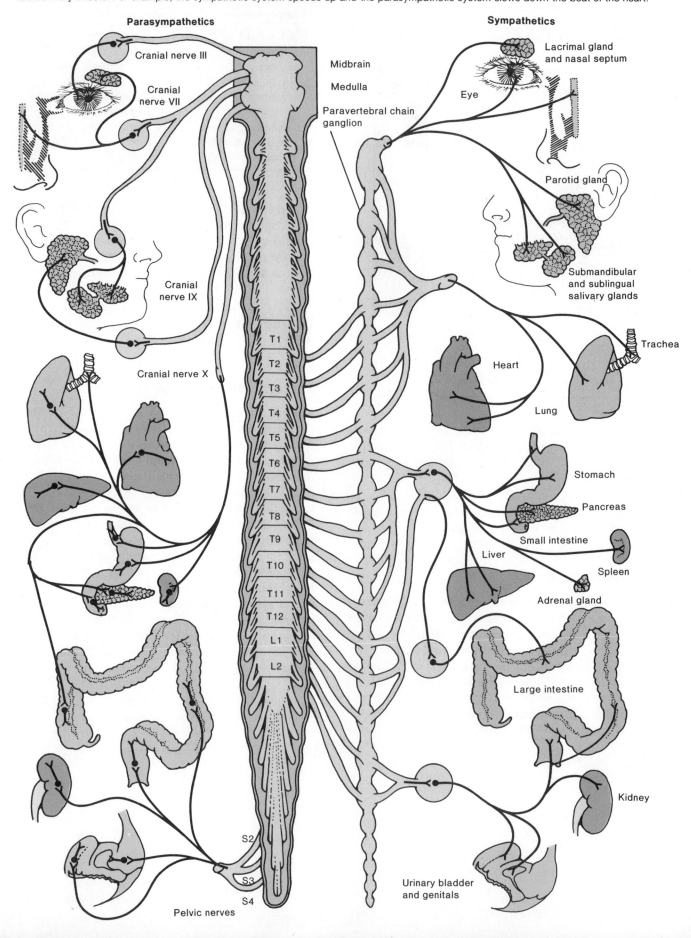

Figure 7.22 Structure and function of the autonomic nervous system. The sympathetic fibers arise from the thoracic and lumbar portion of the cord; the parasympathetic fibers arise from the brain and sacral portion of the cord. Each system innervates the same organs but has contrary effects. For example, the sympathetic system speeds up and the parasympathetic system slows down the beat of the heart.

Parasympathetics

Sympathetics

Cranial nerve III

Cranial nerve VII

Cranial nerve IX

Cranial nerve X

Midbrain

Medulla

Paravertebral chain ganglion

Lacrimal gland and nasal septum

Eye

Parotid gland

Submandibular and sublingual salivary glands

Trachea

Heart

Lung

Stomach

Pancreas

Small intestine

Liver

Spleen

Adrenal gland

Large intestine

Kidney

Urinary bladder and genitals

Pelvic nerves

T1
T2
T3
T4
T5
T6
T7
T8
T9
T10
T11
T12
L1
L2

S2
S3
S4

Summary

I. Nervous System
  A. Divisions of the nervous system. The nervous system is divided into the central nervous system (brain and spinal cord) and the peripheral nervous system (somatic and autonomic systems)
  B. Nerve tissue. Nerve tissue contains neurons and glial cells.
      Although all neurons have the same three parts (axon, dendrite(s), and cell body), there are three different types of neurons (sensory, motor, and interneuron).
  C. Nerve impulse. All neurons transmit the same type of nerve impulse—an electrochemical change that is propagated along the fiber.
  D. Transmission across a synapse. Transmission of a nerve impulse across a synapse is dependent on transmitter substances that change the permeability of the postsynaptic membrane.
  E. Reflex arc. The reflex arc is a main functional unit of the nervous system. It allows us to react to internal and external stimuli.
II. Central Nervous System. The CNS is protected by the meninges and the cerebrospinal fluid.
  A. The spinal cord is the center for reflexes, and takes impulses from the sensory neurons to the brain and from the brain to the motor neurons.
  B. Brain. The medulla oblongata and the hypothalamus both control the internal organs.
      The thalamus receives sensory impulses from all parts of the body and channels them to the cerebrum.
      Consciousness is the province of the cerebrum, the most highly developed portion of the brain, which is responsible for higher mental processes, including the interpretation of sensory input and the initiation of voluntary muscular movements.
      An EEG is a record of the brain's electrical activity and can be used as a diagnostic tool.
      The limbic system is particularly involved in the emotions, and in memory and learning.
III. Peripheral Nervous System
  A. Somatic nervous system. Cranial nerves take impulses to and/or from the brain. Spinal nerves take impulses to and from the spinal cord.
  B. Autonomic system. The autonomic nervous system controls the functioning of internal organs without need of conscious control.
      The sympathetic nervous system brings about those responses we associate with "fight or flight."
      The parasympathetic nervous system brings about those responses we associate with normally restful activities.

Study Questions

1. What are the two main divisions of the nervous system? How are these divisions subdivided?
2. What are the three types of neurons? How are they similar, and how are they different?
3. What does the term *resting potential* mean, and how is it brought about? Describe the two parts of an action potential and the change that may be associated with each part.
4. What is the sodium–potassium pump, and when is it active?
5. What is a neurotransmitter substance; where is it stored; how does it function; and how is it destroyed? Name two well-known neurotransmitters.
6. What is the path of a spinal reflex that involves three neurons?
7. What are the three different meninges and what is their function?
8. What is cerebrospinal fluid? Where is it made and how does it circulate?
9. Describe the anatomy of the spinal cord. What are the functions of the gray and white matter in the spinal cord?
10. What are the various parts of the brain and their functions?
11. What does it mean to say that the cerebral cortex can be mapped? Discuss in relation to the primary motor areas and the primary sensory areas.
12. Give two examples to show that neurological illnesses can be associated with a particular neurotransmitter in the CNS.
13. What are the different cranial nerves and what is the function of each?
14. What is the structure and function of spinal nerves?
15. What is the autonomic nervous system and what are its two major divisions? Give several similarities and differences between these divisions.
16. In what anatomical way are spinal nerves categorized?

Objective Questions

Fill in the blanks.

1. An _____ carries nerve impulses away from the cell body.
2. During the upswing of the action potential, _____ ions are moving to the _____ of the nerve fiber.
3. The space between the axon of one neuron and the dendrite of another is called the _____ .
4. ACh is broken down by the enzyme _____ after it has altered the permeability of the postsynaptic membrane.
5. Motor nerves innervate _____ .
6. The vagus nerve is _____ nerve that controls _____ .
7. In a reflex arc only the _____ is completely within the CNS.
8. The brain and cord are covered by protective layers called _____ .
9. The _____ is that part of the brain that allows us to be conscious.
10. The _____ is the part of the brain responsible for coordination of body movements.

Medical Terminology Reinforcement Exercise

Pronounce the following terms and analyze the meaning by word parts as dissected:

1. neuropathogenesis (nu″ro-path″o-jen′e-sis)—neuro/patho/genesis
2. anesthesia (an″esthe′ze-ah)—an/esthesia
3. encephalomyeloneuropathy (en-sef″ah-lo-mi″ĕ-lo-nu-rop′ah-the)—encephalo/myelo/neuro/pathy
4. hemiplegia (hem″e-ple′je-ah)—hemi/plegia
5. glioblastoma (gli″o-blas-to′mah)—glio/blast/oma
6. subdural hemorrhage (sub-du′ral hem′or-ij)—sub/dural hemo/rrhage
7. cephalometer (sef″ah-lom′ĕ-ter)—cephalo/meter
8. pneumoencephalography (nu″mo-en-sef″ah-log′rah-fe)—pneumo/encephalo/graphy
9. meningoencephalocele (mĕ-ning″go-en-sef″ah-lo-sēl″)—meningo/encephalo/cele
10. neurorrhaphy (nu-ror′ah-fe)—neuro/rraphy
11. ataxiaphasia (ah-tak″se-ah-fa′ze-ah)—a/taxi/a/phasia
12. dysphagia (dis-fa′je-ah)—dys/phagia

8

Senses

Chapter Outline

Learning Objectives

After you have studied this chapter, you should be able to:

1. Categorize receptors according to the system used in the text.
2. Name the five senses of the skin and state the location of these receptors.
3. Name the chemoreceptors and state their location, anatomy, and mechanism of action.
4. Describe the anatomy and function of the accessory organs of the eye.
5. Describe the anatomy of the eye and the function of each part.
6. Describe the receptors for sight, their mechanism of action, and the mechanism for stereoscopic vision.
7. Describe common disorders of sight discussed in the text.
8. Describe the anatomy of the ear and the function of each part.
9. Describe the receptors for balance and hearing, and their mechanism of action.
10. Describe the two types of deafness.

General Receptors

Sense perception is dependent upon receptors, the first component of the reflex arc, which was described in chapter 7. When a receptor (sense organ) is stimulated, it generates nerve impulses that are transmitted to the spinal cord and/or brain, but we are conscious of a sensation only if the impulses reach the cerebrum.

We will divide receptors in two types—general receptors and special receptors.

General receptors are those that are found throughout the body. For example, the skin (fig. 8.1) contains receptors for touch, pressure, pain, and temperature. It is a mosaic of these tiny receptors, as you can determine by passing a metal probe slowly over the skin. At certain points, there will be a feeling of pressure, and at others, a feeling of hot or cold (depending on the temperature of the probe). Certain parts of the skin contain more receptors for a particular sensation; for example, the fingertips have an abundance of touch receptors.

The sense of position and movement of limbs is dependent upon receptors termed **proprioceptors.** There are proprioceptors located in the joints and associated ligaments and tendons that respond to stretching, pressure, and pain. Nerve endings from these receptors are integrated within the cerebellum with those received from other types of receptors so that the person knows the position of body parts.

The receptors in the skin respond to touch, pressure, pain, and temperature; those at the joints (proprioceptors) help us know the location of body parts.

Chemoreceptors

Taste and smell are called the chemical senses because these receptors are sensitive to certain chemical substances in the food we eat and the air we breathe.

There are four types of tastes (bitter, sour, salty, and sweet) and the **taste buds** for each are concentrated on the tongue in particular regions (fig. 8.2). Most taste buds lie along the walls of the papillae, the small elevations visible to the naked eye. They are pockets of cells in the tongue epithelium with microvilli that project out of a pore. Nerve impulses are generated when chemicals bind to the microvilli.

Figure 8.1 Receptors in human skin. *a.* Free nerve endings are pain receptors; Pacinian corpuscles are pressure receptors; and Merkel's disks and Meissner's corpuscles are touch receptors, as are the nerve endings surrounding the hair follicle. *b.* Enlargements of three of these.

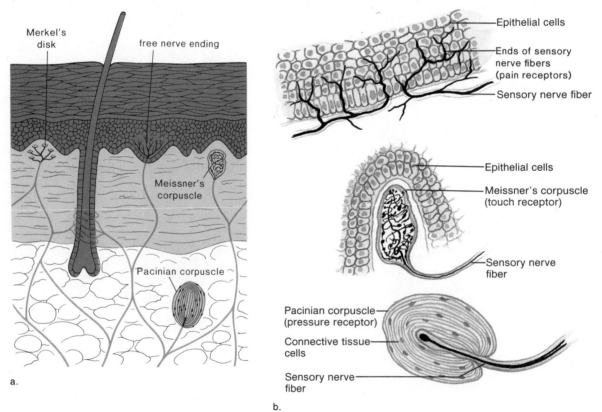

The **olfactory cells** (fig. 8.3) are located high in the roof of the nasal cavity. These cells, which are specialized endings of the fibers that make up the olfactory nerve, lie among supporting epithelial cells. Each cell ends in a tuft of six to eight cilia, and when chemicals bind to these cilia, nerve impulses are generated.

The sense of taste and the sense of smell supplement each other, creating a combined effect when interpreted by the cerebral cortex. For example, when we have a cold, we think that our food has lost its taste, but actually we have lost the ability to sense its smell. This may work in the reverse also. When we smell something, some of the molecules move from the nose down into the mouth region and stimulate the taste buds there. Therefore, part of what we refer to as smell may actually be taste.

The receptors for taste (taste buds) and the receptors for smell (olfactory cells) work together to give us our senses of taste and smell.

Photoreceptor—The Eye

The eyes are located in orbits formed by seven of the skull's bones (frontal, lacrimal, ethmoid, zygomatic, maxilla, sphenoid, and palatine). The supraorbital ridge protects the eye from blows, and the eyebrow diverts sweat around the eye. We will be discussing the accessory organs before discussing the eye itself.

Accessory Organs

Extrinsic Muscles

Within the orbit, the eyes are anchored in place by the **extrinsic muscles** whose contractions move the eyes. There are three pairs of antagonistic extrinsic muscles (fig. 8.4):

superior rectus: rolls eye upward
inferior rectus: rolls eye downward

Figure 8.2 Taste buds. *a.* Elevations, called papillae, indicate the presence of taste buds. The location of those containing taste buds responsive to sweet, sour, salt, and bitter is indicated. *b.* Drawing of a taste bud shows the various cells that make up a taste bud. Sensory cells in the bud end in microvilli that have receptors for the chemicals that exhibit the tastes noted in *a.* When the chemicals combine with the receptors, nerve impulses are generated.

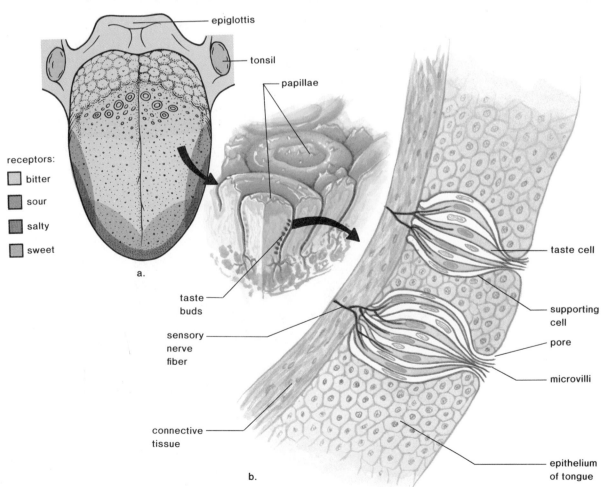

receptors:
bitter
sour
salty
sweet

epiglottis
tonsil
papillae
taste cell
supporting cell
pore
microvilli
epithelium of tongue
taste buds
sensory nerve fiber
connective tissue

a.

b.

Figure 8.3 *a.* Position of olfactory epithelium in a nasal passageway. *b.* The olfactory receptor cells, which have cilia projecting into the nasal cavity, are supported by columnar epithelial cells. When these cells are stimulated by chemicals in the air, nerve impulses begin and are conducted to the brain by olfactory nerve fibers.

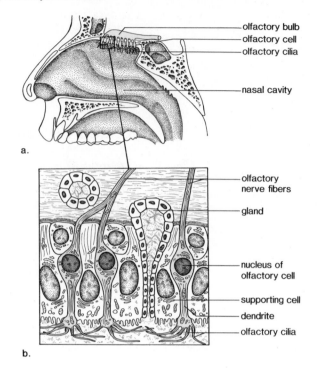

olfactory bulb
olfactory cell
olfactory cilia

nasal cavity

a.

olfactory
nerve fibers

gland

nucleus of
olfactory cell

supporting cell

dendrite

olfactory cilia

b.

Figure 8.4 The extrinsic muscles of the eye.

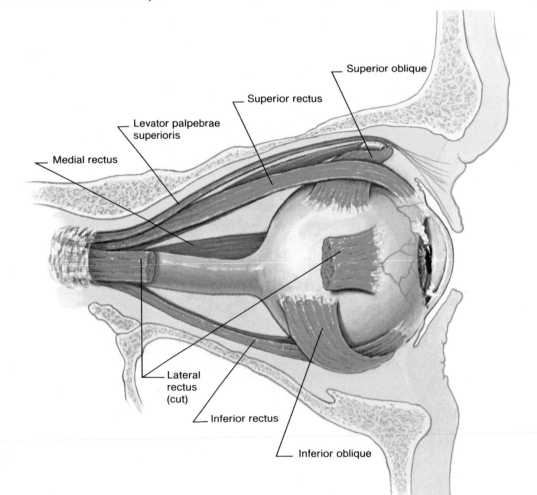

Superior oblique

Superior rectus

Levator palpebrae
superioris

Medial rectus

Lateral
rectus
(cut)

Inferior rectus

Inferior oblique

Figure 8.5 Sagittal section of the eye (anterior portion) and eyelid.

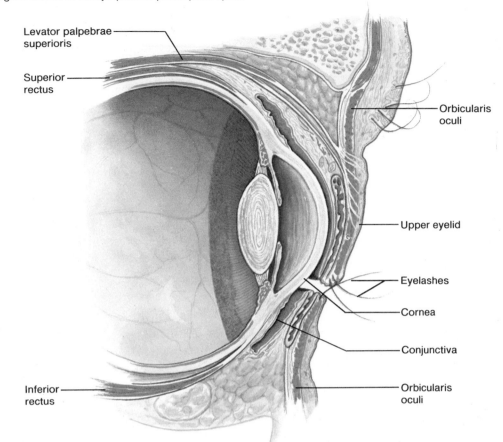

Levator palpebrae superioris

Superior rectus

Orbicularis oculi

Upper eyelid

Eyelashes

Cornea

Conjunctiva

Orbicularis oculi

Inferior rectus

lateral rectus: turns eye outward away from midline

medial rectus: turns eye inward toward midline

superior oblique: rotates eye counterclockwise

inferior oblique: rotates eye clockwise

There are three cranial nerves that control these muscles. They are the oculomotor, abducens, and trochlear. The oculomotor innervates four of these (the superior, inferior, medial rectus, and inferior oblique); the abducens innervates one (the lateral rectus); and the trochlear innervates one (the superior oblique).

Eyelids and Eyelashes

An eyelid (fig. 8.5) has an outer layer of thin skin covering muscle and connective tissue. Lining the inner surface is a transparent mucous membrane, called the conjunctiva. The conjunctiva folds back to cover the anterior of the eye except for the cornea. Therefore, tears cannot enter the orbits. The eyelids have eyelashes and even a particle of grit caught by one lash will cause the eyes to close immediately. There are sebaceous glands associated with each eyelash, and they produce an oily secretion that lubricates the eye. Inflammation of one of the glands is called a *sty*.

The eyelids are operated by the orbicularis oculi, whose contraction closes the lid, and the levator palpebrae superioris, which raises the lid. When a person suffers from *myasthenia gravis,* muscle weakness due to an inability to respond to ACh, the eyelids have to be taped open.

Lacrimal Apparatus

A **lacrimal apparatus** consists of a lacrimal gland and the lacrimal sac along with its ducts (fig. 8.6). The lacrimal gland, which lies in the orbit above the eye, produces tears which flow over the eye with the help of the blinking eyelids. The tears, collected by two small ducts, pass into the lacrimal sac before draining into the nose by way of the nasolacrimal duct.

Structure of the Eye

The eye (fig. 8.7 and table 8.1), an elongated sphere about one inch in diameter, has three layers or coats. The outer **sclera** is a white fibrous layer except for the transparent cornea, the window of the eye. The middle thin, dark brown layer, the **choroid,** contains many blood vessels and absorbs stray light rays. Toward the front,

Figure 8.6 Lacrimal apparatus consists of the lacrimal gland, which produces tears, and the lacrimal sac, which drains them away to enter the nasal cavity.

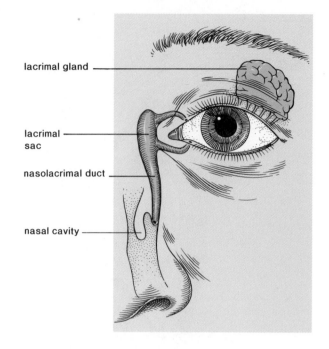

lacrimal gland

lacrimal sac

nasolacrimal duct

nasal cavity

Figure 8.7 Anatomy of the human eye. Notice that the sclera becomes the cornea; the choroid becomes the ciliary body and iris. The ciliary body is thrown into seventy to eighty radiating folds that contain the ciliary muscle and ligaments that hold and adjust the shape of the lens. The retina contains the receptors for sight, and vision is most acute in the fovea centralis where there are only cones. A blind spot occurs where the optic nerve leaves the retina, and there are no receptors for sight.

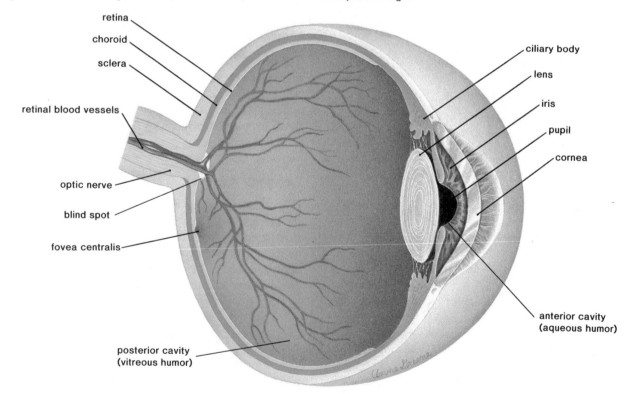

retina

choroid

sclera

retinal blood vessels

optic nerve

blind spot

fovea centralis

posterior cavity
(vitreous humor)

ciliary body

lens

iris

pupil

cornea

anterior cavity
(aqueous humor)

Table 8.1 Name and Function of Parts of the Eye

Part	Function
Lens	Refracts and focuses light rays
Iris	Regulates light entrance
Pupil	Admits light
Choroid	Absorbs stray light
Sclera	Protects eyeball
Cornea	Refracts rays of light
Humors	Refracts rays of light
Ciliary body	Holds lens in place, accommodates
Retina	Contains receptors
Rods	Black-and-white vision
Cones	Color vision
Optic nerve	Transmits impulses
Fovea	Region of cones in retina
Ciliary muscle	Accommodation

the choroid thickens and forms a ring-shaped structure, the ciliary body, containing the **ciliary muscle,** which controls the shape of the lens for near and far vision. Finally, the choroid becomes a thin, circular, muscular diaphragm, the iris, which regulates accommodation, a change in the size of a center hole, the **pupil,** through which light enters the eyeball. The **lens,** attached to the ciliary body by ligaments, divides the cavity of the eye into two chambers. A viscous, gelatinous material, the **vitreous humor,** fills the posterior cavity behind the lens. The anterior cavity between the cornea and the lens is filled with an alkaline, watery solution, secreted by the ciliary body and called the **aqueous humor.**

Retina

The inner layer of the eye, the **retina,** contains the rods and cones, the receptors for sight. The nerve fibers which leave the retina pass in front of the retina, forming the **optic nerve,** which turns to pierce all three layers of the eye. There are no rods and cones where the optic nerve passes through the retina; therefore, this is a **blind spot** (fig. 8.8) where vision is impossible. Because the optic nerve crosses at the *optic chiasma* (fig. 8.9), each occipital lobe receives information about only part of an object. Later, the two halves communicate to arrive at a complete three-dimensional interpretation of the whole object.

The retina contains the rods and cones, which are the receptors for sight. The nerve fibers leaving the retina form the optic nerve, which sends branches to the occipital lobes.

Figure 8.8 *a.* Structure of retina. Rods and cones are located toward the back of the retina, followed by the bipolar cells and the ganglionic cells whose fibers become the optic nerve. *b.* The blind spot where the optic nerve pierces the eyeball is clearly visible in eye examinations.

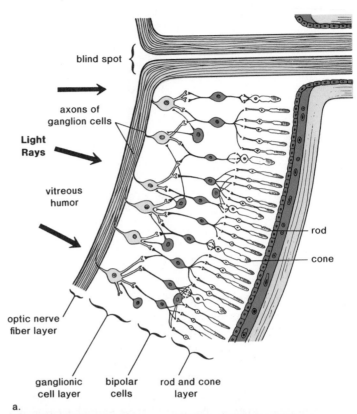

a.

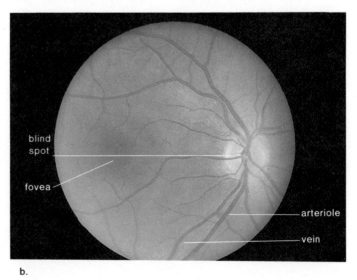

b.

Glaucoma A small amount of aqueous humor is continually produced each day. Normally, it leaves the anterior chamber by way of tiny ducts that are located where the sclera meets the cornea. If these drainage ducts are blocked, pressure rises and compresses the retinal arteries whose capillaries feed nerve fibers located in the retina. With the passage of time, some of these fibers die, and the result is partial or total blindness.

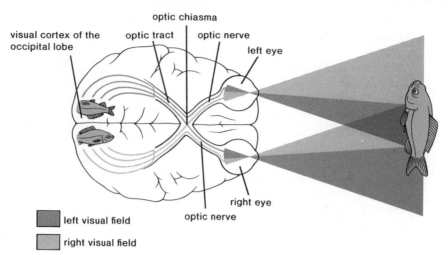

Figure 8.9 Both eyes "see" the entire object, but information from the right half of each retina goes to the right visual cortex and information from the left half of each retina goes to the left visual cortex. When the information is pooled, the brain "sees" the entire object and "sees" it in depth.

Rods

In dim light, the pupils enlarge so that more rays of light can enter the eyes. As the rays of light enter, they strike the rods and cones, but only the 160 million rods located in the periphery of the eyes are sensitive to faint light. The **rods** do not detect fine detail or color, so at night, for example, all objects appear to be blurred and have a shade of gray. Rods do detect even the slightest motion, however, because of their abundance and position in the eyes.

The rods contain *rhodopsin,* a molecule that contains the protein opsin and the pigment retinal. When light strikes rhodopsin, it breaks down to its components and this generates nerve impulses. The more rhodopsin present in the rods, the more sensitive are our eyes to dim light. Therefore, during the time required for adjustment to dim light, when we find it difficult to see, rhodopsin is being formed in the rods. Retinal is a derivative of vitamin A. Vitamin A is abundant in carrots, so the suggestion that we should eat carrots for good vision is not without foundation.

The rods function in dim light. They do not see fine detail or color, but do detect motion.

Cones

The **cones** are located primarily in the fovea centralis, an oval yellowish area with a depression where there are only cone cells. In bright light, the pupils get smaller so that less light enters the eyes. Bright light activates the cones, which detect the fine detail and color of an object. Color vision depends on three kinds of cones, one kind for each of three colors: blue, green, or red light. All the colors we see are believed to be dependent on which of these cones are activated. Complete color blindness is extremely rare. In most instances, a particular type of cone is lacking or deficient in number. The lack of red and green cones is the most common, affecting about 5 percent of the American population. If the eye lacks red cones, the green colors become accentuated, and vice versa (fig. 8.10).

The cones function in bright light. They detect fine detail and are responsible for color vision.

Lens

When we look at an object, light rays are bent (refracted) and focused on the retina. The cornea, vitreous humor, and lens all help in this process. Although the lens remains flat when we view distant objects, it

Figure 8.10 Test plates for color blindness. When looking at the plate on the left, the person with normal color vision will see the number 8 and when looking at the plate on the right, the person with normal color vision will see the number 12. The most common form of color blindness involves an inability to distinguish reds and greens.

These plates have been reproduced from Ishihara's Tests for Colour Blindness published by KANEHARA & CO., LTD., Tokyo, Japan, but tests for color blindness cannot be conducted with this material. For accurate testing, the original plates should be used.

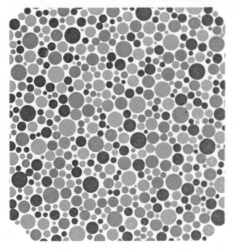 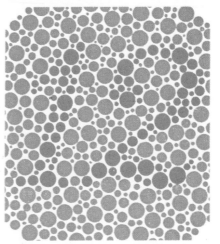

rounds up when we view close objects. This is called **accommodation** (fig. 8.11). The shape of the lens is controlled by the ciliary muscle within the ciliary body. When we view a distant object, the ciliary muscle is relaxed, causing the ligaments attached to the ciliary body to be under tension; therefore, the lens remains relatively flat. When we view a close object, the ciliary muscle contracts, releasing the tension on the ligaments; therefore, the lens rounds up due to its natural elasticity (table 8.2). Since close work requires contraction of the ciliary muscle, it very often causes "eye strain."

The lenses function to focus light rays on the retina. They are flat for distant vision and round up for near vision.

Corrective Lenses The majority of people can see what is designated as a size "20" letter 20 feet away and, therefore, are said to have 20/20 vision. Persons who can see close objects, but cannot see the letters from this distance are said to be *nearsighted*. They often have an elongated eyeball and when they attempt to look at a far object, the image is brought to focus in front of the retina (fig. 8.11c). They need to wear concave lenses that diverge the light rays so that the image can be focused on the retina when viewing a distant object.

Table 8.2	Accommodation	
Object	**Ciliary Muscle**	**Lens**
Near object	Ciliary muscle contracts, ligaments relax	Lens becomes round
Far object	Ciliary muscle relaxes, ligaments under tension	Lens is flattened

Persons who can easily see the optometrist's chart, but cannot see close objects well are *farsighted*. They often have a shortened eyeball, and when they try to see near, the image is focused behind the retina (fig. 8.11d). They need to wear a convex lens to increase the bending of light rays so that the image will be focused on the retina when viewing a close object.

With aging, the lens loses some of its elasticity and is unable to accommodate in order to bring close objects into focus. This necessitates the wearing of *bifocals,* which means that the upper part of the lens is for distant vision and the lower part is for near vision.

Figure 8.11 Accommodation. *a.* When the eye focuses on a far object, the lens is flat because the eye muscles holding the lens are relaxed and the suspensory ligament is taut. *b.* When the eye focuses on a near object, the lens rounds up because the eye muscles contract, causing the suspensory ligament to relax. *c.* In nearsighted individuals, the eyeball is too long. They cannot see far because when viewing a distant object the image is brought to focus in front of the retina. *d.* In farsighted individuals, the eyeball is too short. They cannot see near objects well because when viewing a near object the image is brought to focus behind the retina.

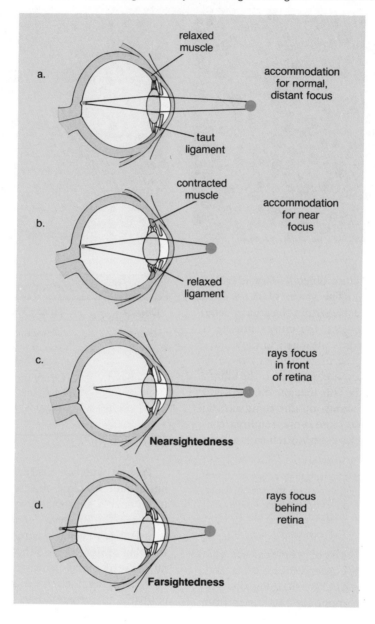

Mechanoreceptor—The Ear

The ear accomplishes two sensory functions: balance and hearing. The sense cells for both of these are located in the inner ear and consist of hair cells with cilia that respond to mechanical stimulation. Each hair cell has from 30 to 150 extensions that are called cilia despite the fact that they contain tightly packed filaments rather than microtubules. When the cilia of any particular hair cell are displaced in a certain direction, the cell generates nerve impulses, which are sent along a cranial nerve to the brain.

Structure of the Ear

Table 8.3 lists the parts of the ear, and figure 8.12 is a drawing of the ear. The ear has three divisions: outer, middle, and inner. The **outer ear** consists of the **pinna** (external flap) and **auditory canal.** The opening of the auditory canal is lined with fine hairs and sweat glands. In the upper wall are ceruminous glands, modified sweat glands that secrete earwax, a substance that helps guard the ear against the entrance of foreign materials such as air pollutants.

Table 8.3 The Ear

	Outer Ear	Middle Ear	Inner Ear	
			Cochlea	Sacs plus Semicircular Canals
Function	Directs sound waves to tympanic membrane	Picks up and amplifies sound waves	Hearing	Maintains equilibrium
Anatomy	Pinna Auditory canal	Tympanic membrane Ossicles	Contains organ of Corti Auditory nerve starts here	Saccule and utricle Semicircular canals
Media	Air	Air (eustachian tube)	Fluid	Fluid

Note: Path of vibration: Sound waves—vibration of tympanic membrane—vibration of hammer, anvil, and stirrup—vibration of oval window—fluid pressure waves in canals of inner ear lead to stimulation of hair cells—bulging of round window.

The **middle ear** begins at the **tympanic membrane** (eardrum) and ends at a bony wall in which are found two small openings covered by membranes. These openings are called the **oval** and **round windows.** Three small bones are found between the tympanic membrane and the oval window. Collectively called the **ossicles,** individually they are the **malleus** (hammer), **incus** (anvil), and **stapes** (stirrup) (fig. 8.12) because their shapes resemble these objects. The malleus adheres to the tympanic membrane, and the stapes touches the oval window. The posterior wall of the middle ear also has an opening that leads to many air spaces within the mastoid process.

The **eustachian tubes** (fig. 8.12) extend from the middle ear to the nasopharynx and permit equalization of air pressure. Chewing gum, yawning, and swallowing in elevators and airplanes help move air through the eustachian tubes upon ascent and descent.

Whereas the outer ear and middle ear contain air, the inner ear is filled with fluid. The **inner ear** (fig. 8.12b), anatomically speaking, has three areas; the first two, called the vestibule and semicircular canals, are concerned with balance; and the third, the cochlea, is concerned with hearing.

The **semicircular canals** are arranged so that there is one in each dimension of space. The base of each canal, called an **ampulla,** is slightly enlarged. Within the ampullae (fig. 8.12b) are little hair cells whose cilia are inserted into a gelatinous medium.

A **vestibule,** or chamber, lies between the semicircular canals and the cochlea. It contains two small sacs called the **utricle** and **saccule.** Within both of these are little hair cells whose cilia protrude into a gelatinous substance. Resting on this substance are calcium carbonate granules, or **otoliths.**

The **cochlea** resembles the shell of a snail because it spirals. Within the tubular cochlea are three canals: the vestibular, the **cochlear canal,** and the tympanic canal (fig. 8.15). Along the length of the basilar membrane, which forms the lower wall of the cochlear canal, are little hair cells whose cilia come into contact with another membrane called the tectorial membrane. The hair cells plus the **tectorial membrane** are called the **organ of Corti.** When this organ sends nerve impulses to the cerebral cortex, it is interpreted as sound.

The outer ear, middle ear, and cochlea are necessary for hearing. The vestibule and semicircular canals are concerned with the sense of balance.

Balance (Equilibrium)

The sense of balance has been divided into two senses: **static equilibrium,** referring to knowledge of movement in one plane, either vertical or horizontal, and **dynamic equilibrium,** referring to knowledge of angular and/or rotational movement.

When the body is still, the otoliths in the utricle (fig. 8.13) and saccule rest on the hair cells. When the head and/or body moves horizontally, or vertically, the granules in the utricle and saccule are displaced. Displacement causes the cilia to bend slightly so that the cell generates nerve impulses that travel by way of a cranial nerve to the brain.

When the body is moving about, the fluid within the semicircular canals moves back and forth. This causes bending of the cilia attached to hair cells within the ampullae (fig. 8.14), and they initiate nerve impulses that travel to the brain. Continuous movement of the fluid in the semicircular canals causes one form of motion sickness.

Figure 8.12 Anatomy of the human ear. *a.* In the middle ear, the malleus, incus, and stapes amplify sound waves. Otosclerosis is a condition in which the stapes become attached to the inner ear and is unable to carry out its normal function. It can be replaced by a plastic piston, and thereafter the individual hears normally because sound waves are transmitted as usual to the cochlea, which contains the receptors for hearing. *b.* Inner ear. The sense organs for balance are in the inner ear: the vestibule contains the utricle and saccule, and the ampullae are at the bases of the semicircular canals. The receptors for hearing are also in the inner ear: the cochlea has been cut to show the location of the organ of Corti.

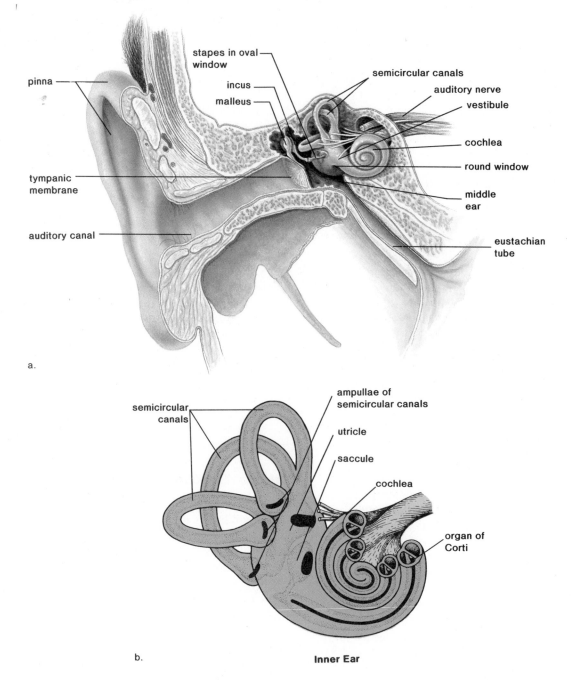

Figure 8.13 Receptor hair cells in the utricle and saccule are involved in our sense of static equilibrium: responsiveness to movement sideways or up and down. *a.* When the head is upright, a gelatinous material and otoliths are balanced directly on the cilia of hair cells. *b.* When the head is bent forward, the material and otoliths shift. The cilia bend, causing nerve impulses to begin.

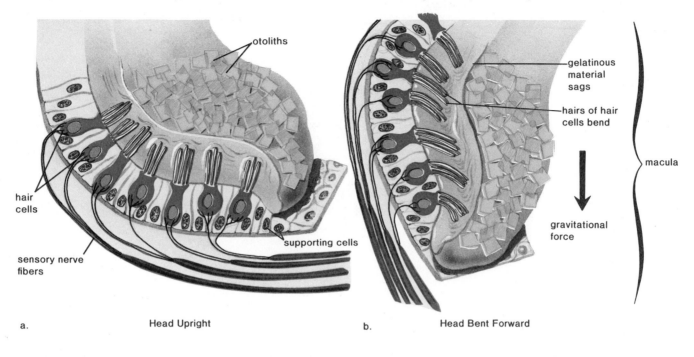

a.   Head Upright   b.   Head Bent Forward

Movement of the otoliths within the utricle and saccule are important for static equilibrium. Movement of fluid within the semicircular canals contributes to our sense of dynamic equilibrium.

Figure 8.14 Receptor hair cells in an ampulla are involved in dynamic equilibrium. Within each ampulla, hair cells are surrounded by gelatinous material. When the body rotates, the fluid within the semicircular canals is displaced, causing the material to move and the cilia to bend.

Hearing

The process of hearing begins when sound waves enter the auditory canal. Just as ripples travel across the surface of a pond, sound travels by the successive vibrations of molecules. Ordinarily, sound waves do not carry much energy, but when a large number of waves strike the eardrum, it moves back and forth (vibrates) ever so slightly. The malleus then takes the pressure from the inner surface of the eardrum and passes it by way of the incus to the stapes in such a way that the pressure is multiplied about twenty times as it moves from the eardrum to the stapes. The stapes strikes the oval window, causing it to vibrate and in this way, the pressure is passed to the fluid within the inner ear.

If the cochlea is unwound, as shown in figure 8.15, the vestibular canal is seen to connect with the tympanic canal; therefore, as the figure indicates, pressure waves move from one canal to the other toward the round window, a membrane that can bulge to absorb the pressure. As a result of the movement of the fluid within the cochlea, the basilar membrane moves up and

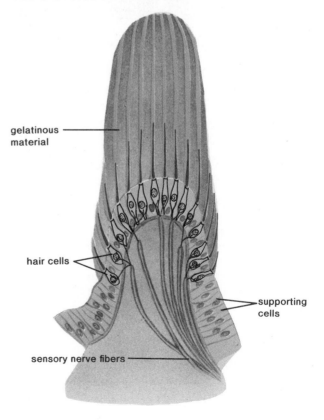

Figure 8.15 Organ of Corti. *a.* Enlarged cross section through the organ of Corti, showing the receptor hair cells from the side. *b.* Cochlea unwound, showing the placement of the organ of Corti along its length. The arrows represent the pressure waves that move from the oval window to the round window. These cause the basilar membrane to vibrate and the cilia of at least a portion of the 15,000 hair cells to bend against the tectorial membrane. The resulting nerve impulses result in hearing.

a. From *Physiology of the Human Body,* 3d edition, by J. R. McClintic. Copyright © 1985 John Wiley & Sons, Inc., New York, NY. Reprinted by permission.

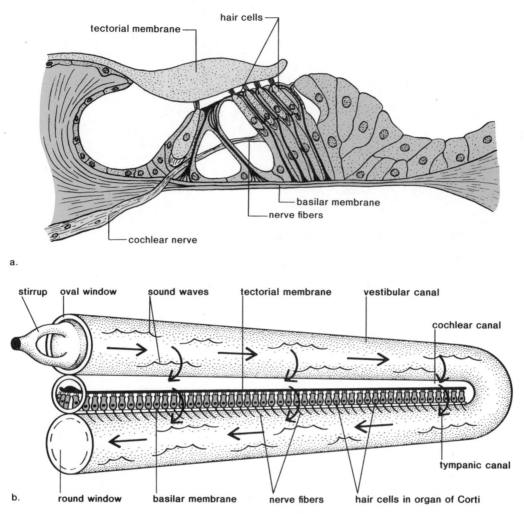

down, and the cilia of the hair cells rub against the tectorial membrane. This bending of the cilia initiates nerve impulses that pass by way of the cochlear nerve to the temporal lobe of the brain, where the impulses are interpreted as a sound.

The sense receptors for sound are hair cells on the basilar membrane (the organ of Corti). When the basilar membrane vibrates, the delicate hairs, which touch the tectorial membrane, bend, and nerve impulses begin and are transmitted in the cochlear nerve to the brain.

Infections and Deafness

Infections of the middle ear (*otitis media*) may occur frequently during childhood. As a precautionary measure to prevent perforation of the tympanic membrane and to keep the auditory tube open, an incision of the tympanic membrane, called *myringotomy* is followed by the insertion of a tiny tube into the membrane. With time, the tubes are sloughed out of the ear.

There are two major types of deafness: conduction and nerve deafness. *Conduction deafness* can be due to a congenital defect, as those that occur when a pregnant woman contracts German measles during the first trimester of pregnancy. (For this reason, every female should be sure to be immunized against rubella before the childbearing years.) Conduction deafness can also be due to repeated infections or to otosclerosis. With *otosclerosis,* the normal bone of the middle ear is replaced by vascular spongy bone. In any case, the ossicles tend to fuse together, restricting their ability to magnify sound waves.

Nerve deafness most often occurs when cilia on the sense receptors within the cochlea have worn away.

Since this may happen with normal aging, old people are more likely to have trouble hearing; however, nerve deafness also occurs when young people listen to loud music amplified to 130 decibels. Because the usual types of hearing aids are not helpful for nerve deafness, it is wise to avoid subjecting the ears to any type of continuous loud noise. Costly cochlear implants that directly stimulate the auditory nerve are available. Those who have these electronic devices report that the speech they hear is like that of a robot.

Summary

I. General Receptors
   A. Skin. Specialized receptors in the human skin respond to temperature, touch, pressure, and pain.
   B. Muscles and joints. Sense receptors in the muscles and joints, called proprioceptors, give us a sense of how our body parts are positioned.
II. Special Senses
   A. Chemoreceptors. The receptors for taste (taste buds) and the receptors for smell (olfactory microvilli) work together to give us our senses of taste and smell.
   B. Photoreceptor—the eye.
      The eye has three layers: the outer sclera, the middle choroid, and the inner retina. Only the retina contains receptors for sight.

The sense receptors for sight are the rods and cones. The rods are responsible for vision in dim light, and the cones are responsible for vision in bright light and color vision. When either is stimulated, nerve impulses begin and are transmitted in the optic nerve to the brain.

The lens, assisted by the cornea and vitreous humor, focuses images on the retina. The shape of the eyeball determines the need for corrective lenses; the inability of the lenses to accommodate as we age also necessitates corrective lens for close vision.
   C. Mechanoreceptor—the ear.
      The outer ear, middle ear, and cochlea are necessary for hearing. The vestibule and semicircular canals are concerned with the sense of balance.

Movement of the otoliths within the utricle and saccule are important for static equilibrium. Movement of fluid within the semicircular canals contributes to our sense of dynamic equilibrium.

The receptors for sound are hair cells on the basilar membrane (the organ of Corti). When the basilar membrane vibrates, the delicate hairs which touch the tectorial membrane bend, and nerve impulses begin and are transmitted in the cochlear nerve to the brain.

Study Questions

1. What is another name for receptor?
2. What type receptors are categorized as general and what type are categorized as special receptors?
3. Discuss the receptors of the skin, viscera, and joints.
4. Discuss the chemoreceptors.
5. Describe the anatomy of the eye, and explain focusing and accommodation.
6. Describe sight in dim light. What chemical reaction is responsible for vision in dim light? Discuss color vision.
7. Relate the need for corrective lenses to two possible shapes of the eye. Discuss bifocals.
8. Describe the anatomy of the ear and how we hear.
9. Describe the role of the utricle, saccule, and semicircular canals in balance.
10. Discuss the two causes of deafness, including why young people frequently suffer loss of hearing.

Objective Questions

Fill in the blanks.
1. The sense organs for position and movement are called _____ .
2. Taste buds and olfactory receptors are termed _____ because they are sensitive to chemicals in the air and food.
3. The receptors for sight, the _____ and _____ , are located in the _____ , the inner layer of the eye.
4. The cones give us _____ vision and work best in _____ light.
5. The lens _____ for viewing close objects.
6. People who are nearsighted cannot see objects that are _____ . A _____ lens will restore this ability.
7. The ossicles are the _____ , _____ , and _____ .
8. The semicircular canals are involved in our sense of _____ .
9. The organ of Corti is located in the _____ canal of the _____ .
10. Vision, hearing, taste, and smell do not occur unless nerve impulses reach the proper portion of the _____ .

Medical Terminology Reinforcement Exercise

Pronounce the following words and analyze the meaning by word parts as dissected:

1. ophthalmologist (of"thal-mol'o-jist)—ophthalmo/log/ist

2. presbyopia (pres"be-o'pe-ah)—presby/opia

3. blepharoptosis (blef"ah-ro-to'sis)—blepharo/ptosis

4. keratoplasty (ker'ah-to-plas"te)—kerato/plasty

5. optometrist (op-tom'ĕ-trist)—opto/metr/ist

6. lacrimator (lak'rĭ-ma"tor)—lacrima/tor

7. otitis media (o-ti'tis)—ot/itis media

8. myringotomy (mir"in-got'o-me)—myringo/tomy

9. tympanocentesis (tim"pah-no-sen-te'sis)—tympano/centesis

10. microtia (mi"kro'she-ah)—micr/otia

9

Endocrine System

Chapter Outline

Learning Objectives

After you have studied this chapter, you should be able to:

1. Define a hormone and name the endocrine glands studied.
2. Explain the anatomical relationship between the posterior pituitary and the hypothalamus, name two hormones produced by the hypothalamus but secreted by the posterior pituitary, and relate the hormone ADH to diabetes insipidus.
3. Name six hormones produced by the anterior pituitary and indicate which of these control other endocrine glands.
4. Discuss the physiological action of GH and relate stature, including that of a dwarf or giant, and the disorder acromegaly to this hormone.
5. Explain the anatomical relationship between the anterior pituitary and the hypothalamus.
6. Draw a diagram indicating the relationship between the hypothalamus, anterior pituitary, and a gland controlled by the anterior pituitary, and show how each of these is controlled by negative feedback.
7. Discuss the macroscopic and microscopic anatomy of the thyroid gland, and the chemistry and physiological function of thyroxin, hypothyroidism, and hyperthyroidism.
8. State the location of the adrenal glands and describe the relationship between the adrenal medulla and adrenal cortex.
9. Discuss the function of the adrenal medulla and its relationship to the nervous system.
10. Name three categories of hormones produced by the adrenal cortex, give an example of each category, and discuss their physiological action.
11. Describe the symptoms of Addison's disease and Cushing's syndrome, relating these to malfunction of the adrenal cortex.
12. State the location of the parathyroid glands and discuss the function of PTH and calcitonin, relating this to osteoporosis and tetany.
13. State the location of the pancreas, describe its microscopic anatomy, and name two hormones produced by the pancreas and discuss their function.
14. Discuss the two types of diabetes mellitus and the diagnosis of diabetes by means of urinalysis, and contrast insulin shock to diabetic coma.

Endocrine System

Like the nervous system, the endocrine system coordinates the functioning of body parts; however, the endocrine system utilizes hormones whose presence or absence affects our metabolism, our appearance, and our behavior. It is now known that hormones are chemical messengers that help regulate cellular activity.

Endocrine Glands

Hormones are produced by glands (fig. 9.1) called endocrine glands that secrete their products internally, placing them directly in the blood. Since these glands do not have ducts for the transport of their secretions, they are sometimes called ductless glands. All hormones are carried throughout the body by the blood, but each one affects only a specific part or parts, appropriately termed target organs.

Table 9.1 lists the major endocrine glands in humans, the hormones produced by each, and the associated disorders that occur when there is an abnormal level of the hormones, either too much or too little. Each of these glands will now be discussed in some detail.

The endocrine glands secrete their hormones into the bloodstream for transport to target organs.

Pituitary Gland

The pituitary gland, which has two portions called the **anterior pituitary** and the **posterior pituitary,** is a small gland, about 1 centimeter in diameter, that lies at the base of the brain.

Posterior Pituitary

The posterior pituitary is connected by means of a stalk to the hypothalamus, the portion of the brain that is concerned with homeostasis. The hormones released by

---

Figure 9.1 Anatomical location of major endocrine glands in the body. The hypothalamus controls the pituitary, which in turn controls the hormonal secretions of the thyroid, adrenal cortex, and sex organs.

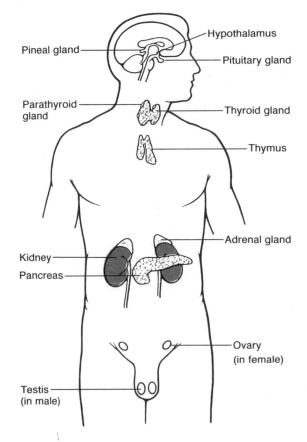

Table 9.1 The Principal Endocrine Glands and Their Hormones

Gland	Hormones	Chief Functions	Disorders Too Much/ Too Little
Hypothalamus	Releasing hormones	Stimulates anterior pituitary	*See* anterior pituitary
Anterior pituitary	Thyroid-stimulating (TSH, thyrotropic)	Stimulates thyroid	*See* thyroid
	Adrenocorticotropic (ACTH)	Stimulates adrenal cortex	*See* adrenal cortex
	Gonadotropic	Stimulates gonads	*See* testes and ovary
	Follicle-stimulating (FSH)	Regulates egg and sperm production	
	Leuteinizing (LH)	Regulates sex hormone production	
	Lactogenic (LTH, prolactin)	Causes milk production	
	Growth (GH, somatotropic)	Promotes growth	Giant, acromegaly/ dwarf
Posterior pituitary	Antidiuretic (ADH, vasopressin)	Causes water retention by kidneys	Diverse^a/diabetes insipidus
	Oxytocin	Causes uterine contraction	
Pineal	Melatonin	Inhibits release of gonadotropins	Sexual immaturity/ sexual maturity
Thymus	Thymosin	Regulates development and function of immune system	Overactive immunity/ underactive immunity
Thyroid	Thyroxin	Increases metabolic rate (cellular respiration)	Exophthalmic goiter/ simple goiter, myxedema, cretinism
	Calcitonin	Decreases plasma level of calcium	Tetany/weak bones
Parathyroid	Parathormone (PTH)	Increases plasma levels of calcium and phosphorus	Weak bones/tetany
Adrenal cortex	Glucocorticoids (cortisol)	Causes gluconeogenesis	
	Mineralocorticoids (aldosterone)	Causes sodium retention; potassium excretion by kidneys	Cushing's syndrome/ Addison's disease
Adrenal medulla	Sex hormones Epinephrine and norepinephrine	Promotes fight or flight	
Pancreas	Insulin	Lowers blood sugar	Shock/diabetes mellitus
	Glucagon	Raises blood sugar	
Testes	Androgens (testosterone)	Promotes secondary male characteristics	Diverse/eunuch
Ovaries	Estrogen (by follicle)	Promotes secondary female characteristics	Diverse/masculinization
	Progesterone (by corpus luteum)		

^aThe word *diverse* in this chart means that the symptoms have not been described as a syndrome in the medical literature.

Figure 9.2 The posterior pituitary is connected to the hypothalamus by a stalk. The hypothalamus produces the hormones (ADH and oxytocin) that are secreted by the posterior pituitary. These are sent to the posterior pituitary by way of nerve fibers.

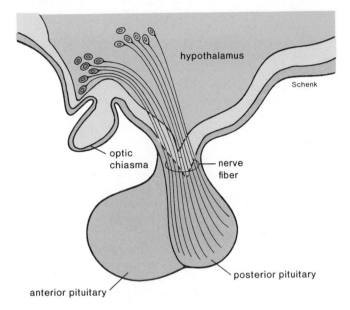

Figure 9.3 Regulation of ADH secretion. When the blood is concentrated, the hypothalamus produces ADH that is released by the posterior pituitary. This acts on the kidneys to retain more water so that the blood is diluted. Thereafter, the hypothalamus does not produce ADH.

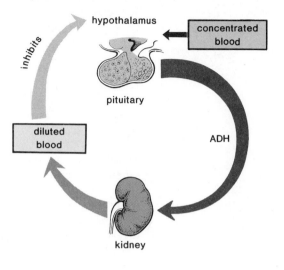

the posterior pituitary are made in nerve cell bodies in the hypothalamus, after which they migrate through axons that terminate in the posterior pituitary (fig. 9.2).

The posterior pituitary releases **antidiuretic hormone (ADH),** sometimes called *vasopressin.* ADH, promotes the reabsorption of water from the collecting duct, a portion of the kidney tubules. It is believed that the hypothalamus contains cells that are sensitive to the amount of water in the blood. When these cells detect that the blood lacks sufficient water, ADH is produced by hypothalamic neurons and is transported by their fibers to the posterior pituitary, where it is released (fig. 9.2). As the blood becomes more dilute, the hormone ceases to be produced and released. Figure 9.3 illustrates how the level of this hormone is controlled by a circular pattern in which the effect of the hormone (dilute blood) acts to shut down the production and release of the hormone. This is an example of control by negative feedback. Negative feedback mechanisms regulate the activities of most hormonal glands.

Inability to produce ADH causes **diabetes insipidus** (watery urine) in which a person produces copious amounts of urine with a resultant loss of salts from the blood. The condition can be corrected by the administration of ADH.

Oxytocin, is another hormone released by the posterior pituitary that is made in the hypothalamus. Oxytocin causes the uterus to contract and may be used to

artificially induce labor. It also stimulates the release of milk from the breast when a baby is nursing.

The hormones of the posterior pituitary, ADH and oxytocin, are produced in the hypothalamus.

Anterior Pituitary

There are tiny blood vessels within a portal system that connect the anterior pituitary to the hypothalamus. The hypothalamus controls the anterior pituitary by producing hypothalamic-releasing hormones, which are transported to the anterior pituitary by the blood within the portal system. Each type of releasing hormone causes the anterior pituitary either to secrete or to stop secreting a specific hormone. The anterior pituitary produces at least six different hormones (fig. 9.4).

Three of the hormones produced by the anterior pituitary have a direct effect on the body. **Growth hormone (GH),** or somatotropin, affects the physical appearance dramatically since it determines the height of the individual (fig. 9.5). If little or no growth hormone is secreted by the anterior pituitary during childhood, the person could become a dwarf of perfect proportions but quite small in stature. If too much growth hormone is secreted, the person could become a giant. Giants usually have poor health, primarily because growth hormone has a secondary effect on blood

Figure 9.4 The anterior pituitary is connected to the hypothalamus only by a portal system. The hypothalamus sends releasing hormones to the anterior pituitary by this circulatory route. The releasing hormones specifically promote or inhibit the secretion of anterior pituitary hormones.

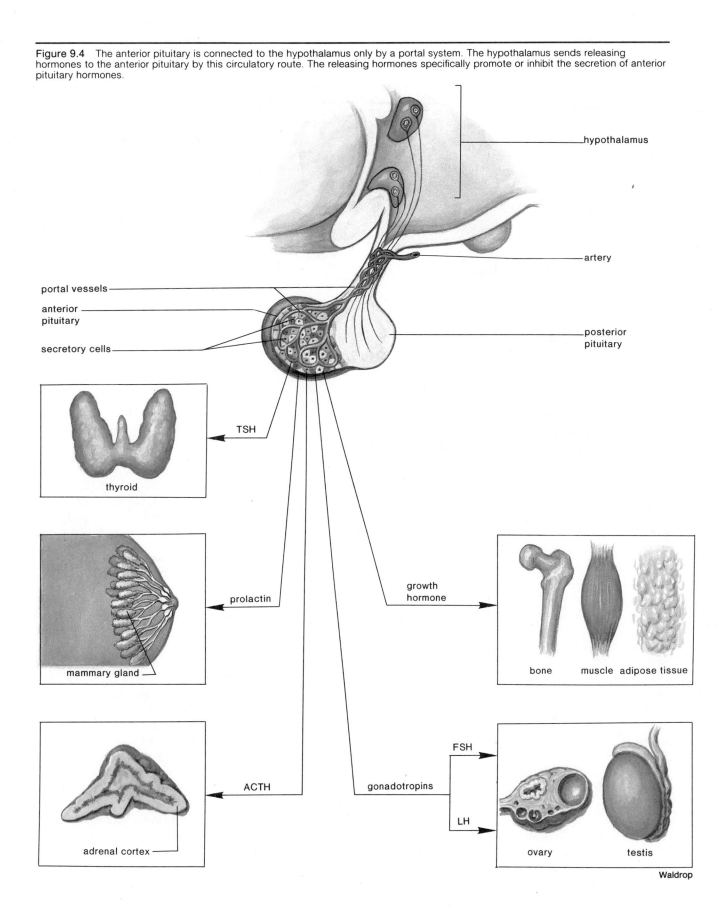

hypothalamus

artery

portal vessels

anterior pituitary

secretory cells

posterior pituitary

TSH

thyroid

prolactin

mammary gland

growth hormone

bone   muscle adipose tissue

ACTH

adrenal cortex

gonadotropins

FSH

LH

ovary

testis

Waldrop

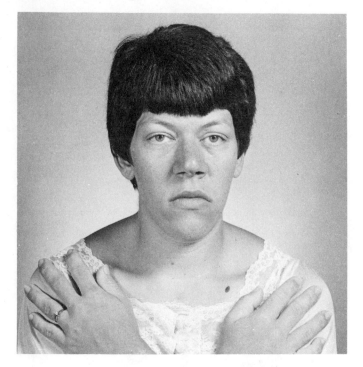

sugar level, promoting an illness called diabetes (sugar) mellitus, which is discussed in one of the following sections on the pancreas.

If the production of GH increases in an adult after full height has been obtained, only certain bones respond. These are the bones of the jaw, eyebrow ridges, nose, fingers, and toes. When these begin to grow, the person takes on a slightly grotesque look with huge fingers and toes, a condition called **acromegaly** (fig. 9.6).

Lactogenic hormone (LTH), also called **prolactin,** is produced in quantity only after childbirth. It causes the mammary glands in the breasts to develop and produce milk.

Melanocyte-stimulating hormone (MSH) causes skin color changes in lower vertebrates, but no one knows what it does in humans. However, it is derived from a molecule that is also the precursor for ACTH and the opioids (endorphins and enkephalins) discussed in the following section.

GH and LTH are two hormones of the anterior pituitary. GH influences the height of children and brings about a condition called acromegaly in adults. LTH promotes milk production after childbirth.

As indicated in figure 9.4 and table 9.1, the anterior pituitary secretes the following hormones, which have an effect on other endocrine glands:

1. **TSH,** thyroid-stimulating hormone
2. **ACTH,** a hormone that stimulates the adrenal cortex
3. **Gonadotropic hormones** (FSH and LH in females, and ICSH and LH in males) that stimulate the gonads, the testes in males and the ovaries in females

TSH causes the thyroid to produce thyroxin; ACTH causes the adrenal cortex to produce cortisol; and gonadotropic hormones cause the gonads to secrete sex hormones. Notice that it is now possible to indicate a three-tiered relationship between the hypothalamus, pituitary, and other endocrine glands. The hypothalamus produces releasing hormones that control the anterior pituitary, and the anterior pituitary produces hormones that control the thyroid, adrenal cortex, and gonads. Figure 9.7 illustrates the feedback mechanism that controls the activity of these glands.

Figure 9.7 Control of hormone secretion. The level of thyroxin in the body is controlled in three ways, as shown: *a.* the level of TSH exerts feedback control over the hypothalamus; *b.* the level of thyroxin exerts feedback control over the anterior pituitary, and *c.* over the hypothalamus. In this way, thyroxin controls its own secretion. Substitution of the appropriate terms would also allow this diagram to illustrate control of cortisol and sex hormone levels.

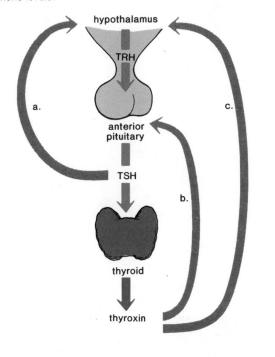

The hypothalamus, anterior pituitary, and the other endocrine glands controlled by the anterior pituitary, are all involved in a self-regulating feedback loop.

Thyroid Gland

The thyroid gland (fig. 9.1) is located in the neck and is attached to the trachea just below the larynx. Internally, the gland is composed of a large number of follicles filled with thyroglobulin, the storage form of thyroxin. The production of both of these requires iodine. Iodine is actively transported into the thyroid gland, where the concentration may become as much as twenty-five times that of the blood. If iodine is lacking in the diet, the thyroid gland enlarges, producing a goiter (fig. 9.8). This is the reason for iodized salt today. The cause of thyroid enlargement becomes clear if we refer to figure 9.7. When there is a low level of thyroxin in the blood, a condition called hypothyroidism, the anterior pituitary is stimulated to produce **TSH.** TSH causes the thyroid to increase in size so that enough

Figure 9.8 Simple goiter. An enlarged thyroid gland is often caused by a lack of iodine in the diet. Without iodine, the thyroid is unable to produce thyroxin and continued anterior pituitary stimulation causes the gland to enlarge.

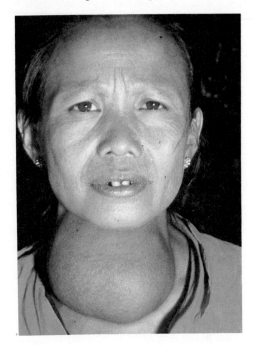

thyroxin usually is produced. In this case, enlargement continues because enough thyroxin is never produced. An enlarged thyroid that produces some thyroxin is called a **simple goiter.**

Activity and Disorders

Thyroxin increases the metabolic rate. It does not have one target organ; instead, it stimulates most of the cells of the body to metabolize at a faster rate. For example, it causes more glucose to be broken down.

If the thyroid fails to develop properly, a condition called **cretinism** results. Cretins (fig. 9.9) are short, stocky persons who have had extreme hypothyroidism since childhood and/or infancy. Thyroxin therapy can initiate growth, but unless treatment is begun within the first two months, mental retardation results. The occurrence of hypothyroidism in adults produces the condition known as **myxedema** (fig. 9.10), which is characterized by lethargy, weight gain, loss of hair, slower pulse rate, decreased body temperature, and thickness and puffiness of the skin. The administration of adequate doses of thyroxin restores normal function and appearance.

Figure 9.9 Cretinism. Cretins are individuals who have suffered from thyroxin insufficiency since birth or early childhood. Skeletal growth is usually inhibited to a greater extent than soft tissue growth; therefore, the child appears short and stocky. Sometimes the tongue becomes so large that it obstructs swallowing and breathing.

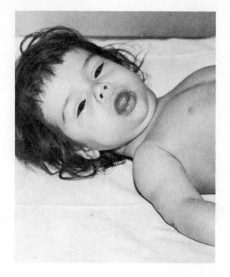

Figure 9.10 Myxedema is caused by thyroid insufficiency in the older adult. An unusual type of edema leads to swelling of the face and bagginess under the eyes.

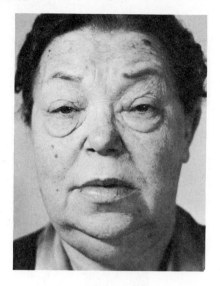

In the case of hyperthyroidism (too much thyroxin), the thyroid gland is enlarged and overactive, causing a goiter to form and the eyes to protrude because of edema in eye socket tissues and swelling of extrinsic eye muscles. This type of goiter is called **exophthalmic goiter** (fig. 9.11). The patient usually becomes hyperactive, nervous, irritable, and suffers from insomnia. Removal or destruction of a portion of the thyroid by means of radioactive iodine is sometimes effective in curing the condition.

Calcitonin

In addition to thyroxin, the thyroid gland also produces the hormone **calcitonin.** This hormone helps regulate the calcium level in the blood and opposes the action of parathyroid hormone. The interaction of these two hormones is also mentioned on page 171.

The anterior pituitary produces TSH, a hormone that promotes the production of thyroxin by the thyroid. Thyroxin, which speeds up metabolism, can affect the body as a whole as exemplified by cretinism and myxedema.

Parathyroid Glands

The parathyroid glands are embedded in the posterior surface of the thyroid gland, as shown in figure 9.12. Many years ago, these four small glands were sometimes removed by mistake during thyroid surgery.

Figure 9.11 Exophthalmic goiter. Protruding eyes occur when an active thyroid gland enlarges.

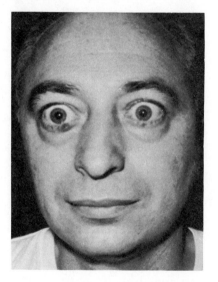

Under the influence of **parathyroid hormone (PTH),** also called parathormone, the calcium (Ca^{++}) level in the blood increases and the phosphate (HPO_4^{-2}) level decreases. The hormone promotes the formation of vitamin D, which in turn stimulates the absorption of calcium from the gut, promotes the retention of calcium by the kidneys, and promotes the demineralization of bone. In other words, PTH promotes the activity of osteoclasts, the bone-resorbing cells. Although this also raises the level of phosphate in the blood, PTH acts on the kidneys to excrete phosphate in the urine. When

Figure 9.12 Parathyroid glands. *a.* These small glands are embedded in the posterior surface of the thyroid gland. Yet the parathyroids and thyroid glands have no anatomical or physiological connection with one another. *b.* Regulation of parathyroid hormone secretion. A low blood level of calcium causes the parathyroids to secrete parathyroid hormone, which causes the kidneys and gut to retain calcium and osteoclasts to break down bone. The end result is an increased level of calcium in the blood. A high blood level of calcium inhibits hormonal secretion of parathyroid hormone.

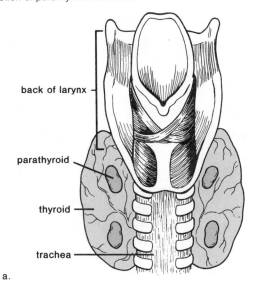

back of larynx

parathyroid

thyroid

trachea

a.

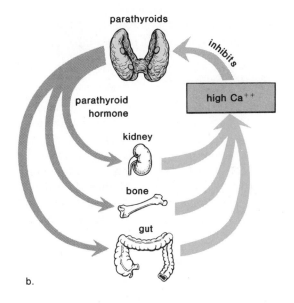

parathyroids

inhibits

parathyroid hormone

high Ca⁺⁺

kidney

bone

gut

b.

a woman stops producing the female sex hormone estrogen following menopause, she is more likely to suffer from osteoporosis, characterized by a thinning of the bones. It is, therefore, reasoned that estrogen makes bones less sensitive to PTH.

If insufficient parathyroid hormone is produced, the level of calcium in the blood drops, resulting in **tetany.** In tetany, the body shakes from continuous muscle contraction. The effect is really brought about by increased excitability of the nerves, which fire spon-

taneously and without rest. Calcium plays an important role in both nervous conduction and muscle contraction.

The level of PTH secretion is controlled by a feedback mechanism involving calcium (fig. 9.12). When the calcium level rises, PTH secretion is inhibited; and when the calcium level lowers, PTH secretion is stimulated.

As mentioned previously, the thyroid secretes calcitonin, which also influences blood calcium level. Although calcitonin has the opposite effect of PTH, particularly on the bones, its action is not believed to be as significant. Still, the two hormones function together to regulate the level of calcium in the blood.

Parathyroid hormone maintains a high blood level of calcium by promoting its absorption in the gut, its reabsorption by the kidneys, and demineralization of bone. These actions are opposed by calcitonin produced by the thyroid.

Adrenal Glands

The adrenal glands, as their name implies (*ad* = near; *renal* = kidneys), lie atop the kidneys (fig. 9.1). Each consists of an outer portion, called the *cortex,* and an inner portion, called the *medulla.* These portions, like the anterior and posterior pituitary, have no connection with one another.

The hypothalamus exerts control over the activity of both portions of the adrenal glands. The hypothalamus can initiate nerve impulses that travel by way of the brain stem, nerve cord, and sympathetic nerve fibers (fig. 9.13) to the adrenal medulla, which then secretes its hormones. The hypothalamus, by means of ACTH-releasing hormone, controls the anterior pituitary's secretion of **ACTH,** which in turn stimulates the adrenal cortex. Stress of all types, including both emotional and physical trauma, prompts the hypothalamus to stimulate the adrenal glands.

The adrenal glands have two parts, an outer cortex and an inner medulla. The adrenal medulla is under nervous control and the cortex is under hormonal control of ACTH, an anterior pituitary hormone.

Adrenal Medulla

The adrenal medulla (fig. 9.13) secretes **epinephrine** (adrenalin) and **norepinephrine** (noradrenalin). The postganglionic fibers of the sympathetic nervous system also secrete norepinephrine. In fact, the adrenal medulla is often considered to be an adjunct to the sympathetic nervous system.

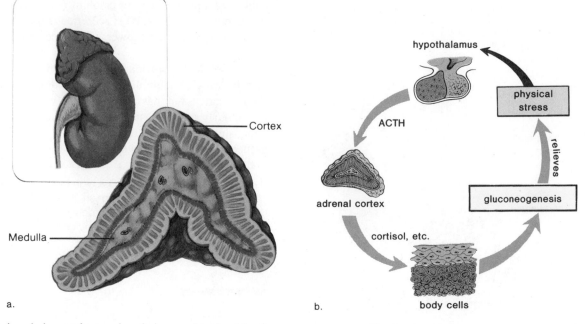

Epinephrine and norepinephrine are involved in the body's immediate response to stress. They bring about all those effects that occur when an individual reacts to an emergency. Blood glucose level rises, the metabolic rate increases, as does breathing and the heart rate. The blood vessels in the intestine constrict, but those in the muscles dilate. Increased circulation to the muscles causes them to have more strength than usual. The individual has a wide-eyed look and is extremely alert.

The adrenal medulla releases epinephrine and norepinephrine into the bloodstream helping us cope with situations that seem to threaten our survival.

Adrenal Cortex

Although the adrenal medulla may be removed with no ill effects, the adrenal cortex is absolutely necessary to life. The two major types of hormones made by the adrenal cortex are the glucocorticoids and the mineralocorticoids. It also secretes a small amount of male and an even smaller amount of female sex hormones. All of these hormones are steroids.

Glucocorticoids

Of the various glucocorticoids, the hormone responsible for the greatest amount of activity, is **cortisol.** The secretion of cortisol helps an individual recover from stress (fig. 9.13). Cortisol also counteracts the inflammatory response (p. 235). During the inflammatory re-

sponse, capillaries become more permeable and fluid leaks out, causing swelling in surrounding tissues. This causes the pain and swelling of joints that accompany arthritis and bursitis. The administration of cortisol aids these conditions because it reduces inflammation.

Mineralocorticoids

The secretion of mineralocorticoids, the most important of which is **aldosterone,** is not under the control of the anterior pituitary. These hormones maintain the level of sodium (Na^+) and potassium (K^+) in the blood, and their primary target organ is the kidney where they promote renal absorption of sodium and renal excretion of potassium. The level of Na^+ and K^+ in the blood are critical for nerve conduction and muscle contraction; in fact, cardiac failure may result from too low a level of potassium.

The level of Na^+ is particularly important to the maintenance of blood pressure, because its concentration indirectly regulates the secretion of aldosterone. When the sodium level is low, the kidneys secrete renin. Renin is an enzyme that leads to the conversion of plasma protein angiotensinogen to angiotensin I, which is converted to angiotensin II as blood passes through the lungs. This substance stimulates the adrenal cortex to release aldosterone (fig. 9.14). This is called the *renin-angiotensin-aldosterone system.* The effect of this system is to raise the blood pressure. First, angiotensin constricts the arteries directly, and second, when aldosterone causes the kidneys to reabsorb sodium, blood volume is raised as water is reabsorbed.

Figure 9.14 If the blood level of sodium is low, the kidneys secrete renin. The increased renin acts, via the increased production of angiotensin I and II, to stimulate aldosterone secretion. Aldosterone promotes reabsorption of sodium by the kidney. When the sodium level in the blood rises, the kidneys stop secreting renin.

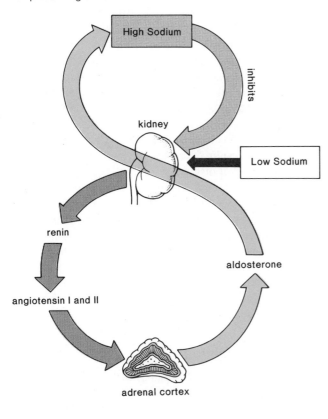

Cortisol, leading to gluconeogenesis, and aldosterone, leading to sodium retention, are two hormones secreted by the adrenal cortex.

Disorders

Addison's Disease When there is a low level of adrenal cortex hormones in the body, the person begins to suffer from Addison's disease. Because of the lack of cortisol, there is a high susceptibility to any kind of stress. Even a mild infection can cause death. Due to the lack of aldosterone, the blood sodium level is low, and the person experiences low blood pressure. In addition, the patient has a peculiar bronzing of the skin (fig. 9.15).

Cushing's Syndrome When there is a high level of adrenal cortex hormones in the body, the person suffers from Cushing's syndrome (fig. 9.16). Cortisol causes a tendency toward diabetes mellitus, a decrease in muscular protein, and an increase in subcutaneous fat. Because of these effects, the person usually develops thin arms and legs and an enlarged trunk. Due to the high

Figure 9.15 Addison's disease. This condition is characterized by a peculiar bronzing of the skin, as seen in the face and the thin skin of the nipples of this patient.

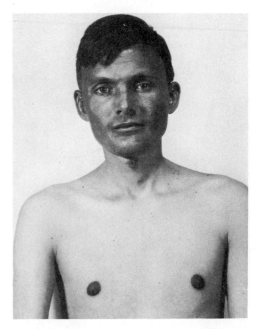

level of sodium in the blood, the patient's blood is basic (p. 202), hypertension occurs, and there is edema of the face, which gives the face a moonlike shape. Masculinization may occur in women due to oversecretion of adrenal male sex hormone.

Addison's disease is due to adrenal cortex hyposecretion, and Cushing's syndrome is due to adrenal cortex hypersecretion.

Sex Organs

The sex organs are the testes in the male and the ovaries in the female. As will be discussed in detail in chapter 16, the testes produce the androgens, which are the male sex hormones, and the ovaries produce estrogen and progesterone, the female sex hormones. The hypothalamus and pituitary gland control the hormonal secretions of these organs in the same manner as described for the thyroid gland in figure 9.7.

The sex hormones control the secondary sex characteristics of the male and female (pp. 279 and 284). Among other traits, males have greater muscle strength than do females. Generally, athletes believe that the intake of anabolic steroids, that is, the male sex hormone testosterone or synthetically related steroids, will cause greater muscle strength. The reading on page 175 discusses the disadvantages of taking these steroids,

Figure 9.16 Cushing's syndrome. Persons with this condition tend to have an enlarged trunk and moonlike face. Masculinization may occur in women due to the excessive male sex hormones in the body.

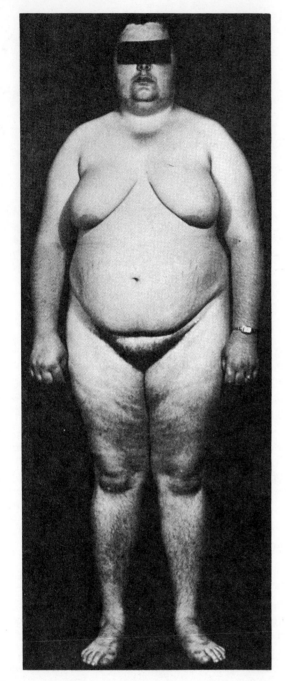

which are considered illegal by the International Olympic Committee. Any Olympic athlete whose urine tests positive for steroids at the time of an event is immediately disqualified from winning a medal.

Androgens (specifically testosterone) are the male sex hormones. Estrogen and progesterone are the female sex hormones.

Pancreas

The pancreas is a long, soft organ that lies transversely in the abdomen (fig. 9.17) between the kidneys and near the duodenum of the small intestine. It is composed of two types of tissues; one of these produces and secretes the digestive juices that go by way of the pancreatic duct to the small intestine, and the other type, called the **islets of Langerhans,** produces and secretes the hormones insulin and glucagon directly into the blood. Insulin is secreted by beta cells and glucagon is secreted by alpha cells (fig. 9.17). Insulin and glucagon are hormones that affect the blood glucose level in opposite directions—**insulin** decreases the level and **glucagon** increases the level of glucose.

Insulin is secreted when there is a high level of glucose in the blood, which usually occurs just after eating. Once the glucose level returns to normal, insulin is not secreted, as illustrated in figure 9.18. Insulin is believed to cause all of the cells of the body to take up glucose. When the liver and muscles take up glucose, they convert to glycogen any glucose not needed immediately. Therefore, insulin promotes the storage of glucose as glycogen.

Diabetes Mellitus

The symptoms of **diabetes mellitus** (sugar diabetes) include the following: sugar in the urine; frequent, copious urination; abnormal thirst; rapid loss of weight; general weakness; drowsiness and fatigue; itching of the genitals and skin; visual disturbances, blurring; and skin disorders, such as boils, carbuncles, and infection.

Many of these symptoms develop because sugar is not being metabolized by the cells. The liver fails to store glucose as glycogen, and all of the cells fail to utilize glucose as an energy source. This means that the blood glucose level rises very high after eating, causing glucose to be excreted in the urine. More water than usual is therefore excreted so that the diabetic is extremely thirsty.

Since carbohydrates are not being metabolized, the body turns to the breakdown of proteins and fat for energy. Unfortunately, the breakdown of these molecules leads to the buildup of acids in the blood (acidosis) and respiratory distress. It is the latter that can eventually cause coma and death of the diabetic. The symptoms that lead to coma (table 9.2) develop slowly.

There are two types of diabetes. In *Type I* diabetes, formerly called juvenile-onset diabetes, the pancreas is not producing insulin. Therefore, the patient must have daily insulin injections. These injections control the diabetic symptoms, but may still cause inconveniences, since either an overdose of insulin or the absence of regular eating can bring on the symptoms of insulin shock

More about . . .

Side Effects of Steroids

*B*eing a steroid user may cost an athlete far more than his or her Olympic medal: a growing body of medical evidence indicates that athletes who take steroids have experienced problems ranging from sterility to loss of libido, and the drug has been implicated in the deaths of young athletes from liver cancer and a type of kidney tumor. Steroid use has also been linked to heart disease. "Athletes who take steroids are playing with dynamite," says Robert Goldman, 29, a former wrestler and weight lifter who is now a research fellow in sports medicine at Chicago Osteopathic Medical Center and who has just published a book on steroid abuse, *Death in the Locker Room* (Icarus). "Any jock who uses these drugs is taking chances not just with his health but with his life."

Anabolic steroids are essentially the male hormone testosterone and its synthetic derivatives. . . .

The great majority of physicians say the drugs upset the body's natural hormonal balance, particularly that involving testosterone, which is present, though in different amounts, in both men and women. Normally, the hypothalamus, the part of the brain that regulates many of the body's functions, "tastes" the testosterone levels; if it finds them too low, it signals the pituitary gland to trigger increased production. When the hypothalamus finds the testosterone levels too high, as it does in the case of steroid abusers, it signals the pituitary to stop production. Problems can also arise in some cases after athletes stop taking the drugs and the hypothalamus fails to get the system started again.

The results can be traumatic. Many men experience atrophy, or shrinking of the testicles, falling sperm counts, temporary infertility and a lessening of sexual desire; some men grow breasts, while others may develop enlargement of the prostate gland, a painful condition not usually found in men under fifty. Women who take too many steroids can develop male sexual characteristics. Some grow hair on their chests and faces and lose hair from their heads; many experience abnormal enlargement of the clitoris. Some cease to ovulate and menstruate, sometimes permanently.

There are several other health risks. Steroids can cause the body to retain fluid, which results in rising blood pressure. This often tempts users to fight "steroid bloat" by taking large doses of diuretics. A postmortem on a young California weight lifter who had a fatal heart attack after using steroids . . . showed that by taking diuretics he had purged himself of electrolytes (p. 202), chemicals that help regulate the heart.

Figure 9.17   Gross and microscopic anatomy of the pancreas. The pancreas lies in the abdomen between the kidneys near the duodenum. As an exocrine gland, it secretes digestive enzymes that enter the duodenum by the common bile duct. As an endocrine gland, it secretes insulin and glucagon into the bloodstream. (Top right) The "alpha" cells of the islets of Langerhans produce glucagon and the "beta" cells produce insulin.

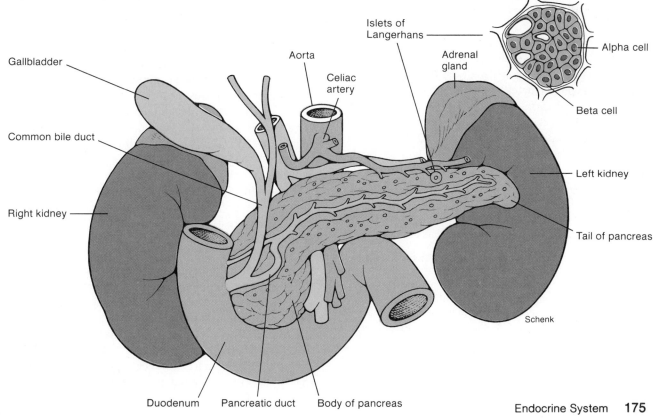

Figure 9.18 Regulation of insulin secretion. In response to a high blood sugar level, the pancreas secretes insulin, which promotes the uptake of glucose in body cells, muscles, and the liver. As a result of a low blood glucose level, the pancreas stops secreting insulin.

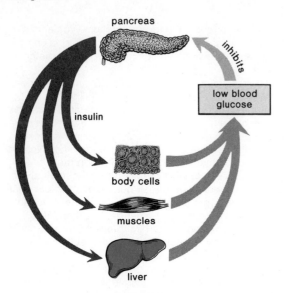

pancreas

inhibits

insulin

low blood glucose

body cells

muscles

liver

Table 9.2 Symptoms of Insulin Shock and Diabetic Coma	
Insulin Shock	**Diabetic Coma**
Sudden onset	Slow, gradual onset
Perspiration, pale skin	Dry, hot skin
Dizziness	No dizziness
Palpitation	No palpitation
Hunger	No hunger
Normal urination	Excessive urination
Normal thirst	Excessive thirst
Shallow breathing	Deep, labored breathing
Normal breath odor	Fruity breath odor
Confusion, disorientation, strange behavior	Drowsiness and great lethargy leading to stupor
Urinary sugar absent or slight	Large amounts of urinary sugar
No acetone in urine	Acetone present in urine

From *How To Live With Diabetes* by Henry Dolger, M.D., and Bernard Seeman of W. W. Norton and Company, Inc. Copyright © 1972, 1965, 1958 by Henry Dolger and Bernard Seeman.

(table 9.2). These symptoms appear because the blood sugar level has decreased below normal levels. Since the brain requires a constant supply of sugar, unconsciousness results. The cure is quite simple: an immediate source of sugar, such as a sugar cube or fruit juice, can counteract insulin shock immediately.

Of the over six million people who now have diabetes in the United States, at least five million have Type II, formerly called maturity-onset diabetes. In this type of diabetes, now known to occur in obese people of any age, the pancreas is producing insulin, but the cells do not respond to it. At first, the cells lack the receptors necessary to detect the presence of insulin, and later the cells are even incapable of taking up glucose. If Type II is left untreated, the results can be as serious as Type I diabetes. Diabetics are prone to blindness, kidney disease, and circulatory disorders, including strokes. Pregnancy carries an increased risk of diabetic coma, and the child of a diabetic is somewhat more likely to be stillborn or to die shortly after birth. It is important, therefore, that Type II diabetes be prevented or at least controlled. The best defense is a nonfattening diet and regular exercise. If that fails, there are oral drugs that make the cells more sensitive to the effects of insulin or stimulate the pancreas to make more of it.

The most common illness due to hormonal imbalance is diabetes mellitus, caused by a lack of insulin or insensitivity of cells to insulin. Insulin lowers blood glucose levels by causing the cells to take it up and the liver to convert it to glycogen.

Other Hormones

There are some other glands in the body that produce hormones. The **pineal gland** is a cone-shaped gland located in the roof of the third ventricle (fig. 9.1). It is smaller than the pituitary gland and gets even smaller as one ages. In the adult, it becomes only a thickened strand of fibrous tissue. The pineal gland secretes the hormone **melatonin,** particularly at night, and melatonin may regulate our daily rhythms such as our sleep pattern. An injection of melatonin can bring on sleep. It also inhibits the secretion of the gonadotropic hormones FSH and LH; therefore, excessive amounts of melatonin inhibit the ovarian and uterine cycles.

In the child, the **thymus gland** may be as large as a fist, but it too regresses in size with maturity. The thymus is the site of production of T cells, a type of lymphocyte that is responsible for cell-mediated immunity (p. 239). It also secretes **thymosin,** which is believed to regulate the development and function of the immune system.

Prostaglandins (PG) are called local hormones because they act on tissues or cells in their immediate vicinity. They are active in very small quantity, and have quite diverse actions affecting nervous system function, blood flow in the kidneys, pregnancy, and the inflammation of arthritis just to mention a few known effects. Sometimes prostaglandins have contrary effects. For example, one type helps prevent blood clots, but another helps blood clots to form. Also, a large dose of PG may have an effect opposite to that of a small dose. Therefore, it has been very difficult to standardize PG therapy and, in most instances, prostaglandin drug therapy is still considered experimental.

Summary

I. Pituitary Gland. The pituitary gland has two parts called the posterior pituitary and the anterior pituitary. The anterior pituitary produces six types of hormones, three of which (TSH, ACTH, gonadotropic hormones) affect other endocrine glands.
   A. Posterior pituitary. The hormones of the posterior pituitary, ADH and oxytocin, are produced in the hypothalamus.
   B. Anterior pituitary. GH (growth hormone) and LTH (lactogenic hormone) are two hormones of the anterior pituitary that do not affect other glands. GH influences the height of children and brings about a condition called acromegaly in adults. LTH promotes milk production after childbirth.

   The hypothalamus, anterior pituitary, and the other endocrine glands controlled by the anterior pituitary, are all involved in a self-regulating feedback loop.

II. Thyroid Gland
   A. Activity and disorders. The anterior pituitary produces TSH (thyroid-stimulating hormone), a hormone that promotes the production of thyroxin by the thyroid. Thyroxin, which speeds up metabolism, can affect the body as a whole as exemplified by cretinism and myxedema.

III. Parathyroid Glands. Parathyroid hormone maintains a high blood level of calcium by promoting its absorption in the gut, its reabsorption by the kidneys, and demineralization of bone. These actions are opposed by calcitonin produced by the thyroid.

IV. Adrenal Glands. The adrenal glands have two parts, an outer cortex and an inner medulla. The adrenal medulla is under nervous control and the cortex is under hormonal control of ACTH (adrenocorticotropic hormone, an anterior pituitary hormone).
   A. Adrenal medulla. The adrenal medulla releases adrenalin and noradrenalin into the blood-

stream helping us cope with situations that seem to threaten our survival.
   B. Adrenal cortex. Cortisol, leading to gluconeogenesis, and aldosterone, leading to sodium retention, are two hormones secreted by the adrenal cortex.

   Addison's disease is due to adrenal cortex hyposecretion, and Cushing's syndrome is due to adrenal cortex hypersecretion.

V. Sex Organs. Androgens (specifically testosterone) are the male sex hormones. Estrogen and progesterone are the female sex hormones.

VI. Pancreas
   A. Diabetes mellitus. The most common illness due to hormonal imbalance is diabetes mellitus, caused by a lack of insulin or insensitivity of cells to insulin. Insulin lowers blood glucose levels by causing the cells to take it up and the liver to convert it to glycogen.

Study Questions

1. What is a hormone and how do hormones work?
2. Define endocrine gland and target organ.
3. How does the hypothalamus control the posterior pituitary? The anterior pituitary?
4. Discuss two hormones secreted by the anterior pituitary that have an effect on the body proper rather than on other glands.
5. For each of the following endocrine glands, name the hormone(s) secreted, the effect of the hormone(s), and the medical illnesses, if any, that result from too much or too little of each hormone: posterior pituitary, thyroid, parathyroids, adrenal cortex, adrenal medulla, pancreas.
6. Give the anatomical location of each of the endocrine glands listed in question 5.
7. Draw a diagram to describe the action and control of ADH, thyroxin, glucocorticoids (e.g., cortisol), aldosterone, parathyroid hormone, and insulin.

Objective Questions

1. The hypothalamus _____ the hormones _____ and _____ , released by the posterior pituitary.
2. The _____ secreted by the hypothalamus control the anterior pituitary.
3. Generally, hormone production is self-regulated by a _____ mechanism.
4. Growth hormone is produced by the _____ pituitary.
5. Simple goiter occurs when the thyroid is producing _____ (too much or too little) _____ .
6. ACTH, produced by the anterior pituitary, stimulates the _____ of the adrenal glands.
7. An overproductive adrenal cortex results in the condition called _____ .
8. Parathyroid hormone increases the level of _____ in the blood.
9. Type I diabetes mellitus is due to a malfunctioning _____ , but Type II diabetes is due to malfunctioning _____ .
10. Prostaglandins are not carried in the _____ as are hormones that are secreted by the endocrine glands.

Medical Terminology Reinforcement Exercise

Pronounce the following terms and analyze the meaning by word parts as dissected:

1. antidiuretic (an″ti-di″u-ret′ik)—anti/di/uret/ic

2. hypophysectomy (hi-pof″ĭ-sek′ to-me)—hypophys/ectomy

3. gonadotropic (gon″ah-do-trop′ik)—gonado/tropic

4. hyperglycemia (hi″per-gli-se′me-ah)—hyper/glyc/emia

5. hypokalemia (hi″po-kal″e-e′me-ah)—hypo/kal/emia

6. acromegaly (ak″ro-meg′ah-le)—acro/megaly

7. lactogenic (lak″to-jen′ik)—lacto/gen/ic

8. adrenopathy (ad″ren-op′ah-the)—adreno/pathy

9. adenomalacia (ad″ĕ-no-mah-la′she-ah)—adeno/malacia

10. parathyroidectomy (par″ah-thi″roi-dek′ to-me)—parathyroid/ectomy

IV

Processing and Transporting

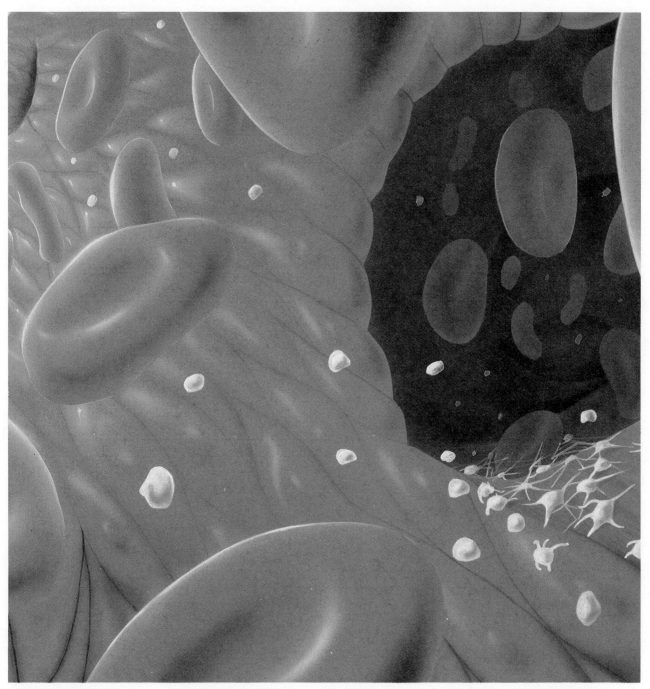

Artist's representation of the interior of a blood vessel. The large saucerlike red blood cells contain hemoglobin, the respiratory pigment that accounts for the red color. The white particles are platelets that initiate the process of blood clotting should this blood vessel suffer an injury.

Digestion

Learning Objectives

After you have studied this chapter, you should be able to:

1. Trace the path of food during digestion, and describe the general structure and function of each organ mentioned.
2. Name the major digestive enzymes and give their preferred pHs.
3. Name the accessory organs of digestion and describe their contribution to the digestive process.
4. Name six functions of the liver.
5. Name factors that can affect the flow of digestive juices.
6. Name and describe disorders of the tract and associated glands.

Digestive System

The digestive system begins with the mouth and ends with the anus (table 10.1 and fig. 10.1). The functions of the digestive system are to ingest the food, digest it to small molecules that can cross cell membranes, absorb these nutrient molecules, and eliminate nondigestible wastes.

Too often, we are inclined to think that since we eat meat (protein), potatoes (carbohydrate), and butter (fat), these are the substances that nourish our bodies. Instead, it is the amino acids from the protein, the sugars from the carbohydrate, and the glycerol and fatty acids from the fat that actually enter the blood and are transported about the body to nourish our cells. Any component of food, such as roughage that is incapable of being digested to small molecules, leaves the gut as waste material.

Mouth

The **mouth** receives the food. Most people enjoy eating because of the combined sensations of smelling and tasting food. The olfactory receptors, located in the nose, are responsible for smelling; tasting is, of course, a function of the taste buds, located on the tongue.

The tongue is striated muscle. Its intrinsic muscles change its shape; it is moved about by extrinsic muscles that insert onto it. The tongue is covered by mucous membrane; a fold of mucous membrane (lingual frenulum) on the underside of thetongue attaches it to the floor of the mouth. Besides moving the food around in the mouth, the tongue is necessary for producing speech.

The teeth chew the food into pieces convenient to swallow. During the first two years of life, the twenty deciduous or baby teeth appear. Eventually, these will be replaced by the adult teeth. Normally, adults have thirty-two teeth (fig. 10.2). One-half of each jaw has teeth of four different types: two chisel-shaped *incisors* for biting; one pointed *canine* for tearing; two fairly flat *premolars* for grinding; and three, more flattened *molars* for crushing. The last molars, called the wisdom teeth, may fail to erupt or, if they do, they are sometimes crooked and useless. Oftentimes, the extraction of wisdom teeth is recommended.

Each tooth (fig. 10.3) has a crown and a root. The crown has a layer of *enamel*, an extremely hard outer covering of calcium compounds; *dentin*, a thick layer of bonelike material; and an inner pulp, which contains the nerves and blood vessels. Dentin and pulp are also in the root. Tooth decay, or *caries*, commonly called cavities, occurs when bacteria within the mouth break down sugar and give off acids that corrode the teeth. It has been found that fluoride treatments, particularly

Table 10.1	Path of Food	
Organ	Special Features	Functions
Mouth	Teeth, tongue	Chewing of food
	Tongue	Formation of bolus
		Digestion of starch
Esophagus		Passageway
Stomach		Storage of food
		Acidity kills bacteria
	Gastric glands	Digestion of protein
Small intestine	Intestinal glands	Digestion of all foods and
	Villi	Absorption of nutrients
Large intestine		Absorption of water
		Storage of nondigestible remains
Anus		Defecation

in children, can make the enamel stronger and more resistant to decay. Gum disease is more apt to occur as one ages. Inflammation of the gums (*gingivitis*) may spread to the periodontal membrane (fig. 10.3) that lines the tooth socket. Then, the individual has *periodontitis,* characterized by a loss of bone and loosening of the teeth, with the possibility that these teeth may have to be pulled. Daily brushing and flossing of teeth along with stimulation of the gums has been found to help prevent these conditions.

The roof of the mouth separates the nasal passages from the mouth cavity. The roof has two parts: an anterior **hard palate** and a posterior **soft palate.** The hard palate contains several cranial bones, but the soft palate is merely muscular. The soft palate ends in the *uvula,* a suspended process often mistaken by the layperson for the tonsils, but as figure 10.4 shows, the **tonsils** are otherwise placed. The *lingual tonsils* are located on the dorsal surface of the base of the tongue; the *pharyngeal tonsils,* also called the adenoids, are located in the posterior wall of the nasopharynx; and the *palatine tonsils* are found in the posterior lateral portion of the oral cavity. The tonsils play a minor role in protecting the body from disease-causing organisms as discussed in chapter 14.

There are three pairs of **salivary glands** (fig. 10.1) which send saliva by way of ducts to the mouth. The *parotid glands* lie at the sides of the face immediately

Figure 10.1 The human digestive system. The liver is drawn smaller than normal size and moved back to show the gallbladder, and to expose the stomach and duodenum.

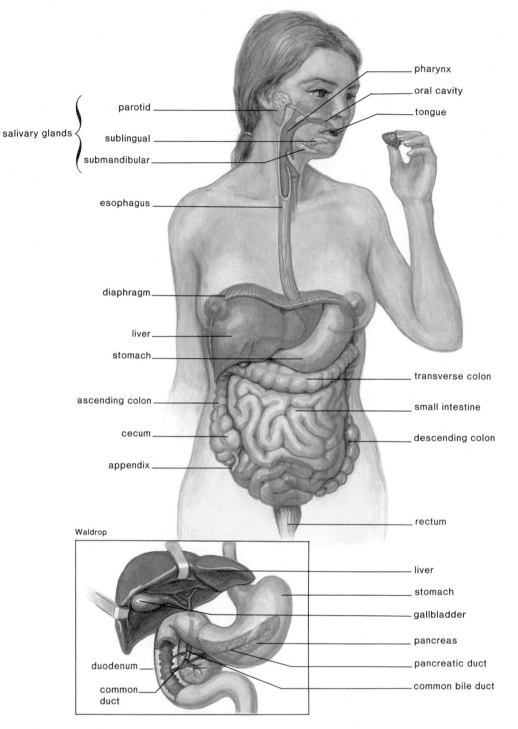

pharynx

oral cavity

tongue

parotid

salivary glands

sublingual

submandibular

esophagus

diaphragm

liver

stomach

transverse colon

ascending colon

small intestine

cecum

descending colon

appendix

rectum

Waldrop

liver

stomach

gallbladder

pancreas

duodenum

pancreatic duct

common duct

common bile duct

Figure 10.2 Diagram of the mouth showing the adult teeth. The sizes and shapes of teeth correlate with their functions.

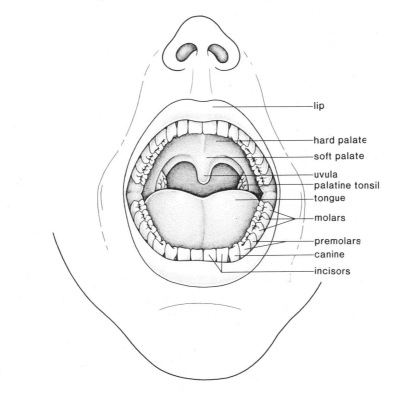

- lip
- hard palate
- soft palate
- uvula
- palatine tonsil
- tongue
- molars
- premolars
- canine
- incisors

Figure 10.3 Longitudinal section of a canine tooth. A tooth contains nerves and blood vessels within the pulp.

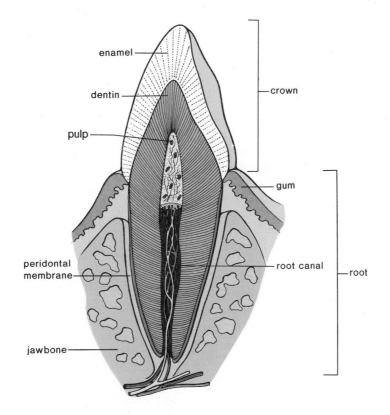

- enamel
- dentin
- pulp
- crown
- gum
- peridontal membrane
- root canal
- root
- jawbone

Figure 10.4 When food is swallowed, the soft palate covers the nasopharyngeal openings, and the epiglottis covers the glottis so that the food bolus must pass down the esophagus. Therefore, one does not breathe during swallowing.

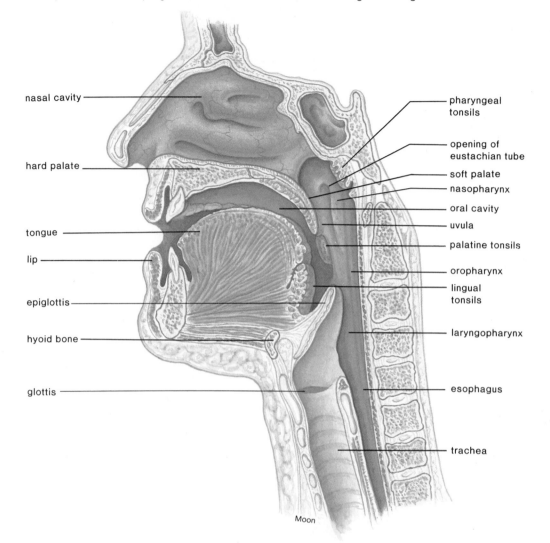

nasal cavity

hard palate

tongue

lip

epiglottis

hyoid bone

glottis

pharyngeal tonsils

opening of eustachian tube

soft palate

nasopharynx

oral cavity

uvula

palatine tonsils

oropharynx

lingual tonsils

laryngopharynx

esophagus

trachea

Moon

below and in front of the ears. They become swollen when a person has the mumps, a viral infection most often seen in children. Each parotid gland has a duct that opens on the inner surface of the cheek just at the location of the second upper molar. The *sublingual glands* lie beneath the tongue, and the *submandibular glands* lie beneath the lower jaw. The ducts from these glands open into the mouth under the tongue. You can locate all these openings if you use your tongue to feel for small flaps on the inside of the cheek and under the tongue. An enzyme in saliva begins the process of digesting the food. Specifically, it acts on starch, a carbohydrate.

The tongue mixes the chewed food with saliva and then forms it into a mass called a **bolus,** in preparation for swallowing.

The salivary glands send saliva into the mouth where the teeth chew the food, and the tongue forms a bolus for swallowing.

Pharynx

Swallowing, movement of food into the esophagus, occurs in the **pharynx** (fig. 10.4), a region that opens into the nose (called the *nasopharynx*), the mouth (called the *oropharynx*), and the larynx (called the *laryngopharynx*). Swallowing is a reflex action, which means the action is usually performed automatically and does not require conscious thought. During swallowing, food normally enters the esophagus, a long muscular tube that extends to the stomach, because the nasal and laryngeal passages are blocked. The nasopharyngeal openings are covered when the soft palate moves back. The opening to the larynx (voice box) at the top of the trachea, called the **glottis**, is covered when the trachea moves up under a flap of tissue, called the epiglottis. This is easy to observe in the up-and-down movement of the Adam's apple, a part of the larynx, when a person eats. Breathing does not occur when one swallows. Why not?

The air passage and food passage cross in the pharynx. When one swallows, the air passage is usually blocked off and food must enter the esophagus.

Esophagus

After swallowing occurs, the **esophagus** conducts the bolus through the thoracic cavity. The wall of the esophagus is representative of the gut in general (fig. 3.14). A mucous membrane layer lines the lumen (space within the tube). This is followed by a submucosal layer of connective tissue that contains nerve and blood vessels; a smooth muscle layer having both a longitudinal and circular muscle; and finally, a serous membrane layer.

A rhythmical contraction of the esophageal wall, called **peristalsis** (fig. 10.5) pushes the food along. Occasionally, peristalsis begins even though there is no food in the esophagus. This produces the sensation of a lump in the throat.

The esophagus stretches in three smooth curves from the back of the pharynx to just below the diaphragm, where it meets the stomach at an angle. The entrance of the esophagus into the stomach is marked by the presence of a constrictor, called the lower esophageal sphincter, although it is not as developed as a true sphincter. **Sphincters** are muscles that encircle tubes and act as valves; tubes close when sphincters contract, and they open when sphincters relax. When food is swallowed, the sphincter relaxes, allowing the bolus to pass into the stomach. Normally, this sphincter prevents the acid contents of the stomach from entering the esophagus. Heartburn, which feels like a burning pain rising up into the throat, occurs when some of the contents of the stomach do escape into the esophagus. When vomiting occurs, a reverse peristaltic wave causes the constrictor to relax and the contents of the stomach are propelled upward through the esophagus.

Stomach

The **stomach** (fig. 10.6) is a thick-walled, J-shaped organ that lies in the upper left quadrant of the peritoneal cavity beneath the diaphragm. The stomach is continuous with the esophagus superiorly and the duodenum of the small intestine inferiorly. The stomach stores food and starts digestion of protein. The wall of the stomach has three layers of muscle and contains deep folds called **rugae** (see fig. 1.1, p. 14), which disappear as the stomach fills. The muscular wall of the stomach churns, mixing the food with gastric secretions. When food leaves the stomach, it is a pasty material called acid **chyme.**

The mucosal lining of the stomach contains millions of microscopic digestive glands called **gastric glands**

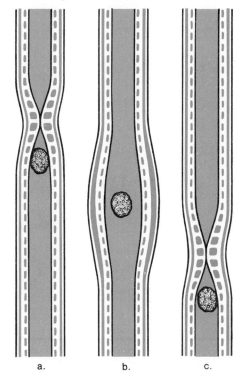

Figure 10.5   Peristalsis in the digestive tract. Rhythmic waves of muscle contraction move material along the digestive tract. The three drawings show how a peristaltic wave moves through a single section of gut over time (a to c).

a.   b.   c.

(the word gastric always refers to the stomach). The gastric glands produce gastric juice, which contains a digestive enzyme and hydrochloric acid (HCl). The acidity of the stomach is beneficial in that it kills most bacteria present in food. Although HCl does not digest food, it does break down the connective tissue of meat and it activates gastric enzymes present in gastric juice. Normally, the wall of the stomach is protected by a thick layer of mucus, but if, by chance, HCl does penetrate this mucus, autodigestion of the wall can begin and an ulcer results. An ulcer is an open sore in the wall caused by the gradual disintegration of tissues. It is believed that the most frequent cause of an ulcer is oversecretion of gastric juice due to too much nervous stimulation. Persons under stress tend to have a greater incidence of ulcers.

Normally, the stomach empties in about two to six hours. Acid chyme leaves the stomach and enters the small intestine by way of the *pyloric sphincter*. The sphincter repeatedly opens and closes allowing acid chyme to enter the small intestine in small squirts only. This assures that digestion in the small intestine will proceed at a slow and thorough rate.

The stomach expands and stores food. While food is in the stomach, it churns, mixing food with the acid gastric juices.

Figure 10.6   Gastric glands in the wall of the stomach produce gastric juice rich in pepsin, an enzyme that digests protein.

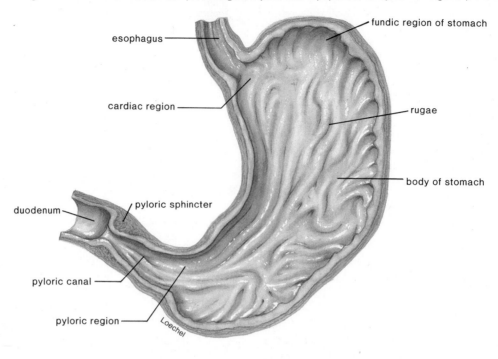

Figure 10.7   Regions of the small intestine. The duodenum is attached to the stomach, the jejunum is the largest portion of the small intestine, and the ileum is attached to the large intestine.

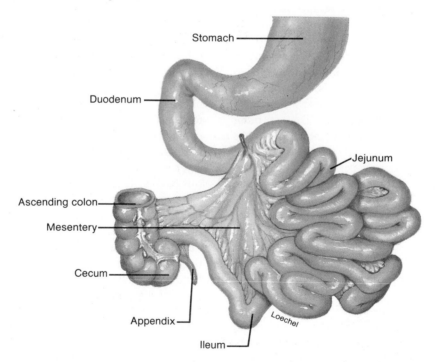

Small Intestine

The **small intestine** (fig. 10.7) gets its name from its small diameter (compared to that of the large intestine); but perhaps it should be called the long intestine because ordinarily it averages about 3.0 m (10 ft) in length compared to the large intestine, which is about 1.5 m (5 ft) long. The small intestine is found in the central and lower portion of the abdominal cavity where it is supported by a fan-shaped mesentery. The small intestine receives secretions from the liver and pancreas, chemically and mechanically breaks down acid chyme; absorbs nutrient molecules; and transports undigested material to the large intestine.

Figure 10.8 Anatomy of intestinal lining. *a.* The products of digestion are absorbed by villi, fingerlike projections of the intestinal wall, *b.* each of which contains blood vessels and a lacteal.

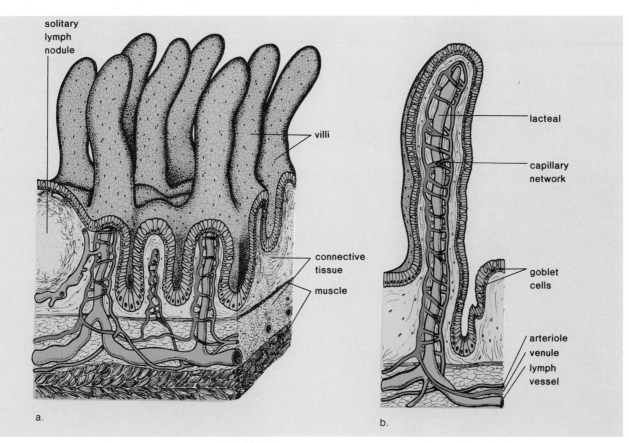

The small intestine is divided into three regions: the **duodenum** is only the first 25 cm (10 in), the **jejunum** is the next 1 m (3 ft), and, finally, the **ileum** is 2 m (6–7 ft). The ileum enters the large intestine through the ileocecal valve at the location of the cecum.

The wall of the small intestine contains fingerlike projections called **villi** (fig. 10.8), and the villi themselves have microvilli, which are visible microscopically. Cells of the villi produce intestinal digestive juices which contain enzymes for finishing the digesting of food to small molecules that can cross cell membranes. The huge number of villi, which increase the surface area for absorption, give the wall a soft, velvety appearance. Each villus has an outer layer of columnar epithelium and contains blood vessels and a small lymph vessel called a lacteal. The lymphatic system is an adjunct to the circulatory system, as discussed in chapter 13.

Absorption of nutrient molecules across the wall of each villus continues until all small molecules have been absorbed. Therefore, absorption is an active process involving active transport of molecules across cell membranes and requiring an expenditure of cellular energy. Sugars and amino acids cross the columnar epithelial cells to enter the blood, the components of fats rejoin before entering the lacteals.

The small intestine is 3 m long and has three divisions (duodenum, ileum, and jejunum). Its walls have fingerlike projections called villi where nutrient molecules are absorbed into the circulatory and lymphatic systems.

Large Intestine

The **large intestine** (fig. 10.9), which includes the cecum, colon, rectum, and anal canal, is larger in diameter than the small intestine (6.5 cm compared to 2.5 cm). The large intestine begins in the lower right quadrant of the peritoneal cavity. The **cecum,** which lies inferior to this point, has a small projection called the **vermiform appendix** (vermiform means wormlike). Superior to this point, the large intestine is termed the **ascending colon.** At the level of the liver, the large intestine bends sharply and becomes the **transverse colon.** At the left abdominal wall, the large intestine bends again to become the **descending colon.** In the pelvic region, the large intestine turns medially to give an S-shaped bend known as the **sigmoid colon.** The last 20 cm of the large intestine, the **rectum,** ends in the anal canal, which opens at the anus.

Figure 10.9   The large intestine. Notice that there are four colons: the ascending colon, the transverse colon, the descending colon, and the sigmoid colon.

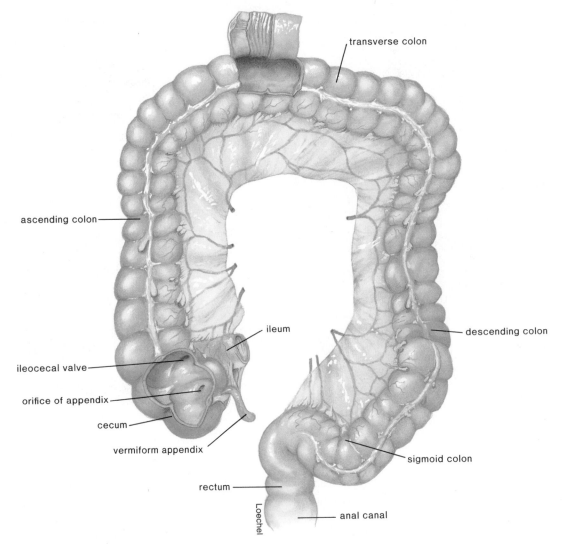

The large intestine absorbs water and electrolytes (p. 202). It also prepares and stores nondigestible material until it is defecated at the anus. In addition to nondigestible remains, feces also contains certain excretory substances such as bile pigments and heavy metals, and large quantities of bacteria, particularly *Escherichia coli.*

The large intestine normally contains a large population of bacteria, particularly *Escherichia coli,* that live off any substances that were not digested earlier. When they break this material down, they give off odorous molecules that cause the characteristic odor of feces. Some of the vitamins (vitamin K and some B complex vitamins), amino acids, and other growth factors produced by these bacteria are absorbed by the gut lining. In this way, *E. coli* and other bacteria perform a service for us.

Water is considered unsafe for swimming when the coliform count reaches a certain level. This is not because *E. coli* normally causes disease, but because

a high count is an indication of the amount of fecal material that has entered the water. The more fecal material present, the greater the possibility that pathogenic, or disease-causing, organisms are also present.

Diarrhea and Constipation

Two common everyday complaints associated with the large intestine are diarrhea and constipation.

The major causes of *diarrhea* are infection of the lower tract and nervous stimulation. In the case of infection, such as food poisoning caused by eating contaminated food, the intestinal wall becomes irritated and peristalsis increases. Lack of absorption of water is a protective measure, and the diarrhea that results serves to rid the body of the infectious organisms. In nervous diarrhea, the nervous system stimulates the intestinal wall and diarrhea results. Loss of water due to diarrhea may lead to dehydration, a serious condition in which the body tissues lose their normal water content.

When a person is *constipated,* the feces are dry and hard. One cause of this condition is that socialized persons have learned to inhibit defecation to the point that the normal reflexes are often ignored. Two components of the diet can help to prevent constipation: water and roughage. A proper intake of water prevents the drying out of the feces. Dietary inclusion of roughage (fiber), or nondigestible plant substances, provides the bulk needed for elimination and may even help protect one from colon cancer. It's believed that the less time that feces are in contact with the membrane of the colon, the better.

The frequent use of laxatives is certainly discouraged, but if it should be necessary to take a laxative, a bulk laxative is the most natural because, like roughage, it produces a soft mass of cellulose in the colon. Lubricants like mineral oil make the colon slippery and saline laxatives like milk of magnesia act osmotically; they prevent water from exiting or even cause water to enter the colon depending on the dosage. Some laxatives are irritants; they increase peristalsis to the degree that the contents of the colon are expelled.

Chronic constipation is associated with the development of hemorrhoids, discussed on page 218.

The large intestine does not produce digestive enzymes; it does absorb water and some electrolytes. In diarrhea, too little water has been absorbed; in constipation, too much water has been absorbed.

Other Conditions

There are other, usually more serious medical conditions associated with the large intestine.

Appendicitis In humans, the appendix is vestigial, meaning that the organ is underdeveloped (in other animals, it is developed and serves as a location for bacterial digestion of cellulose). Unfortunately, the appendix can become infected. When this happens, the individual has *appendicitis,* a very painful condition, in which the fluid content of the appendix may rise to the point that it bursts. The appendix should be removed before this occurs because it may lead to a generalized infection of the peritoneal membrane of the abdominal cavity.

Diverticulosis This condition is characterized by the presence of diverticula, or saclike pouches of the colon. For the most part, these pouches cause no problems. But about 15 percent of persons develop an inflammation known as diverticulitis. The symptoms of diverticulitis are very like those of appendicitis—crampy or steady pain with local tenderness. Fever, loss of appetite, nausea, and vomiting may also occur. Today, high-fiber diets are being recommended to prevent the development of these conditions, and cancer of the colon.

Colostomy The colon is subject to the development of *polyps,* small growths that generally appear on epithelial tissue, such as the epithelial tissue that lines the digestive tract. Whether polyps are benign or cancerous, they can be individually removed along with a portion of the colon if necessary. Should it be necessary to remove the last portion of the rectum and the anal canal, then the intestine is sometimes attached to the abdominal wall, and the digestive remains are collected in a plastic bag fastened around the opening. Recently, the use of metal staples has permitted surgeons to join the colon to a piece of rectum that formerly was considered too short.

The colon is subject to many disorders. Three of these are appendicitis, diverticulosis, and polyps.

Accessory Organs

The pancreas and liver are the accessory organs of digestion. Figure 10.10 shows the anatomy of the ducts which conduct pancreatic juices from the pancreas and bile from the liver to the duodenum.

Pancreas

The **pancreas** lies deep in the peritoneal cavity, resting on the posterior abdominal wall. It is an elongated and somewhat flattened organ that has both an endocrine function (p. 174) and an exocrine function. We are now interested in its exocrine function—most of its cells produce pancreatic juice, which contains sodium bicarbonate ($NaHCO_3$) and digestive enzymes for carbohydrate, protein, and fat. In other words, the pancreas secretes enzymes for the digestion of all types of food. They travel by way of the pancreatic duct and common duct to the duodenum of the small intestine (fig. 10.1). Regulation of pancreatic secretion is discussed in the reading on page 190.

Liver

The **liver,** which is the largest gland in the body, lies mainly in the right upper quadrant of the peritoneal cavity, under the diaphragm. There are two main lobes, the right lobe and smaller left one which crosses the midline to lie above the stomach. These two lobes are further divided into lobules at the center of which are small blood vessels draining into the hepatic vein. This takes blood into the inferior vena cava, the major vein draining blood from the lower half of the body. Between the lobules lie the portal canals, which contain three structures: (1) the branch of the hepatic artery, (2) the branch of the hepatic portal vein, and (3) the channel to collect bile.

More about . . .

Control of Digestive Gland Secretion

The study of the control of digestive gland secretion began in the late 1800s. At that time, Ivan Pavlov showed that dogs would salivate at the ringing of a bell because they had learned to associate the sound of the bell with being fed. Pavlov's experiments demonstrated that even the thought of food can bring about the secretion of digestive juices. Certainly if food is present in the mouth, stomach, and small intestine, digestive secretion occurs. This is attributable to a simple reflex occurrence. The presence of food sets off nerve impulses that travel to the brain. Thereafter, the brain stimulates the digestive glands to secrete.

In this century, investigators have discovered that specific control of digestive secretions is achieved by hormones. A *hormone* is a substance that is produced by one set of cells but affects a different set of cells, called the target cells. Hormones are transported by the bloodstream. For example, when a person has eaten a meal particularly rich in protein, the hormone **gastrin,** produced by the lower part of the stomach, enters the bloodstream (see figure) and soon reaches the upper part of the stomach, where it causes the gastric glands to secrete more gastric juice.

Experimental evidence has also shown that the duodenal wall produces hormones, the most important of which are **secretin** and **CCK** (cholecystokinin). Acid, especially HCl present in the acid chyme, stimulates the release of secretin, while partially digested protein and fat stimulate the release of CCK. These hormones enter the bloodstream (see figure) and signal the pancreas and the gallbladder to send secretions to the duodenum.

Still another hormone has only recently been discovered. **GIP** (gastric inhibitory peptide) produced by the small intestine apparently works in opposition to gastrin because it inhibits gastric acid secretion. This is not surprising because very often the body has hormones that have opposite effects.

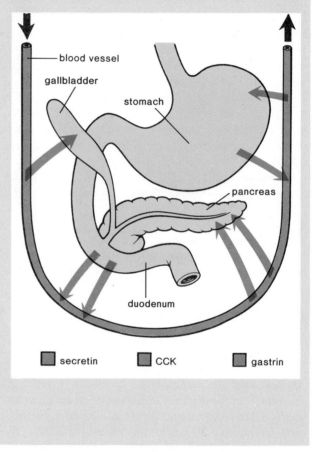

Hormonal control of digestive gland secretions. Especially after eating a protein-rich meal, gastrin produced by the lower part of the stomach enters the bloodstream and, thereafter, stimulates the upper part of the stomach to produce more digestive juices. Acid chyme from the stomach causes the duodenum to secrete secretin and CCK. Secretin stimulates the pancreas, and CCK stimulates the gallbladder to release bile.

blood vessel
gallbladder
stomach
pancreas
duodenum

■ secretin   ■ CCK   ■ gastrin

Bile **Bile** is a yellowish green fluid because it contains the bile pigments bilirubin and biliverdin, which come from the breakdown of hemoglobin—the pigment found in red blood cells. Bile also contains bile salts which emulsify fat once bile reaches the duodenum of the small intestine. When fat is emulsified, it breaks up into droplets that can be acted upon by a digestive enzyme from the pancreas. Emulsification is a process that can be witnessed by adding oil to water in a test tube. The oil has no tendency to mix with the water, but if a liquid detergent is added and the contents of the tube are shaken, the oil does break up and disperse into the water.

About 2¼ pints of bile are produced by the liver each day. This is sent by way of bile ducts to the **gallbladder** where it is stored. The gallbladder is a pear-shaped muscular sac attached to the ventral surface of the liver (fig. 10.10). Here, water is reabsorbed so that bile becomes a thick, mucouslike material. Bile leaves the gallbladder by the common bile duct and proceeds to the duodenum by the common duct (fig. 10.1).

The liver acts, in some ways, as the gatekeeper to the blood. Once nutrient molecules have been absorbed by the small intestine, they enter the *hepatic portal vein* and pass through the blood vessels of the liver before entering the hepatic vein. As the blood passes through the liver, it removes poisonous substances and works to keep the contents of the blood constant. In particular, we may note that the glucose level of the blood is always about 0.1 percent, even though we eat intermittently. Any excess glucose that is present in the hepatic portal vein is removed and stored by the liver as glycogen:

$$\text{glucose} \longrightarrow \text{glycogen} + H_2O$$

Figure 10.10   *a.* Both the liver and the pancreas send secretions to the duodenum. *b.* The liver contains lobules, one of which is shown here in detail.

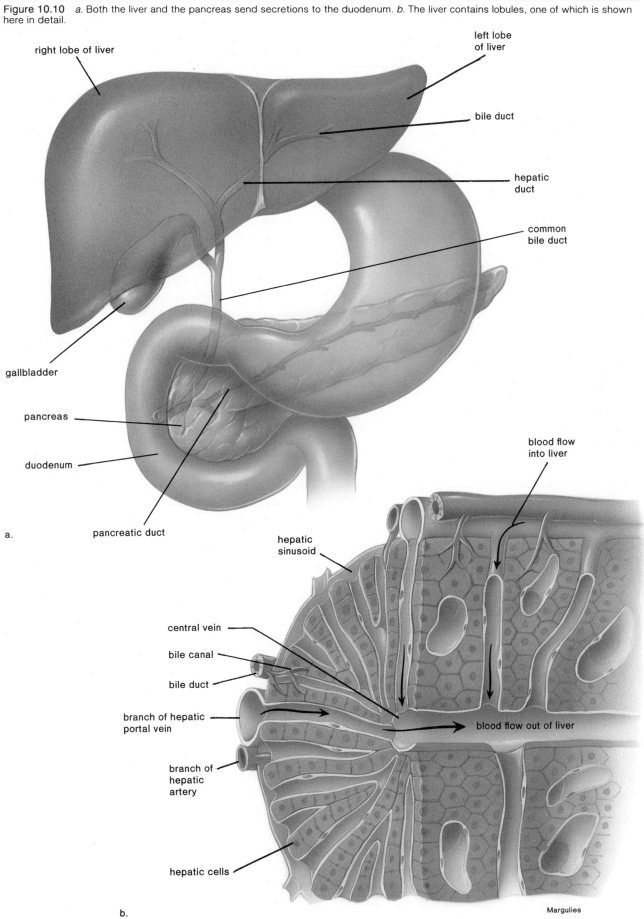

right lobe of liver

left lobe of liver

bile duct

hepatic duct

common bile duct

gallbladder

pancreas

duodenum

pancreatic duct

a.

blood flow into liver

hepatic sinusoid

central vein

bile canal

bile duct

branch of hepatic portal vein

blood flow out of liver

branch of hepatic artery

hepatic cells

b.

Margulies

Between eating periods, glycogen is broken down to glucose, which enters the hepatic vein, and in this way, the glucose content of the blood remains constant.

If, by chance, the supply of glycogen or glucose runs short, the liver will convert amino acids to glucose molecules. In the process, ammonia is given off and converted to urea. **Urea** is the common nitrogen waste product of humans; and after its formation in the liver, it is transported to the kidneys for excretion.

The liver also makes blood proteins from amino acids. These proteins serve important functions within the blood itself.

Altogether, we have mentioned the following functions of the liver:

1. Destruction of old red blood cells and conversion of hemoglobin to the breakdown products in bile (bilirubin and biliverdin).
2. Production of bile, which is stored in the gallbladder before entering the small intestine, where it emulsifies fats.
3. Storage of glucose as glycogen after eating and the breakdown of glycogen to glucose between eating to maintain the glucose concentration of the blood constant.
4. Production of urea from the breakdown of amino acids.
5. Production of the blood proteins.
6. Detoxification of the blood by removing poisonous substances and metabolizing them.

There are two accessory organs of digestion which send secretions to the duodenum via ducts. The pancreas produces pancreatic juice, which contains bicarbonate and digestive enzymes for carbohydrate, protein, and fat. The liver produces bile, which is stored in the gallbladder.

Liver Disorders

Jaundice When a person is jaundiced, there is a yellowish tint to the skin due to an abnormally large amount of bilirubin in the blood. Hemolytic jaundice is due to red blood cell destruction in such quantity that excess bilirubin spills over into the bloodstream.

In obstructive jaundice, there is an obstruction of the bile duct or damage to the liver cells, and this causes an increased amount of bilirubin to enter the bloodstream. Obstruction of the bile duct can be due to the formation of gallstones when materials precipitate out of the bile. The stones may be so numerous that passage of bile along the bile duct is blocked. Liver damage can be due to viral hepatitis. Hepatitis is transmitted by unsanitary food or water. In recent years, persons have been known to acquire the disease after eating shellfish from polluted waters. It is also acquired through blood transfusions, kidney dialysis, sexual intercourse, and injection with inadequately sterilized needles.

Cirrhosis With this condition, the liver first becomes fatty and then liver tissue is replaced by inactive fibrous scar tissue. Cirrhosis is common among alcoholics, in which case it is most likely caused by the need for the liver to detoxify alcohol and break it down.

Chemical Digestion

We have previously mentioned that the various digestive juices contain enzymes that digest particular types of food (table 10.2). We will now consider each of these enzymes as we discuss the digestion of a ham sandwich, which contains starch, a carbohydrate (in the bread), protein (in the ham), and fat (butter on the bread).

In the mouth, salivary amylase is an enzyme present in saliva, which acts on starch. The end product of this reaction (maltose), however, is not small enough to cross cell membranes to any degree; therefore, further digestion is required before absorption is possible.

In the stomach, pepsin is an enzyme present in gastric juices which acts on protein. The end product of this reaction (peptides) is, again, too large to cross cell membranes; therefore, further digestion is required before absorption is possible.

Pancreatic juice, which enters the small intestine at the duodenum, contains digestive enzymes for all types of food. Pancreatic amylase, like salivary amylase, acts on starch. Trypsin, like pepsin, acts on protein. And the enzyme lipase acts on fat droplets after fat has been emulsified by bile salts. The end products of lipase reaction are small enough to cross the cells of the villi where absorption takes place. Glycerol and fatty acids are the end products of fat breakdown. Glycerol and fatty acids enter the cells of the villi, and within these cells, they rejoin to form fat, which enters the lacteals (fig. 10.8), a branch of the lymphatic system.

The intestinal juice contains enzymes that complete the digestion of starch and of protein to small molecules that can cross the cells of the villi. Glucose is the end product of starch breakdown, and amino acids are the end product of protein breakdown. Glucose and amino acids enter the blood capillaries of the villi.

As mentioned previously, these blood capillaries join to form venules and veins which empty into the hepatic portal vein, a vessel that goes to the liver.

Table 10.2 Comparison of Enzymes

Enzyme	Source	Optimum pH*	Type of Food Digested	Product
Salivary amylase	Salivary glands	Neutral	Starch	Maltose
Pepsin	Stomach	Acid	Protein	Peptides
Pancreatic amylase	Pancreas	Basic	Starch	Maltose
Lipase	Pancreas	Basic	Fat	Glycerol; fatty acids
Trypsin	Pancreas	Basic	Protein	Peptides
Nucleases	Pancreas	Basic	RNA, DNA	Nucleotides
Peptidases	Intestine	Basic	Peptides	Amino acids
Maltase	Intestine	Basic	Maltose	Glucose

*pH is discussed on page 202 and in the appendix.

Digestive enzymes present in digestive juices break down food to the nutrient molecules: glucose, amino acids, fatty acids, and glycerol. The first two are absorbed into the blood capillaries of the villi and the last two reform to give fat, which enters the lacteals.

Fate of Nutrients

The blood circulatory system distributes nutrients to the tissue where they are utilized by the body's cells. Glucose is utilized by mitochondria to produce a constant supply of ATP for the cell. In other words, glucose is the immediate energy source for the body. The brain, in particular, needs a constant source of glucose because it utilizes no alternate source.

The liver is able to alter ingested fats to suit the body's needs, except it is unable to produce the fatty acid, linoleic acid. Since this is required for construction of cell membranes, it is considered an **essential fatty acid.** Essential molecules must be present in our food because the body is unable to manufacture them. Fats can be metabolized into their components, which are then used as an energy source if glucose is not available. Therefore, fats are said to be a long-term energy source. When adipose tissue cells store fats, we increase in weight. Cells do have the capability of changing excess sugar molecules into fats for storage and that is why carbohydrate foods can also contribute to weight gain.

Amino acids from protein digestion are used by the cells to construct their own proteins, including the cells' enzymes that carry out metabolism. Protein formation requires twenty different types of amino acids. Of these, nine are required in the diet because the body is unable to produce them. These are termed the **essential amino acids.** The body produces the other amino acids by simply transforming one type into another type. Some protein sources, such as meat, are complete in the sense that they provide all the different types of amino acids. Vegetables do supply us with amino acids, but they are incomplete sources because at least one of the essential amino acids is absent. It is possible, however, to combine foods in order to acquire all of the essential amino acids.

Vitamins and Minerals

Vitamins are organic compounds (other than carbohydrates, fats and proteins) that the body is unable to produce and, therefore, they must be present in the diet. Although vitamins are an important part of a balanced diet, they are required only in very small amounts. These are portions of coenzymes, or enzyme helpers. Coenzymes are needed in only small amounts because each one can be used over and over again.

In addition to vitamins, various **minerals** are also required by the body. Minerals are divided into macronutrients which are needed in gram amounts per day and micronutrients (trace elements) which are needed in only microgram amounts per day. The macronutrients sodium, magnesium, phosphorus, chlorine, potassium, and calcium serve as constituents of cells and body fluids and as structural components of tissues. For example, calcium is needed for the construction of bones and teeth and also for nerve conduction and muscle contraction. The micronutrients seem to have very specific functions. For example, iron is needed for the production of hemoglobin, and iodine is used in the production of thyroxin, a hormone produced by the thyroid gland. As research continues, more and more elements have been added to the list of those considered essential. During the past three decades, molybdenum, selenium, chromium, nickel, vanadium, silicon, and even arsenic have been found to be essential to good health in very small amounts.

Summary

I. Digestive System

  A. The salivary glands send saliva into the mouth where the teeth chew the food, and the tongue forms a bolus for swallowing.

  B. The air passage and food passage cross in the pharynx. When one swallows, the air passage is usually blocked off and food must enter the esophagus.

  C. The stomach expands and stores food. While food is in the stomach, it churns, mixing food with the acid gastric juices.

  D. The small intestine is 3 m long and has three divisions (duodenum, ileum, and jejunum). Its walls have fingerlike projections called villi where nutrient molecules are absorbed into the circulatory and lymphatic systems.

  E. The large intestine consists of the cecum; colons (ascending, transverse, descending, and sigmoid); and the rectum, which ends at the anus.

    The large intestine does not produce digestive enzymes; it does absorb water and some electrolytes. In diarrhea, too little water has been absorbed; in constipation, too much water has been absorbed.

    The colon is subject to many disorders such as appendicitis, diverticulosis, and cancer.

  F. There are two accessory organs of digestion which send secretions to the duodenum via ducts. The pancreas produces pancreatic juice, which contains digestive enzymes for carbohydrate, protein, and fat.

  G. The liver produces bile, which is stored in the gallbladder. The liver receives blood from the small intestine by way of the hepatic portal vein. The liver has numerous important functions, and any malfunction of the liver is a matter of considerable concern.

II. Chemical Digestion

  A. Digestive enzymes are present in digestive juices and break down food to the nutrient molecules: glucose, amino acids, fatty acids, and glycerol. Glucose and amino acids are absorbed into the blood capillaries of the villi. Fatty acids and glycerol reform to give fat, which enters the lacteals.

Study Questions

1. List the parts of the digestive tract, anatomically describe them, and state the contribution of each to the digestive process.
2. List the accessory glands, and describe the part that they play in the digestion of food.
3. List six functions of the liver. How does the liver maintain a constant glucose level in the blood?
4. What is jaundice? Cirrhosis of the liver?
5. What is the common intestinal bacterium? What do these bacteria do for us?
6. What are gastrin, secretin, and CCK? Where are they produced? What are their functions?
7. Discuss the digestion of starch, protein, and fat, listing all of the steps that occur to bring about digestion of each of these.
8. Describe how the stomach and duodenum are protected from the action of acid and digestive enzymes. What condition develops when this protection fails?
9. Describe liver disorders and how they might be prevented.
10. Describe three major disorders associated with the large intestine.

Objective Questions

Fill in the blanks.

1. In the mouth, salivary _____ digests starch.
2. When swallowing, the _____ covers the opening to the larynx.
3. The _____ takes food to the stomach where _____ is primarily digested.
4. The gastric juices are _____ and, therefore, they usually destroy any bacteria in the food.
5. The gallbladder stores _____ , a substance that _____ fat.
6. The pancreas sends digestive juices to the _____ , the first part of the small intestine.
7. Pancreatic juice contains _____ for digesting protein, _____ for digesting fat, and _____ for digesting starch.
8. The products of digestion are absorbed into the cells of the _____ , fingerlike projections of the intestinal wall.
9. After eating, the liver stores glucose as _____ .
10. The large intestine has four _____ and the _____ which opens at the _____ .

Medical Terminology Reinforcement Exercise

Pronounce the following words.
Analyze meaning by word parts as
dissected:

1. stomatoglossitis (sto″mah-to-glos-si′tis)—stomato/gloss/itis

2. glossopharyngeal (glos″o-fah-rin′je-al)—glosso/pharyngeal

3. esophagectasia (ĕ-sof″ah-jek-ta′se-ah)—esophag/ectasia

4. gastroenteritis (gas″tro-en-ter-i′tis)—gastro/enter/itis

5. colostomy (ko-los′to-me)—col/ostomy

6. sublingual (sub-ling′gwal)—sub/lingu/al

7. gingivoperiodontitis (jin″ji-vo-per″e-o-don-ti′tis)—gingivo/peri/odont/itis

8. dentalgia (den-tal′je-ah)—dent/algia

9. pyloromyotomy (pi-lo″ro-mi-ot′o-me)—pyloro/myo/tomy

10. cholangiogram (ko-lan′je-o-gram)—chol/angio/gram

11. cholecystolithotripsy (ko″le-sis″to-lith′o-trip″se)—chole/cysto/litho/tripsy

12. proctosigmoidoscopy (prok″to-sig″moi-dos′ko-pe)—procto/sigmoid/oscopy

11

Blood

Chapter Outline

Learning Objectives

After you have studied this chapter, you should be able to:

1. Describe, in general, the composition of blood.
2. Describe the structure and function of red blood cells and their life cycle.
3. List four types of anemia and the causes of each type.
4. Describe the structure and function of neutrophils and lymphocytes.
5. Distinguish between leukopenia, leukocytosis, and leukemia.
6. State the significance of the following blood tests: hematocrit, hemoglobin determination, and differential white blood cell count.
7. Describe the structure and function of platelets.
8. Discuss the transport function of blood and describe capillary exchange within the tissues.
9. Describe the blood clotting process and how it is associated with thromboembolism.
10. Describe the ABO and Rh systems of blood typing, and how blood is typed.
11. Distinguish between myocardial infarction, cerebrovascular accident, angina pectoris, and hemolytic disease of the newborn.

Composition of Blood

An adult male of average size (70 kilograms or 154 pounds) has a blood volume of about 5 liters (5.2 quarts). If a sample is transferred to a test tube and prevented from clotting, it separates into two layers (fig. 11.1). The lower layer takes up about 45 percent of the volume of whole blood and is composed mainly of red blood cells; therefore, this percentage is called the **hematocrit (HCT).** The upper layer, called **plasma,** contains a variety of inorganic and organic substances dissolved or suspended in water. Plasma accounts for about 55 percent of the volume of whole blood.

Red blood cells (erythrocytes) are one of the **formed elements** in blood. The other formed elements are white blood cells (leukocytes) and blood platelets (thrombocytes). All of the components of blood are listed in table 11.1.

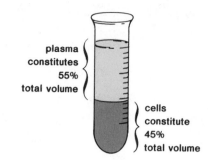

Figure 11.1 Volume relationship of plasma and formed elements (cells) in blood. Red blood cells are by far the most prevalent blood cell, and this accounts for the color of blood.

Table 11.1 Components of Blood

Blood	Function	Source
I. FORMED ELEMENTS		
Red blood cells	Transport oxygen	Bone marrow
Platelets	Clotting	Bone marrow
White blood cells	Fight infection	Bone marrow and lymphoid tissue
II. PLASMA[1]		
Water	Maintains blood volume and transports molecules	Absorbed from intestine
Plasma proteins	All maintain blood osmotic pressure and pH	
Albumin	Transport	Liver
Fibrinogen	Clotting	Liver
Gamma Globulins	Fight infection	Lymphocytes
Gases		
Oxygen	Cellular respiration	Lungs
Carbon dioxide	End product of metabolism	Tissues
Nutrients		
Fats, glucose, amino acids, etc.	Food for cells	Absorbed from intestinal villi
Electrolytes[2]	Maintain blood osmotic pressure	Absorbed from intestinal villi
Ions (Na^+, HCO_3^-, K^+)	and pH; aid metabolism	
Wastes		
Urea and ammonia	End products of metabolism	Tissues
Hormones, vitamins, etc.	Aid metabolism	Varied

[1]Plasma is 90–92% water, 7–8% plasma proteins, not quite 1% salts, and all other components are present in even smaller amounts.
[2]See page 202.

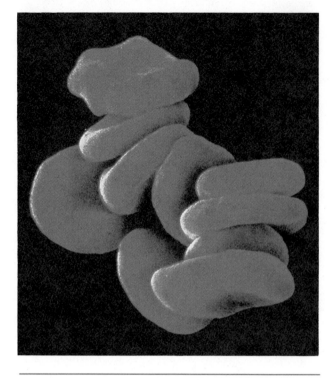

Blood is a liquid tissue; the liquid portion is termed plasma, and the solid portion consists of the formed elements.

Red Blood Cells (Erythrocytes)

The **red blood cells (RBCs)** are small biconcave, disk-shaped cells without nuclei (fig. 11.2). Red cells are continuously manufactured in the red bone marrow of the skull, ribs, vertebrae, and ends of the long bones. Originally, they have nuclei, but they pass through several developmental stages during which time they lose the nucleus and acquire hemoglobin.

Hemoglobin is a carrier for oxygen. It forms a loose association with oxygen in the cool, neutral conditions of the lungs, and readily gives it up under the warm and more acidic conditions of the tissues. **Oxyhemoglobin** (hemoglobin + oxygen) is a bright red color. **Reduced hemoglobin** (hemoglobin − oxygen) is a dark purple color. This accounts for the difference in color between the systemic arteries and the systemic veins.

The number of red blood cells in a cubic millimeter (mm^3) of blood is called the **red blood cell count.** Males usually have 4.6 to 6.2 million cells per mm^3, and females have 4.2 to 5.4 million cells per mm^3. Each cell contains about 200 million hemoglobin molecules. If this much hemoglobin were suspended within the plasma rather than being enclosed within the cells, the blood would be so thick that the heart would have difficulty pumping it.

Red cells live only about 120 days and are destroyed chiefly in the liver and spleen, where they are engulfed by large phagocytic cells. When red cells are broken down, the hemoglobin is released and broken down into **heme,** an iron-containing portion, and **globin,** a protein. The iron is recovered and returned to the red bone marrow for reuse. The rest of the heme undergoes further chemical degradation and is excreted by the liver in the bile as bile pigments. The bile pigments are primarily responsible for the color of feces.

Red blood cells are made in the bone marrow and are broken down in the liver and spleen.

Anemia

When there is an insufficient number of red cells or the cells do not have enough hemoglobin, the individual suffers from **anemia** and has a tired, run-down feeling. In *iron deficiency anemia,* the common type, the hemoglobin count is low. Normally, the hemoglobin blood value is 12 to 17 g/100 ml of whole blood. It may be that the diet does not contain enough iron. Certain foods, such as raisins and liver, are rich in iron and the inclusion of these in the diet can help prevent this type of anemia.

In another type of anemia, called *pernicious anemia,* the digestive tract is unable to absorb enough vitamin B_{12}. This vitamin is essential to the proper formation of red cells and without it, immature red cells tend to accumulate in the bone marrow in large quantities. A special diet and administration of vitamin B_{12} by injection is an effective treatment for pernicious anemia.

In *aplastic anemia,* the red bone marrow has been damaged due to radiation or chemicals, and not enough red cells are produced. In *hemolytic anemia,* there is an increased rate of red blood cell destruction.

In *sickle-cell anemia,* a genetic disease, the red blood cells are sickle-shaped (fig. 11.3). Such cells tend to rupture and wear out easily as they pass through the narrow capillaries, leading to the symptoms of anemia. Sickle-cell anemia is usually seen in blacks because the sickle-shaped cells are a protection against malaria, a disease prevalent in parts of Africa. The parasite that causes malaria can't infect sickle-shaped red blood cells.

Anemia results when the blood has too few red blood cells and/or not enough hemoglobin.

White Blood Cells (Leukocytes)

White blood cells (WBCs), or leukocytes, differ from red blood cells in that they are usually larger; have a nucleus; lack hemoglobin; and, without staining, would appear to be white in color. With staining, white cells

Figure 11.3 Individuals with sickle-cell anemia have red blood cells that are sickle shaped. These cells tend to suffer wear and tear as they pass through the narrow capillary walls.

characteristically appear bluish. White cells are less numerous than red cells. The **white blood cell count** is usually between 5,000 and 10,000 per mm³.

There are several different types of white blood cells (fig. 11.4 and table 11.2), but it is possible to divide them into the granular leukocytes (granulocytes) and the agranular leukocytes (agranulocytes). The granular leukocytes have granules in the cytoplasm and a many-lobed nucleus joined by nuclear threads; therefore, they are called polymorphonuclear. They are formed and mature in the red bone marrow. The agranular leukocytes do not have granules and have a circular, or indented, nucleus. They are produced in the lymphoid tissue of the red bone marrow, and certain of the lymphocytes mature in the thymus. Agranular leukocytes are stored in the spleen, lymph nodes, tonsils, and other lymphoid organs.

White blood cells are divided into the granular leukocytes and the agranular leukocytes.

White blood cells are involved in defending the body against infectious diseases. Infection fighting by WBCs is primarily dependent on the neutrophils, which comprise 60 to 70 percent of all leukocytes, and the lymphocytes, which make up 25 to 30 percent of the leukocytes (fig. 11.5).

Neutrophils are *amoeboid;* they slip through the capillary wall by a process called *diapedesis* and then self-propel themselves by a slow extension of the cytoplasm. They are also phagocytic; when they come in contact with bacteria, they engulf them. Certain **lymphocytes** (called B cells, p. 237) secrete immunoglobulins, antibodies, which combine with foreign substances (antigens) to inactivate them. Neutrophils and lymphocytes may be contrasted in the following manner:

Neutrophils
granules in cytoplasm
polymorphonuclear
produced in bone marrow
phagocytic

Lymphocytes
no granules in cytoplasm
mononuclear
produced in lymphoid tissue
make antibodies

The role of neutrophils and other white blood cells in the *inflammatory response* (p. 235), and the role of monocytes and lymphocytes in the development of *immunity* (p. 235) are discussed in chapter 13.

The neutrophils, which are amoeboid and engulf invaders, may be contrasted to lymphocytes, which produce antibodies that combine with antigens.

Diseases

Certain viral illnesses, like influenza, measles, and mumps, cause the white blood cell count to decrease. **Leukopenia** is a total white cell count below 5,000 per mm³. Others, like appendicitis, cause the white cell count to increase dramatically. **Leukocytosis** is a white blood cell count above 10,000 per mm³.

Often illnesses cause an increase in a particular type of white cell. For this reason, a **differential white blood cell count,** involving the microscopic examination of a blood sample and the counting of each type of white cell to a total of 100 cells, may be done as part of the diagnostic procedure. For example, the characteristic finding in the viral disease **mononucleosis** is a great

Figure 11.4 The formed elements in blood are the cells and platelets, which are fragments of cells. Erythrocytes are red blood cells. Leukocytes (both granular and agranular) are white blood cells. Thrombocytes are platelets.

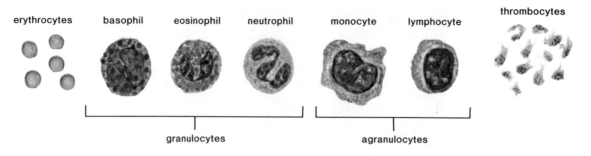

erythrocytes   basophil   eosinophil   neutrophil   monocyte   lymphocyte   thrombocytes

granulocytes   agranulocytes

Table 11.2 Formed Elements of the Blood

Component	Description	Number present	Function
Erythrocyte (red blood cell)	Biconcave disk without nucleus; contains hemoglobin; survives 100–120 days	4,000,000 to 6,000,000/mm³	Transports oxygen and carbon dioxide
Leukocytes (white blood cells)		5,000 to 10,000/mm³	Aid in defense against infections by microorganisms
Granular leukocytes	About twice the size of red blood cells; cytoplasmic granules present; survive 12 hours to 3 days		
1. Neutrophil	Nucleus with 2–5 lobes; cytoplasmic granules stain slightly pink	54%–62% of white cells present	Phagocytic
2. Eosinophil	Nucleus bilobed; cytoplasmic granules stain red in eosin stain	1%–3% of white cells present	Helps to detoxify foreign substances; secretes enzymes that break down clots
3. Basophil	Nucleus lobed; cytoplasmic granules stain blue in hematoxylin stain	Less than 1% of white cells present	Releases anticoagulant heparin
Agranular leukocytes	Cytoplasmic granules absent; survive 100–300 days		
1. Monocyte	2–3 times larger than red blood cell; nuclear shape varies from round to lobed	3%–9% of white cells present	Phagocytic
2. Lymphocyte	Only slightly larger than red blood cell; nucleus nearly fills cell	25%–33% of white cells present	Provides specific immune response (including antibodies)
Thrombocyte (platelet)	Cytoplasmic fragment; survives 5–9 days	130,000 to 360,000/mm³	Clotting

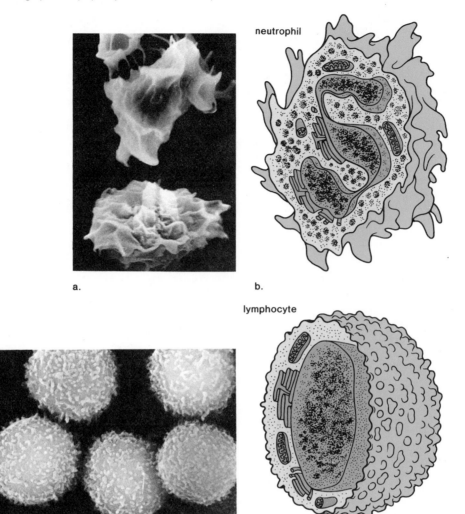

neutrophil

a.

b.

lymphocyte

c.

d.

number of lymphocytes that are larger than mature lymphocytes and stain more darkly. This condition takes its name from the fact that lymphocytes are mononuclear.

Leukemia **Leukemia** is a form of cancer characterized by uncontrolled production of abnormal white blood cells. These cells accumulate in the bone marrow, lymph nodes, spleen, and liver so that these organs are unable to function properly. Largely as a result of combined chemotherapy and radiation, patients with leukemia are subject to severe anemia due to a low red cell count, clotting difficulties due to a low platelet count, and development of infections due to the presence of nonfunctioning white cells.

The white blood cell count and the differential white blood cell count are useful in diagnosing illnesses.

Platelets (Thrombocytes)

Platelets (fig. 11.4) result from fragmentation of certain large cells, called megakaryocytes, in the red bone marrow. The platelet count in blood is 130,000 to 360,000 per mm^3.

Platelets do not have nuclei and are about half the size of red blood cells. They live only about ten days and are involved in repairing damaged blood vessels and in initiating the process of blood clotting as discussed on page 203.

Homeostasis and Body Fluids

The average individual is about 60 percent water by weight. About two-thirds of this water is inside the cells (called intracellular) and the rest is distributed in the plasma, tissue fluid, lymph, and cerebrospinal fluid (called extracellular). Maintenance of the normal content of water in all body fluids is important for continued good health. The osmolarity of the blood is constantly monitored within the hypothalamus and this determines whether we feel thirsty and take a drink of water, and whether the kidneys retain more water. You'll recall that the hypothalamus produces the hormone ADH which is released by the posterior pituitary when more water is to be retained by the kidneys (p. 164).

The osmolarity of the blood is dependent upon the concentration of substances within the blood—particularly electrolytes. **Electrolytes** are compounds and molecules that are able to ionize and, thus, carry an electric current. The most common electrolytes in the plasma are sodium (Na^+), potassium (K^+) and bicarbonate (HCO_3^-). Na^+ and K^+ are termed cations because they are positively charged and HCO_3^- is termed an anion because it is negatively charged.

Sodium. The movement of Na^+ across an axon membrane, you'll recall, is necessary to the formation of the nerve impulse and muscle contraction. The concentration of Na^+ in the blood is also the best indicator of its osmolarity. When there is a reduction in Na^+ in the plasma, the renin-angiotensinogen-aldosterone system is activated to reabsorb Na^+ and blood volume is subsequently raised (p. 172).

Potassium. The movement of K^+ across an axon membrane is also necessary to the formation of the nerve impulse and muscle contraction. Abnormally low K^+ concentrations in the blood, as might happen if diuretics are abused, can lead to cardiac arrest.

Bicarbonate ion. HCO_3^- is the way in which carbon dioxide is carried in the blood. The bicarbonate ion has a very important function in that it helps maintain the pH of the blood as discussed in the following paragraphs.

Other ions. There are many other ions in the plasma. For example, calcium ions (Ca^{++}) and phosphate ions (HPO_4^-) are important to bone formation and cellular metabolism. Their absorption from the gut and excretion by the kidneys is regulated by hormones already discussed in chapter 9.

Hydrogen Ion (H^+) Concentration

The H^+ concentration of body fluids is important because proteins such as cellular enzymes only function properly when this concentration is maintained at about 7. Pure water has an H^+ concentration of about 7 and pure water dissociates to give an equal number of H^+ and OH^- ions:

$$H-O-H \longrightarrow H^+ + OH^-$$

Acids increase the H^+ concentration and bases increase the OH^- concentration. When a base adds hydroxide ions (OH^-) to pure water, it decreases the relative number of H^+ ions present.

Because the body is so sensitive to the H^+ concentration, the pH scale is used to indicate the pH of body fluids. Pure water has a pH of exactly 7. Acidity increases as we go from 7 to 0, and basicity increases as we move from 7 to 14. For every higher pH unit, there is a tenfold decrease in H^+ concentration. For more information about the pH scale, see p. 317 of the appendix.

The pH of the blood stays at just about 7.4 because the blood is buffered. A **buffer** is a chemical or combination of chemicals that can take up excess H^+ or excess OH^-. One of the most important buffers in the blood is carbonic acid (H_2CO_3) and the bicarbonate ion (HCO_3^-):

$$H_2CO_3 \rightleftharpoons H^+ + HCO_3^-$$

If the pH of the blood rises (less acidity), carbonic acid dissociates to release H^+. If the pH of the blood decreases (more acidity), the bicarbonate ion combines with it to give carbonic acid. Proteins also help buffer the blood because they are charged in such a way that they can combine with either H^+ or OH^-.

Plasma

Plasma is the liquid portion of blood (fig. 11.1). Plasma is approximately 92 percent water, but still contains many different types of molecules including the plasma proteins. Two types of plasma proteins have a very special function. The **gamma globulins**[1] are antibodies that help fight infection. The protein **fibrinogen** is converted to fibrin threads when blood clotting occurs.

Plasma proteins, along with electrolytes, create an osmotic pressure that draws water from the tissues into the blood. This function of the plasma proteins is particularly associated with albumin, the smallest and most plentiful of the plasma proteins.

1. When globulins undergo electrophoresis (are put in an electrical field), they separate into major components called alpha globulin, beta globulin, and gamma globulin. Almost all circulating antibodies are found in the gamma globulin fraction.

There are several different plasma proteins, each with specific functions.

Two Functions of Blood

Blood has many functions, two of which we will be discussing here.

Transport

The blood transports oxygen from the lungs and nutrients from the intestine to the capillaries, where they enter tissue fluid. Here, blood also takes up carbon dioxide and nitrogen waste (i.e., ammonia) given off by the cells and transports them away. Carbon dioxide exits the blood at the lungs, and ammonia exits at the liver, where it is converted to urea, a substance that later travels by way of the bloodstream to the kidneys and is excreted. Figure 11.6 diagrams the major transport functions of blood, indicating the manner in which these functions help keep the internal environment relatively constant. The reading on the opposite page gives further information.

Homeostasis is possible only because blood brings nutrients to the cells and removes their wastes.

Capillary Exchange within the Tissues

When arterial blood enters the tissue capillaries (fig. 11.7), it is bright red because the red cells are carrying oxygen. It is also rich in nutrients dissolved in the plasma. At this end of the capillary, water along with oxygen and nutrients (glucose and amino acids) exit from the capillary. This is a filtration process facilitated by blood pressure (p. 36).

At the venous end of the capillary, water along with the waste molecules carbon dioxide and ammonia are drawn into a capillary due to osmotic pressure. The blood that leaves a capillary is deep purple in color because the red cells contain reduced hemoglobin that is not carrying oxygen. Ammonia is dissolved in the plasma and carbon dioxide is carried as the bicarbonate ion (HCO_3^-).

Oxygen and nutrient molecules (e.g., glucose and amino acids) exit a capillary near the arterial end; waste molecules (e.g., carbon dioxide and ammonia) enter a capillary near the venous end.

This method of retrieving fluid is not completely effective. There is always some fluid that is left and not picked up at the venous end of the capillary. This excess tissue fluid enters the lymphatic vessels (fig. 13.2). Lymph is tissue fluid contained within lymphatic vessels. Lymph is returned to systemic venous blood where

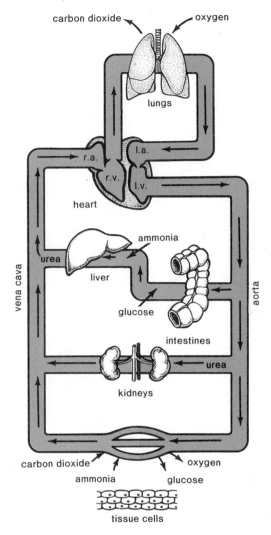

Figure 11.6 Diagram illustrating the transport function of blood. Oxygen is transported from the lungs to the tissues, and carbon dioxide is transported from the tissues to the lungs. Ammonia is transported from the tissues to the liver where it is converted to urea, a molecule excreted by the kidneys. Glucose is absorbed by the gut and may be temporarily stored in the liver as glycogen before it is transported to the tissues.

the major lymphatic vessels enter subclavian veins (p. 231).

Blood Clotting

When a blood vessel is cut, it immediately constricts so that the flow of blood to the area is reduced. Then, the platelets start to stick to collagen fibers in the connective tissue layer beneath the endothelium. As more and more platelets congregate, they form a **platelet plug** that can fill the break if it is small enough.

Another response to injury of a blood vessel is formation of a blood clot. There are at least twelve *clotting factors* in the blood that participate in the

Figure 11.7 Diagram of a capillary illustrating the exchanges that take place and the forces that aid the process. At the arterial end of a capillary, the blood pressure is higher than the osmotic pressure, and therefore, water, oxygen, and glucose tend to leave the bloodstream. At the venous end of a capillary, the osmotic pressure is higher than the blood pressure, and therefore, water, ammonia, and carbon dioxide tend to enter the bloodstream. Notice that the red blood cells and plasma proteins are too large to exit a capillary.

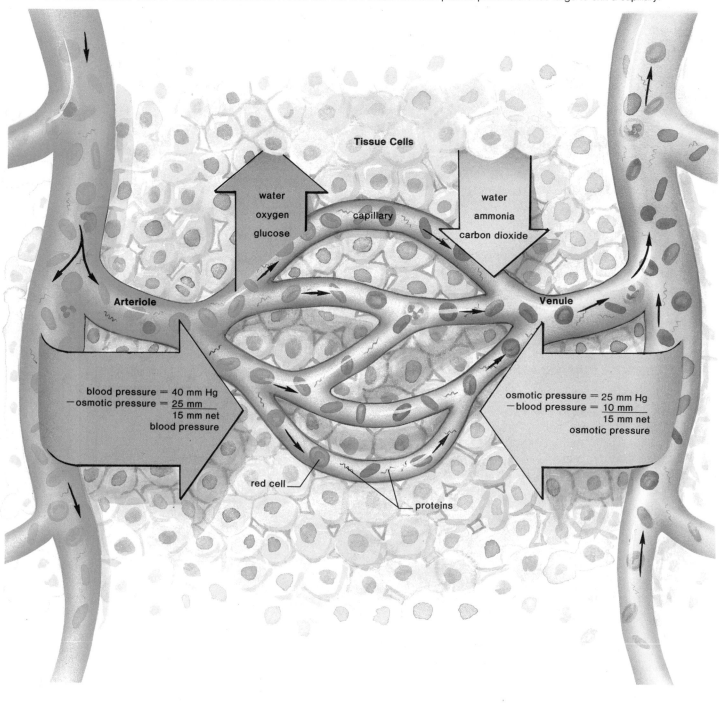

formation of a blood clot. We are going to be discussing the role played by platelets, **prothrombin,** and **fibrinogen.** Prothrombin and fibrinogen are plasma proteins produced by the liver. Vitamin K is necessary to the production of prothrombin and if, by chance, this vitamin is missing from the diet, hemorrhagic disorders develop.

The clotting process (fig. 11.8) consists of a cascade of enzymatic reactions in which each reaction leads to the next and so forth. It is possible to summarize the main events in the following manner. The clotting process is initiated when platelets and damaged tissue release prothrombin activator. **Prothrombin activator** is

Figure 11.8   Injury to a blood vessel causes blood vessel constriction, platelet plug formation, and blood clotting. Blood clotting requires the steps shown.

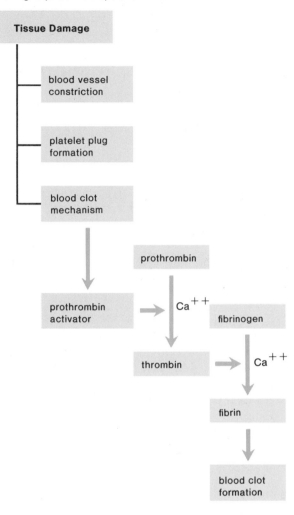

Figure 11.9   When blood clots, serum is squeezed out as a solid plug is formed. In a blood vessel, this plug helps prevent further blood loss.

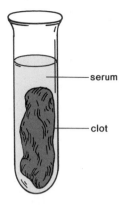

Table 11.3	Body Fluids
Name	**Composition**
Blood	Formed elements and plasma
Plasma	Liquid portion of blood
Serum	Plasma minus fibrinogen
Tissue fluid	Plasma minus proteins
Lymph	Tissue fluid within lymph vessels

an enzyme that converts prothrombin to thrombin. This reaction requires the presence of Ca^{++}. **Thrombin,** in turn, acts as an enzyme that brings about a change in fibrinogen so that it forms long threads of fibrin. Once formed, **fibrin** threads wind around the platelet plug and provide a framework for the clot. Red blood cells are also trapped within the fibrin threads and their presence makes a clot appear red.

A blood clot consists of red blood cells entangled within fibrin threads.

If blood is placed in a test tube and allowed to clot, a yellowish fluid rises above the clotted material (fig. 11.9). This fluid is called **serum,** and it contains all of the components of plasma except fibrinogen. Since we have now used a number of different terms to refer to portions of the blood, table 11.3 reviews these terms for you.

In the body, many clots disappear with time. An enzyme called *plasmin* destroys the fibrin network and restores the fluidity of plasma. This is a protective measure because the presence of a blood clot can lead to a *heart attack* (myocardial infarction).

Thromboembolism

A blood clot that remains stationary in a blood vessel is a **thrombus,** but when it dislodges and moves along with the blood, it is called an **embolus.** If **thromboembolism** is not treated, complications can arise as discussed in the following paragraph.

Thromboembolism is especially associated with **atherosclerosis,** an accumulation of soft masses of fatty materials, particularly cholesterol, beneath the inner linings of the arteries. Such deposits are called **plaque,** and as they develop, they tend to protrude into the vessel and interfere with the flow of blood. This is why it is now recommended that diet be used to control the cholesterol blood level when necessary.

If the coronary artery is partially blocked (occluded) by the presence of atherosclerosis or a blood clot, the individual may suffer from **ischemic heart disease.** Although enough oxygen may normally reach the heart, the individual experiences insufficiency during exercise or stress. At that time, the individual may suffer **angina pectoris,** characterized by a radiating pain in the left arm. If total blockage occurs, a portion of the heart may die due to lack of oxygen. Dead tissue is known as an infarct and, therefore, the individual has undergone a **myocardial infarction,** commonly called a heart attack.

On occasion, it is a cerebral artery that is occluded by an embolus, and in that case, the person suffers a **cerebrovascular accident** or *stroke.* Strokes are also associated with hypertension and occur when a blood vessel bursts.

Thromboembolism and/or atherosclerosis can lead to a myocardial infarction or cerebrovascular accident.

There are two surgical procedures associated with occluded coronary blood vessels. In *thrombolytic therapy,* a plastic tube can be threaded into an artery of an arm or leg and guided through a major blood vessel toward the heart. Once it reaches a blockage, it is possible to inflate a balloon attached to the end of the tube to break up the clot or inject streptokinase to dissolve the clot. In the *coronary bypass operation,* a segment from another blood vessel such as a large vein in the leg is stitched from the aorta to the coronary artery past the point of obstruction. Now blood can flow normally again from the aorta to the heart.

Blood Typing

ABO Grouping

Two antigens that may be present on the red cells have been designated as A and B. They are termed antigens because they are foreign substances to a possible recipient. As table 11.4 shows, an individual may have one of these antigens (i.e., type A or type B), or both (type AB), or neither (type O). Therefore, ABO blood type is dependent on which of these antigens are present on the red cells. As you can see, type O blood is most common in the United States.

In the ABO blood grouping system, there are four types of blood: A, B, AB, and O. Type O blood has neither the A nor the B antigen on the red cells; the other types of blood are designated by the antigens present on the red cells.

Within the plasma of an individual, there are antibodies to the antigens that are *not* present on that individual's red cells. For example, type A blood has an

Table 11.4 Blood Groups

Type	Antigen	Anti-body	% U.S. Black[a]	% U.S. Caucasian[a]
A	A	anti-B	25	41
B	B	anti-A	20	7
AB	A,B	None	4	2
O	None	anti-A anti-B	51	50

[a]Blood type frequency for other races is not available.

antibody called anti-B in the plasma. Type AB blood has neither anti-A nor anti-B antibodies because both antigens are on the red cells. This is reasonable because antibodies are molecules that combine with antigens and if the same antigen and antibody are present, **agglutination,** or clumping, of red cells will occur. Agglutination of red cells can cause the blood to stop circulating in small blood vessels and this leads to organ damage. It is also followed by hemolysis, which brings about death of the individual.

For a recipient to receive blood from a donor, the recipient's plasma must not have an antibody that would cause the donor's cells to agglutinate. For this reason, it is important to determine each person's blood type. Figure 11.10 demonstrates a way to use the antibodies derived from plasma to determine the type of blood. If clumping occurs after a sample of blood is exposed to a particular antibody, the person has that type of blood.

Rh System

Another important antigen in matching blood types is the **Rh factor.** Persons with this particular antigen on the red cells are Rh positive (Rh$^+$); those without it are Rh negative (Rh$^-$). Rh negative individuals do not normally have antibodies to the Rh factor, but they make them when exposed to the Rh factor. It is possible to use anti-Rh antibodies for blood type testing. When Rh positive blood is mixed with anti-Rh antibodies, agglutination occurs.

The designation of blood type usually also includes whether the person has the Rh factor (Rh$^+$) or does not have the Rh factor (Rh$^-$) on the red cells.

During pregnancy (fig. 11.11), if the mother is Rh negative and the father is Rh positive, the child may be Rh positive. The Rh positive red cells may begin leaking across into the mother's circulatory system as placental tissues normally break down before and at birth. This causes the mother to produce anti-Rh antibodies. If the mother becomes pregnant with another Rh positive baby, anti-Rh antibodies may cross the pla-

Figure 11.10 The standard test to determine ABO and Rh blood type consists of putting a drop of anti-A antibodies, anti-B antibodies, and anti-Rh antibodies on a slide. To each of these a drop of the person's blood is added. *a*. If agglutination occurs, as seen at lower left, the person has this antigen on the red blood cells. *b*. Several results are possible.

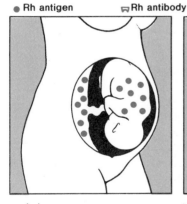
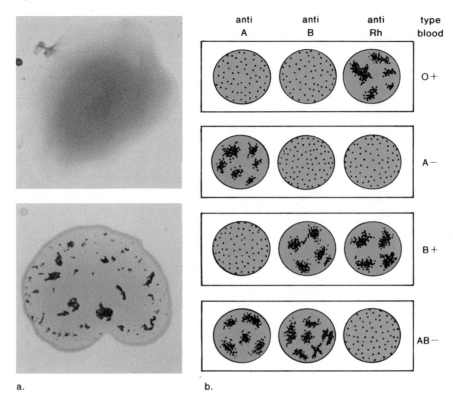

a.

b.

Figure 11.11 Diagram describing the development of hemolytic disease of the newborn. *a*. Baby's red blood cells carry Rh antigen. *b*. Some of these cells escape into mother's system. *c*. Mother begins manufacturing anti-Rh antibodies. *d*. During a subsequent pregnancy, mother's anti-Rh antibodies cross placenta to destroy the new baby's red blood cells.

● Rh antigen   ⎍ Rh antibody

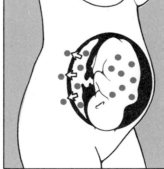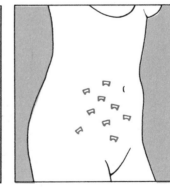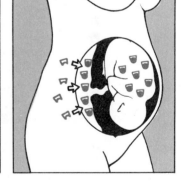

a. during pregnancy   **b. before delivery**   **c. months and years later**   **d. subsequent pregnancy**

centa and cause destruction of the child's red cells. This is called hemolytic disease of the newborn.

The Rh problem has been solved by giving Rh negative women an Rh immune globulin injection either midway through the first pregnancy or no later than 72 hours after giving birth to an Rh positive child. This injection contains anti-Rh antibodies that attack any

of the baby's red cells in the mother's blood before these cells can stimulate the mother to produce her own antibodies.

The possibility of hemolytic disease of the newborn exists when the mother is Rh⁻ and the father is Rh⁺.

Summary

I. Composition of Blood. Blood is a liquid tissue; the liquid portion is termed plasma, and the solid portion consists of the formed elements.
  A. Red blood cells are made in the bone marrow and are broken down in the liver and spleen.

  Illness (anemia) results when the blood has too few red cells and/or not enough hemoglobin.
  B. White blood cells are divided into the granular leukocytes and the agranular leukocytes.

  The neutrophils, which are amoeboid and engulf invaders, may be contrasted to the lymphocytes, which produce antibodies that combine with antigens.

  The white blood cell count and the differential white blood cell count are useful in diagnosing illnesses.
  C. Platelets are fragments of cells that are involved in vessel repair and blood clotting.
  D. Plasma proteins have specific functions. Gamma globulins help fight infection; fibrinogen is necessary to blood clotting; and albumin, in particular, helps maintain blood volume.

II. Two Functions of Blood
  A. Transport. Homeostasis is only possible because blood brings nutrients to the cells and removes their wastes.

  Oxygen and nutrient molecules (e.g., glucose and amino acids) exit a capillary near the arterial end; waste molecules (e.g., carbon dioxide and ammonia) enter a capillary near the venous end.
  B. Blood clotting. A blood clot consists of red blood cells entangled within fibrin threads.

Thromboembolism and/or atherosclerosis can lead to a myocardial infarction or cerebrovascular accident.

Surgical treatment is available for persons who have partially or wholly occluded coronary blood vessels.

III. Blood Typing
  A. In the ABO blood grouping system, there are four types of blood: A, B, AB, and O. Type O blood has no antigens on the red cells; the other types of blood are designated by the antigens present on the red cells.
  B. The designation of blood type usually also includes whether the person has the Rh factor (Rh⁺) or does not have the Rh factor (Rh⁻) on the red cells.

  The possibility of hemolytic disease of the newborn exists when the mother is Rh⁻ and the father is Rh⁺.

Study Questions

1. Define blood, plasma, tissue fluid, lymph, and serum.
2. Name the formed elements, describe, in general, the structure and function of each.
3. Describe the life cycles of a red blood cell. What factor controls the quantity of red blood cells in the body?
4. Contrast the neutrophils to the lymphocytes in at least four ways.
5. Name three plasma proteins and give a function for each.
6. Draw a diagram of a capillary, illustrating the exchanges that occur in the tissues. What forces operate to facilitate exchange of molecules across the capillary wall?
7. Name the events that take place after a blood vessel is injured. Which substances are present in the blood at all times and which appear during the clotting process?
8. What are the four ABO blood types in humans? For each type, tell what antibodies are in the plasma.

Objective Questions

I. Fill in the blanks.
  1. The liquid part of blood is called _____ .
  2. Red blood cells carry _____ , and white blood cells _____ _____ .
  3. Hemoglobin that is carrying oxygen is called _____ .
  4. Human red blood cells lack a _____ and only live about _____ days.
  5. When a blood clot occurs, fibrinogen has been converted to _____ threads.
  6. The most common granular leukocyte is the _____ , a phagocytic white blood cell.
  7. Lymphocytes are made in _____ tissue and produce _____ that react with antigens.
  8. At a capillary, _____ , _____ _____ , and _____ leave the arterial end, and _____ and _____ enter the venous end.
  9. AB blood has the antigens _____ and _____ on the red cells and _____ antibodies in the plasma.
  10. Hemolytic disease of the newborn can occur when the mother is _____ and the father is _____ .

II. Matching
  For questions 11–14, match the items in the key to the descriptions below.
  Key: a. hematocrit
        b. red blood cell count
        c. white blood cell count
        d. hemoglobin
  11. 5,000 to 10,000 per mm³
  12. 4.6 to 5.2 million per mm³ in males
  13. 45 percent red blood cells and 55 percent plasma
  14. 12 to 17 g/100 ml

  For questions 15–18, match the items in the key to the descriptions below.

Key: a. iron deficiency anemia
      b. pernicious anemia
      c. aplastic anemia
      d. sickle-cell anemia
15. a genetic disease
16. lack of vitamin B$_{12}$
17. faulty diet
18. X-ray treatment

For questions 19–25, match the items in the key to the descriptions below.
Key: a. polycythemia
      b. leukopenia
      c. leukocytosis
      d. thromboembolism
      e. atherosclerosis
      f. myocardial infarction
      g. cerebrovascular accident

19. overabundance of red blood cells
20. stroke
21. low white blood cell count
22. heart attack
23. high white blood cell count
24. blood clots in vessels
25. plaque on walls of blood vessels

Medical Terminology Reinforcement Exercise

Pronounce the following terms and analyze the meaning by word parts as dissected:

1. hematemesis (hem″ah-tem′ĕ-sis)—hemat/emesis

2. erythrocytometry (ĕ-rith″ro-si-tom′ĕ-tre)—erythro/cyto/metry

3. leukocytogenesis (loo″ko-si″to-jen′ĕ-sis)—leuko/cyto/genesis

4. thrombocytopenia (throm″bo-si″to-pe′ne-ah)—thrombo/cyto/penia

5. hemophobia (he″mo-fo′be-ah)—hemo/phobia

6. afibrinogenemia (ah-fi″brin-o jĕ-ne′me-ah) a/fibrino/gen/emia

7. polycythemia (pol″e-si-the′me-ah)—poly/cyt/hemia

8. lymphosarcoma (lim″fo-sar-ko′mah)—lympho/sarcoma

9. phagocytosis (fag″o-si-to′sis)—phago/cyt/osis

10. phlebotomy (flĕ-bot′o-me)—phleb/otomy

12

Circulation

Chapter Outline

Learning Objectives

After you have studied this chapter, you should be able to:

1. Name the parts of the heart and trace the path of blood through the heart.
2. Describe the heartbeat and mechanisms for controlling the heartbeat.
3. Label and explain a normal electrocardiogram.
4. Name and describe the structure and function of blood vessels.
5. Name the parts of the circulatory system, and trace the path of blood in general and specifically to any organ in the body.
6. Describe the location, operation, and function of valves in the vessels and in the heart.
7. Describe the factors that control the flow of blood in the arteries, capillaries, and veins.

Heart

Humans have a circulatory system which brings nutrients and oxygen to the cells and takes away their wastes. At the center of this system is the heart, which pumps the blood.

The **heart** is a cone-shaped, muscular organ about the size of a fist. It is located between the lungs, directly behind the sternum, and is tilted so that the apex is directed to the left. The heart lies within the double layered **pericardial sac (pericardium)** (fig. 12.1). The outer layer of the sac is the *parietal pericardial membrane* and the inner layer is the *visceral pericardial membrane*. The potential space between the two membranes is the *pericardial cavity.* A thin liquid film lies between the two pericardial membranes and this lubricates them as the heart beats. The major portion of the heart, called the **myocardium,** consists largely of cardiac muscle tissue. The muscle fibers within the myocardium are branched and joined to one another so tightly that, prior to studies with the electron microscope, it was thought that they formed one continuous muscle. Now it is known that there are individual fibers, but they are bound end to end at intercalated disks, areas of folded cell membrane between the cells (fig. 3.11). The inner surface of the heart is lined with the *endocardium,* which consists of a layer of connective tissue and then endothelial tissue.

Internally, the heart has a right and left side, separated by the interventricular **septum** (fig. 12.2). The heart has four chambers: two upper, thin-walled **atria** (singular, atrium), which are sometimes called auricles, and two lower, thick-walled **ventricles.** The atria are much smaller than the strong, muscular ventricles.

The heart also has valves that direct the flow of blood and prevent a backflow. The valves that lie between the atria and ventricles are called the **atrioventricular valves.** The valves are supported by strong fibrous strings called *chordae tendineae.* The chordae tendineae are attached to the *papillary muscles* which project inward from the ventricular walls. They support the valves and prevent them from inverting. The atrioventricular valve on the right side is called the *tricuspid valve* because it has three leaflets, or flaps; and the valve on the left side is called the bicuspid, or *mitral valve* because it has two flaps. There are also **semilunar valves,** which resemble half moons, between the ventricles and their attached vessels. The semilunar valve on the right side of the heart is the pulmonary semilunar valve and that on the left side is the aortic semilunar valve.

Humans have a four-chambered heart (two atria and two ventricles), in which the right side is separated from the left by a septum.

Figure 12.1   The heart lies in the thoracic cavity, between the lungs and surrounded by the pericardial sac.

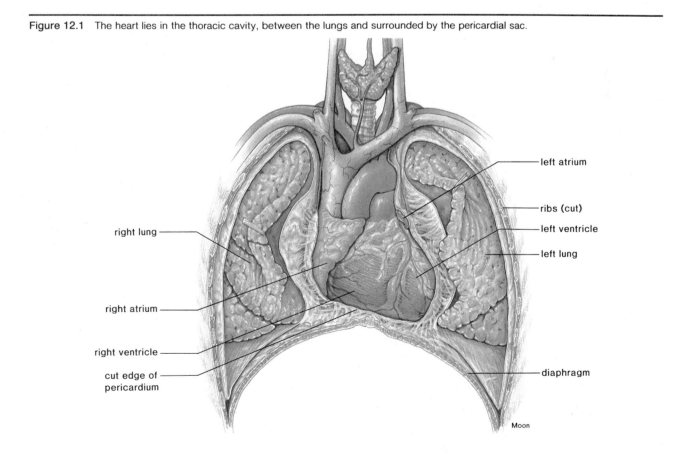

right lung

right atrium

right ventricle

cut edge of pericardium

left atrium

ribs (cut)

left ventricle

left lung

diaphragm

Moon

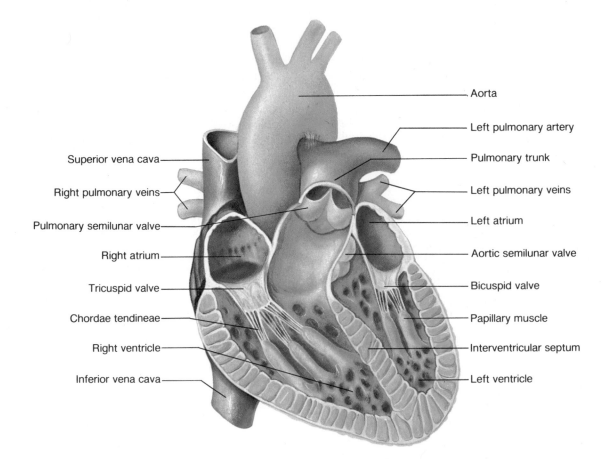

Double Pump

The right side of the heart sends blood through the lungs, and the left side sends blood throughout the body (fig. 12.3). Therefore, there are actually two circular paths (circuits) of the blood: (1) from the heart to the lungs and back to the heart, and (2) from the heart to the body and back to the heart. The right side of the heart is a pump for the first of these circuits, and the left side of the heart is a pump for the second; thus, the heart is a double pump. Since the left ventricle has the harder job because it pumps blood to all of the body, its walls are much thicker than those of the right ventricle.

Path of Blood in the Heart

It is possible to trace the path of blood through the heart in the following manner (figs. 12.2 and 12.3). Deoxygenated blood (low in oxygen and high in carbon dioxide) enters the right atrium from the **superior** and **inferior venae cavae,** the largest veins in the body. Contraction of the right atrium forces the blood through the tricuspid valve to the right ventricle. The right ventricle pumps it through the pulmonary semilunar valve, which allows blood to enter the pulmonary trunk. The

Figure 12.3 Diagram of pulmonary and systemic circuits. The blue-colored vessels carry deoxygenated blood, while the red-colored vessels carry oxygenated blood. Notice that the blood cannot move from the right side of the heart to the left side without passing through the lungs.

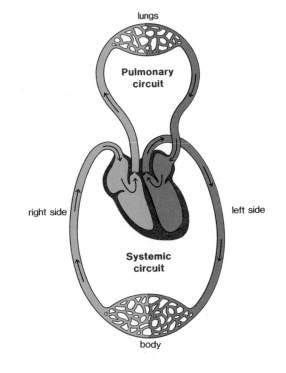

Figure 12.4 Stages in the cardiac cycle. *a*. When the heart is relaxed, both atria and ventricles are filling with blood. *b*. When the atria contract, the ventricles are relaxed and filling with blood. *c*. When the ventricles contract, the atrioventricular valves are closed, the semilunar valves are open, and blood is pumped into the pulmonary artery and aorta.

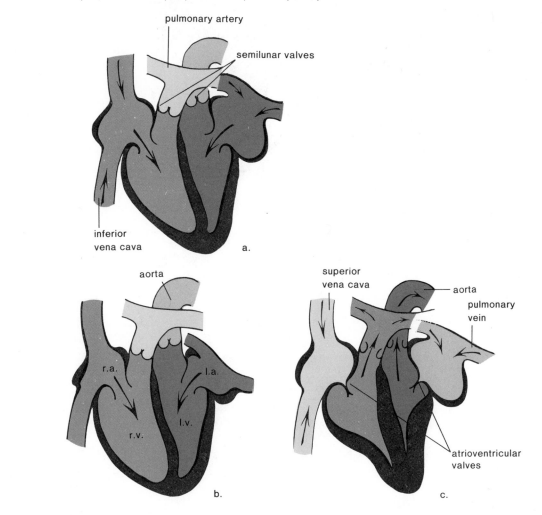

pulmonary trunk divides into the **pulmonary arteries,** which take blood to the lungs. From the lungs, oxygenated blood (high in oxygen and low in carbon dioxide) enters the left atrium from the **pulmonary veins.** Contraction of the left atrium forces blood through the bicuspid valve into the left ventricle. The left ventricle pumps it through the aortic semilunar valve into the **aorta,** the largest artery in the body. The aorta sends blood to all body tissues. Notice that deoxygenated blood never mixes with oxygenated blood and that blood must pass through the lungs before entering the left side of the heart.

The right side of the heart pumps blood to the lungs, and the left side pumps blood to the tissues.

Heartbeat

Cardiac Cycle

From this description of the path of blood through the heart, it might seem that the right and left side of the heart beat independently of one another, but actually they contract together. First, the two atria contract simultaneously; then the two ventricles contract at the same time. The word **systole** refers to contraction of heart muscle, and the word **diastole** refers to relaxation of heart muscle; therefore, atrial systole is followed by ventricular systole. The heart contracts, or beats, about seventy times a minute and each heartbeat lasts about 0.85 second. Each heartbeat, or *cardiac cycle* (fig. 12.4), consists of the following elements:

Time	Atria	Ventricles
0.15 sec.	Systole	Diastole
0.30 sec.	Diastole	Systole
0.40 sec.	Diastole	Diastole

This shows that while the atria contract, the ventricles relax, and vice versa; and that all chambers rest at the same time for 0.40 second. The short systole of the atria is appropriate since the atria send blood only into the ventricles. It is the muscular ventricles that actually

Figure 12.5   Heart valves. *a.* Close-up view of closed semilunar valves. *b.* Drawing showing the relative positions of all the valves.

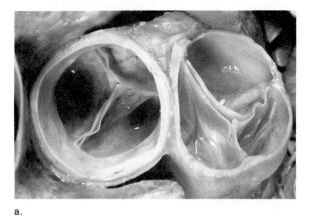

a.

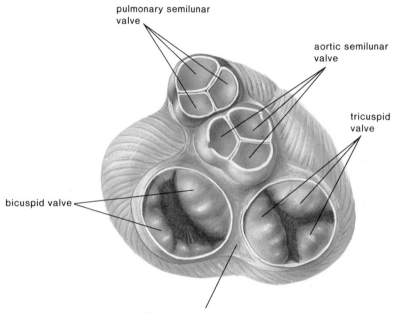

pulmonary semilunar valve

aortic semilunar valve

tricuspid valve

bicuspid valve

fibrous connective tissue

b.

pump blood out into the circulatory system proper. When the word *systole* is used alone, it usually refers to the left ventricular systole.

The heartbeat is divided into three phases: first, the atria contract and then, the ventricles contract. (When the atria are in systole, the ventricles are in diastole, and vice versa.) Finally, all chambers are in diastole.

Heart Sounds

When the heart beats, the familiar lub-DUPP sound may be heard as the valves of the heart close. The lub is caused by vibrations of the heart when the atrio-

ventricular valves close, and the DUPP is heard when vibrations occur due to the closing of the semilunar valves (fig. 12.5). Heart murmurs, or a slight slush sound after the lub, are often due to ineffective valves that allow blood to pass back into the atria after the atrioventricular valves have closed. Rheumatic fever resulting from a strep infection is one cause of a faulty valve, particularly the mitral valve. *Mitral stenosis* is a narrowing of the opening of the mitral valve. If operative procedures are unable to open and/or restructure the valve, it may be replaced by an artificial valve.

The heart sounds are due to the closing of the heart valves.

Table 12.1 Valves of the Heart

Valve	Location	Function	Valve	Location	Function
Tricuspid valve	Right atrioventricular valve	Prevents blood from moving from right ventricle into right atrium	Bicuspid (mitral) valve	Left atrioventricular valve	Prevents blood from moving from left ventricle into left atrium
Pulmonary semilunar valve	Entrance to pulmonary trunk	Prevents blood from moving from pulmonary trunk into right ventricle	Aortic semilunar valve	Entrance to aorta	Prevents blood from moving from aorta into left ventricle

From John W. Hole, Jr., *Human Anatomy and Physiology*, 5th ed. Copyright © 1990 Wm. C. Brown Publishers, Dubuque, Iowa. All Rights Reserved. Reprinted by permission.

Cardiac Conduction System

The beat of the heart is *intrinsic,* meaning the heart will beat independently of any nervous stimulation. In fact, it is possible to remove the heart of a small animal, such as a frog's heart, and watch it undergo contraction in a petri dish. The reason for this lies in the fact that there is a unique type of tissue called nodal tissue, with both muscular and nervous characteristics, located in two regions of the heart. The first of these, the **SA (sinoatrial) node,** is found in the upper dorsal wall of the right atrium; the other, the **AV (atrioventricular) node,** is found in the base of the right atrium very near the septum (fig. 12.6). The SA node, or the pacemaker, initiates the heartbeat and automatically sends out an excitation impulse every 0.85 second to cause the atria to contract. After the impulse reaches the AV node, it passes into a group of large fibers called the *AV bundle (bundle of His)* and, thereafter, spreads out by way of the **Purkinje fibers.** These fibers lie within the myocardium and signal the ventricles to contract.

The SA node is called the **pacemaker** because it usually keeps the heartbeat regular. If a person is suffering from heart block, the heart is not beating regularly due to a fault in the cardiac conduction system. To correct heart block, it is possible to implant in the body an artificial pacemaker that automatically gives an electric shock to the heart every 0.85 second. This causes the heart to beat regularly again.

On occasion, the heart may beat slower or faster than usual. A condition in which there are fewer than 60 heartbeats per minute is called *bradycardia* and one in which there are more than 100 heartbeats per minute is called *tachycardia.* The heart is in *fibrillation* when it beats rapidly but the contractions are uncoordinated. This is apt to occur when a person is suffering from a heart attack. The heart can sometimes be defibrillated by briefly applying a strong electrical current to the chest.

Electrocardiogram (EKG) With the contraction of any muscle, including the myocardium, electrolyte changes occur that can be detected by electrical recording devices. Therefore, it is possible to study the heartbeat by recording voltage changes that occur when the heart contracts. (Voltage, which in this case is measured in millivolts, is the difference in polarity between two electrodes attached to the body.) The record that results is called an **electrocardiogram** (fig. 12.6b), which clearly shows an atrial phase and a ventricular phase. The first wave in the electrocardiogram, called the P wave, represents the excitation and contraction of the atria. The second wave, or the QRS wave, occurs during ventricular excitation and contraction. The third, or T, wave is caused by the recovery of the ventricles. An examination of the electrocardiogram indicates whether the heartbeat has a normal or irregular pattern.

The conduction system of the heart includes the SA node, the AV node, and the Purkinje fibers. The EKG can be used to determine if the conduction system and, therefore, the beat of the heart is regular.

Control of the Heartbeat

The rate of the heartbeat is also under nervous control. There is a cardiac center in the medulla oblongata (p. 133) of the brain, which can alter the beat of the heart by way of the autonomic nervous system. The parasympathetic system causes the heartbeat to slow down and the sympathetic system increases the heartbeat. Various factors, such as the relative need for oxygen or the blood pressure level, determine which of these systems becomes activated.

Other factors, too, influence the heartbeat rate. A cold temperature slows the heartbeat rate, which is why the body temperature of persons undergoing open-heart surgery is lowered. The correct plasma concentration

Figure 12.6 Control of the heart cycle. *a.* The SA node sends out a stimulus that causes the atria to contract. When this stimulus reaches the AV node, it signals the ventricles to contract by way of the Purkinje fibers. *b.* A normal EKG indicates that the heart is functioning properly. The P wave indicates that the atria have contracted; the QRS wave indicates that the ventricles have contracted; and the T wave indicates that the ventricles are recovering from contraction.

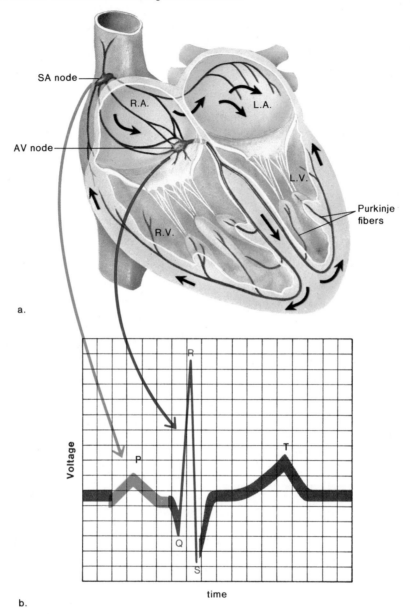

of electrolytes like potassium (K$^+$) and calcium (Ca^{++}) are important to a regular heartbeat.

The heart rate is regulated largely by the autonomic nervous system.

Vascular System

Blood Vessels

The blood vessels are of three types: arteries, capillaries, and veins (fig. 12.7).

Arteries and Arterioles

Arteries, which take blood away from the heart, have thick walls (fig. 12.8) because, in addition to an inner endothelium layer (tunica intima) and an outer connective tissue layer (tunica externa), they have a thick middle layer (tunica media) of elastic and muscle fibers. The elastic fibers enable an artery to expand and accommodate the sudden increase in blood volume that results after each heartbeat. Arterial walls are sometimes so thick that the walls themselves are supplied with blood vessels. The **arterioles** are small arteries just visible to the naked eye. The middle layer of these vessels has some elastic tissue, but is composed mostly of smooth muscle whose fibers encircle the arteriole. The

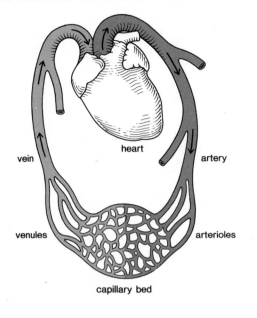

Figure 12.7 Diagram illustrating the path of blood. Blood leaving the heart moves from an artery to arterioles to capillaries to venules and then returns to the heart by way of a vein. Thus, arteries are vessels that take blood away from the heart, and veins are vessels that return blood to the heart.

vein

heart

artery

venules

arterioles

capillary bed

contraction of the smooth muscle cells is under involuntary control by the autonomic nervous system. If the muscle fibers contract, the lumen of the arteriole gets smaller; if the fibers relax, the lumen of the arteriole enlarges. Whether arterioles are constricted or dilated affects blood pressure. The greater the number of vessels dilated, the lower the blood pressure.

Capillaries

Arterioles branch into small vessels called **capillaries**. Each one is an extremely narrow, microscopic tube with a wall composed of only one layer of endothelial cells. *Capillary beds* (a network of many capillaries) are present in all regions of the body; consequently, a cut to any body tissue draws blood. The capillaries are an important part of the circulatory system because an exchange of nutrient and waste molecules takes place across their thin walls. Oxygen and glucose diffuse out of a capillary into the tissue fluid that surrounds cells, and carbon dioxide and ammonia diffuse into the capillary (fig. 11.7). Since it is the capillaries that serve the needs of the cells, the heart and other vessels of the circulatory system can be considered a means by which blood is conducted to and from the capillaries.

Figure 12.8 A comparison of artery and vein structure shows that arteries have strong walls, while veins have weak walls. This is largely due to the difference in size of the tunica media, which is composed of smooth muscle and connective tissue.

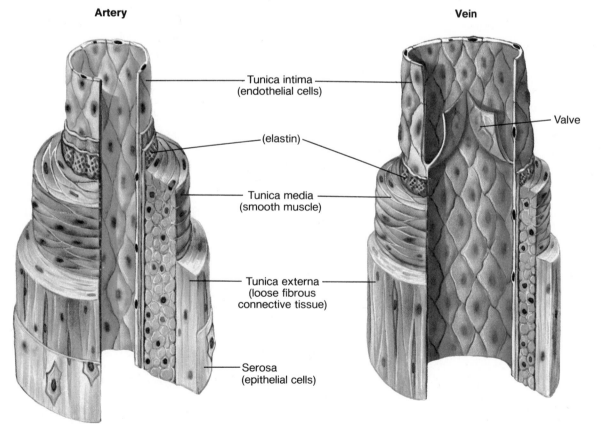

Artery

Vein

Tunica intima
(endothelial cells)

Valve

(elastin)

Tunica media
(smooth muscle)

Tunica externa
(loose fibrous
connective tissue)

Serosa
(epithelial cells)

Figure 12.9 Anatomy of a capillary bed. Capillary beds form a matrix of vessels that lie between an arteriole and a venule. *a.* Sphincter muscles are found at the junctions between an arteriole and capillaries. When these are contracted, the capillary bed is closed. Blood moves from the arteriole to the venule by way of a shunt. *b.* When a capillary bed is open, blood moves freely in the matrix of vessels making up the bed. If all capillary beds were open at the same time, an individual would suffer very severe low blood pressure.

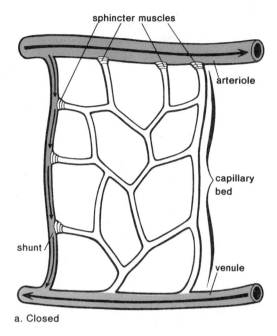

a. Closed

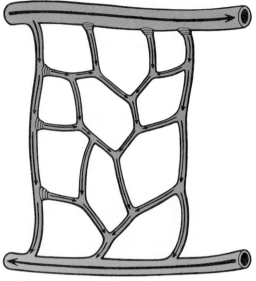

b. Open

Not all capillary beds (fig. 12.9) are open or in use at the same time. After eating, the capillary beds of the digestive tract are usually open; during muscular exercise, the capillary beds of the skeletal muscles are open. Most capillary beds have a shunt that allows blood to move directly from the arteriole to the venule when the capillary bed is closed. There are sphincter muscles that encircle the entrance to each capillary. These are constricted, preventing blood from entering the capillaries, when the bed is closed and relaxed when the bed is open. As would be expected, the larger the number of capillary beds open, the lower the blood pressure.

Veins

Veins and smaller vessels called **venules** take blood from the capillary beds to the heart. First, the venules drain the blood from the capillaries and then join together to form a vein. The wall of a vein is much thinner than that of an artery because the middle layer of muscle and elastic fibers is poorly developed (fig. 12.8). Within some veins, especially in the major veins of the arms and legs, there are **valves** (fig. 12.19) that allow blood to flow only toward the heart when they are open and prevent the backward flow of blood when they are closed.

At any given time, more than half of the total blood volume is found in the veins and venules. If a loss of blood occurs, for example, due to hemorrhaging, sympathetic nervous stimulation causes the veins to constrict, providing more blood to the rest of the body. In this way, the veins act as a blood reservoir.

Arteries and arterioles carry blood away from the heart, veins and venules carry blood to the heart, and capillaries join arterioles to venules.

Varicose Veins and Phlebitis *Varicose veins* are abnormal and irregular dilations in superficial (near the surface) veins, particularly those in the lower legs. Varicose veins in the rectum, however, are commonly called piles, or more properly, *hemorrhoids*. Varicose veins develop when the valves of the veins become weak and ineffective due to a backward pressure of the blood. The problem can be aggravated when venous blood flow is obstructed by crossing the legs or by sitting in a chair so that its edge presses against the back of the knees.

Phlebitis, or inflammation of a vein, is a more serious condition, particularly when a deep vein is involved. Blood in the inflamed vessel may clot, in which case *thromboembolism* (p. 205) occurs. An embolus that originates in a systemic vein may eventually come to rest in a pulmonary arteriole, blocking circulation through the lungs. This condition, termed *pulmonary embolism,* can result in death.

Veins have weak walls, and this occasionally leads to medical disorders.

Path of Circulation

The vascular system, which is diagrammatically represented in figure 12.10, can be divided into two circuits: the **pulmonary circuit,** which circulates blood

Figure 12.10   Blood vessels in the pulmonary and systemic circulatory circuits. The veins (blue-colored) carry deoxygenated blood, and the arteries (red-colored) carry oxygenated blood; the arrows indicate the flow of blood. Lymph vessels collect excess tissue fluid and return it to the subclavian veins in the shoulders.

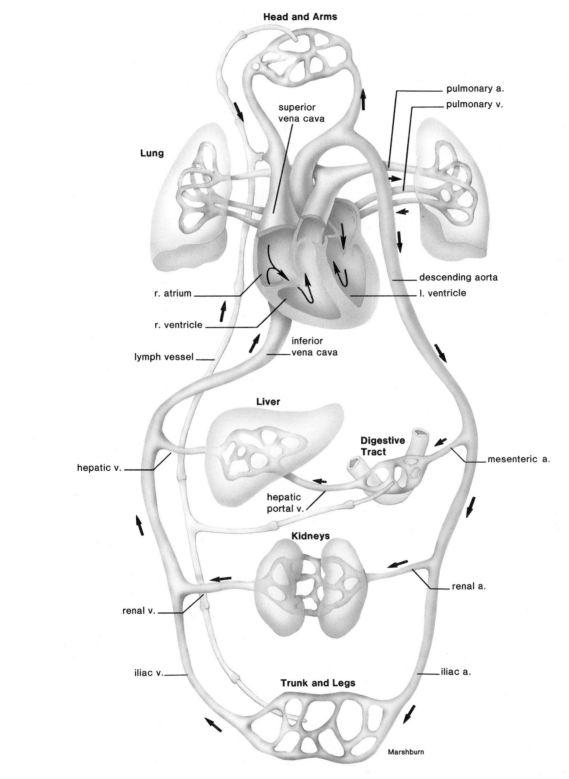

Head and Arms

pulmonary a.

pulmonary v.

superior
vena cava

Lung

descending aorta

r. atrium

l. ventricle

r. ventricle

inferior
vena cava

lymph vessel

Liver

**Digestive
Tract**

hepatic v.

mesenteric a.

hepatic
portal v.

Kidneys

renal a.

renal v.

iliac v.

iliac a.

Trunk and Legs

Marshburn

through the lungs, and the **systemic circuit,** which serves the needs of the body's tissues.

Pulmonary Circuit

The path of blood through the lungs can be traced as follows. Blood from all regions of the body first collects in the right atrium and then passes into the right ventricle, which pumps it into the pulmonary trunk. The pulmonary trunk divides into the **pulmonary arteries,** which divide up into the arterioles of the lungs. The arterioles take blood to the pulmonary capillaries, where carbon dioxide and oxygen are exchanged. The blood then enters the pulmonary venules and flows through the **pulmonary veins** back to the left atrium. Since the blood in the pulmonary arteries is deoxygenated but the blood in the pulmonary veins is oxygenated, it is not correct to say that all arteries carry blood that is high in oxygen and all veins carry blood that is low in oxygen. It is just the reverse in the pulmonary system.

The pulmonary arteries take deoxygenated blood to the lungs, and the pulmonary veins return oxygenated blood to the heart.

Systemic Circuit

The systemic circuit includes all of the other arteries and veins of the body. The largest artery in the systemic circuit is the **aorta,** and the largest veins are the **superior** (anterior) and **inferior** (posterior) venae cavae. The superior vena cava collects blood from the head, chest, and arms, and the inferior vena cava collects blood from the lower body regions. Both enter the right atrium. The aorta and the venae cava serve as the major pathways for blood in the systemic system.

The path of systemic blood to any organ in the body begins in the left ventricle, which pumps blood into the aorta. Branches from the aorta go to the major body regions and organs. To trace the path of blood to any organ in the body, you need only mention the aorta, the proper branch of the aorta, the organ, and the returning vein to the vena cava. In many instances, the artery and vein that serve the same organ have the same name. For example, the path of blood to the kidneys may be traced as follows: left ventricle–aorta–renal artery–renal arterioles, capillaries, venules–renal vein–inferior vena cava–right atrium. In the systemic circuit, unlike the pulmonary system, arteries contain oxygenated blood and appear a bright red, but veins contain deoxygenated blood and appear a purplish color.

The systemic circuit takes blood from the left ventricle of the heart to the right atrium of the heart. It serves the body proper.

The Major Systemic Arteries If you examine the path of the aorta (fig. 12.11) after it leaves the heart, you can tell why it is divided into the *ascending aorta,* the *aortic arch,* and the *descending aorta.* There are three major arteries that branch off the aortic arch: the **brachiocephalic** artery, the **left common carotid** artery, and the **left subclavian.** The brachiocephalic soon divides into the **right common carotid** and the **right subclavian** arteries. Therefore, it is possible to easily make out that these blood vessels serve the head and arms:

Head: right and left common carotids

Arms: right and left subclavian

The descending aorta is divided into the *thoracic aorta,* which gives off branches to the organs within the thoracic cavity, and the *abdominal aorta,* which gives off branches to the organs in the abdominal cavity. These blood vessels are depicted in figure 12.11 and listed in table 12.2.

The descending aorta ends when it divides into the **common iliac** arteries that branch into the **internal iliac** artery and the **external iliac** artery. The internal iliac serves the pelvic organs and the external iliac serves the legs.

All the arteries in the systemic system can eventually be traced from the aorta.

Major Systemic Veins Figure 12.12 shows the major veins of the body. The external and internal jugular veins drain blood from the brain, head, and neck. The external jugular veins enter the subclavian veins and they, along the internal jugular veins, enter the brachiocephalic veins. These vessels merge to give rise to the superior vena cava.

In the abdominal cavity, we have to remember that the hepatic portal vein receives blood from the abdominal viscera (fig. 12.10). The hepatic portal vein enters the liver. Emerging from the liver, the hepatic veins enter the inferior vena cava.

In the pelvic region, veins from the various organs enter the internal iliacs while the veins from the legs enter the external iliac veins. The internal and external iliacs become the common iliac veins that merge to produce the inferior vena cava.

Figure 12.11   Major systemic arteries (a. = artery).

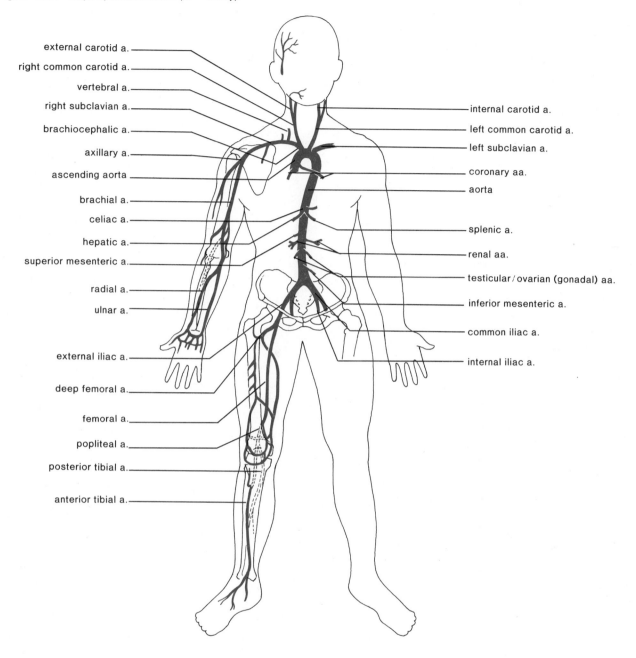

external carotid a.

right common carotid a.

vertebral a.

right subclavian a.

brachiocephalic a.

axillary a.

ascending aorta

brachial a.

celiac a.

hepatic a.

superior mesenteric a.

radial a.

ulnar a.

external iliac a.

deep femoral a.

femoral a.

popliteal a.

posterior tibial a.

anterior tibial a.

internal carotid a.

left common carotid a.

left subclavian a.

coronary aa.

aorta

splenic a.

renal aa.

testicular / ovarian (gonadal) aa.

inferior mesenteric a.

common iliac a.

internal iliac a.

Figure 12.12 Major systemic veins. Left leg shows superficial veins. Right leg shows deep veins (v. = vein).

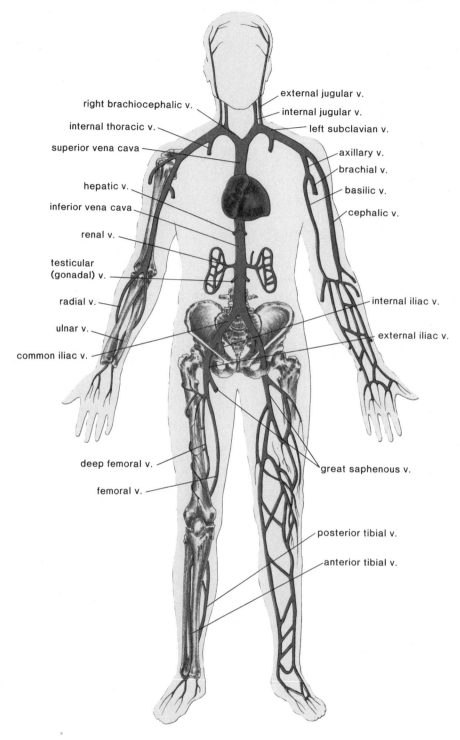

right brachiocephalic v.
internal thoracic v.
superior vena cava
hepatic v.
inferior vena cava
renal v.
testicular (gonadal) v.
radial v.
ulnar v.
common iliac v.

external jugular v.
internal jugular v.
left subclavian v.
axillary v.
brachial v.
basilic v.
cephalic v.

internal iliac v.
external iliac v.

deep femoral v.
femoral v.

great saphenous v.

posterior tibial v.
anterior tibial v.

Table 12.2 Aorta and Its Principal Branches

Portion of Aorta	Major Branch	Regions Supplied
Ascending aorta	Coronary arteries	Heart
Arch of aorta	Brachiocephalic	
	Right common carotid	Right side of head
	Right subclavian	Right arm
	Left common carotid	Left side of head
	Left subclavian	Left arm
Descending aorta		
Thoracic aorta	Bronchial artery	Bronchi
	Esophageal artery	Esophagus
	Intercostal	Thoracic wall
Abdominal aorta	Celiac artery	Stomach, spleen, and liver
	Superior mesenteric	Small and large intestine (ascending and transverse colons)
	Renal artery	Kidney
	Gonadal artery	Ovary or testis
	Inferior mesenteric	Lower digestive system (transverse and descending colons, and rectum)
	Common iliac	Pelvic organs and legs

All the veins in the systemic system can eventually be traced to the venae cavae.

Vital Systemic Circulatory Routes

Blood Supply to the Brain The brain is supplied with arterial blood in arteries (vertebral and internal carotids) that give off branches which join to form a circle in the region of the pituitary gland. This circle is called the **circle of Willis** (fig. 12.13). The value of having the blood vessels join in this way is that if one becomes blocked, there are still three other routes by which the brain can receive blood.

The circulation to the brain includes the circle of Willis, a feature that protects the brain from reduced blood supply.

Blood Supply to the Heart The **coronary arteries** (fig. 12.14), which are a part of the systemic circuit, are extremely important arteries because they serve the heart muscle itself. (The heart is not nourished by the blood in its chambers.) The coronary arteries arise from the aorta just above the aortic semilunar valve. They lie on the exterior surface of the heart, where they branch off in various directions into smaller arteries and then arterioles. The coronary capillary beds join to form venules. The venules converge into the cardiac veins, which empty into the right atrium. Although the coronary arteries receive blood under high pressure, they have a very small diameter and may become blocked, as discussed on page 205.

The circulation to the heart is dependent upon the proper functioning of the coronary arteries.

Blood Supply to the Liver The body has an important portal system, the **hepatic portal system** (fig. 12.15) that takes blood from the stomach, intestines, and other organs to the liver. A portal system is one that begins and ends in capillaries; in this case, the first set of capillaries occurs at the villi of the small intestine, and the second occurs in the liver. Blood passes from the capillaries of the villi into venules and then into veins that join to form the hepatic portal vein, a vessel that enters the liver. Emerging from the liver, the hepatic veins enter the inferior vena cava.

The hepatic portal system takes blood from the stomach and intestines to the liver.

Features of the Circulatory System

When the left ventricle contracts, the blood is sent out into the aorta under pressure.

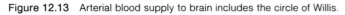

Figure 12.13 Arterial blood supply to brain includes the circle of Willis.

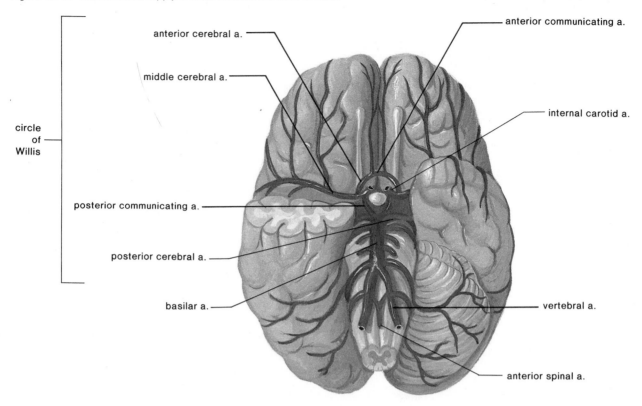

circle of Willis

anterior cerebral a.

middle cerebral a.

posterior communicating a.

posterior cerebral a.

basilar a.

anterior communicating a.

internal carotid a.

vertebral a.

anterior spinal a.

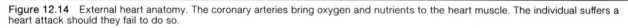

Figure 12.14 External heart anatomy. The coronary arteries bring oxygen and nutrients to the heart muscle. The individual suffers a heart attack should they fail to do so.

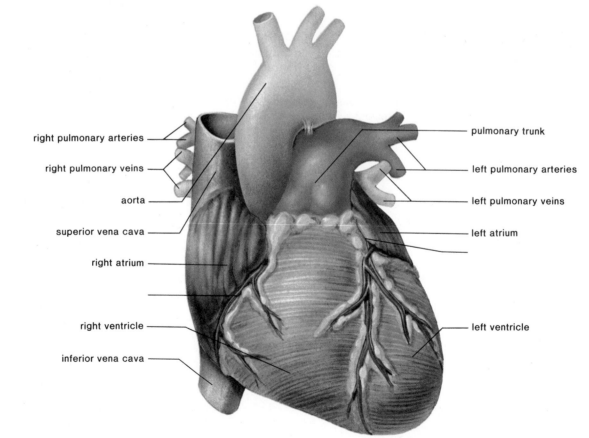

right pulmonary arteries

right pulmonary veins

aorta

superior vena cava

right atrium

right ventricle

inferior vena cava

pulmonary trunk

left pulmonary arteries

left pulmonary veins

left atrium

left ventricle

Figure 12.15 Hepatic portal system. The hepatic portal vein drains blood from vessels serving the intestines and stomach before it enters the liver.

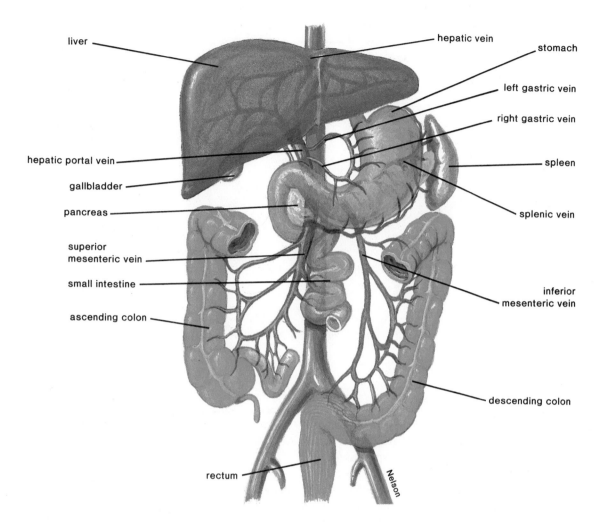

liver

hepatic vein

stomach

left gastric vein

right gastric vein

hepatic portal vein

gallbladder

pancreas

spleen

splenic vein

superior mesenteric vein

small intestine

inferior mesenteric vein

ascending colon

descending colon

rectum

Nelson

Figure 12.16 Sites where the pulse can be taken are shown (a. = artery).

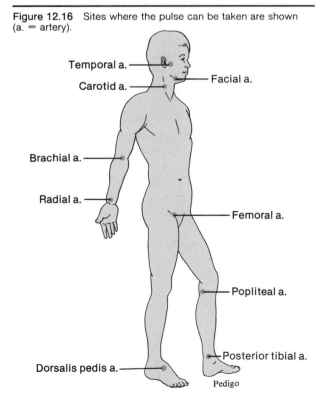

Temporal a.

Carotid a.

Facial a.

Brachial a.

Radial a.

Femoral a.

Popliteal a.

Posterior tibial a.

Dorsalis pedis a.

Pedigo

Pulse

The surge of blood entering the arteries causes their elastic walls to swell, but then they almost immediately recoil. This alternate expanding and recoiling of an arterial wall can be felt as a **pulse** in any artery that runs close to the surface. It is customary to feel the pulse by placing several fingers on the radial artery, which lies near the outer border of the palm side of the wrist. There are several other locations where the pulse can be obtained (fig. 12.16). The pulse rate indicates the rate of the heartbeat because the arterial walls pulse whenever the left ventricle contracts.

The pulse rate indicates the heartbeat rate.

Blood Pressure

Blood pressure is the pressure of the blood against the wall of a blood vessel.

Figure 12.17 Regulation of blood pressure involves these aspects.

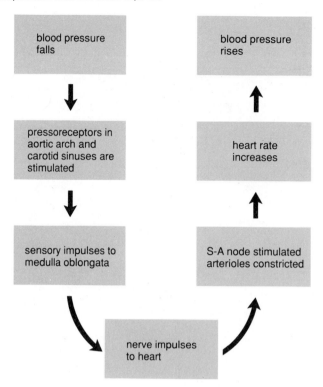

Regulation of Blood Pressure

Two factors affect blood pressure and these are cardiac output and peripheral resistance. *Cardiac output* is the amount of blood leaving the left ventricle per minute. Obviously, when the heart is beating rapidly, it is putting out more blood than when it is beating slowly. *Peripheral resistance* primarily refers to the degree of constriction of the arterioles. When the arterioles are constricted, there is more peripheral resistance than when they are dilated.

There are pressoreceptors (baroreceptors) in the aortic arch and also in the carotid sinuses (enlarged areas of the carotid arteries). If the blood pressure falls as it might when we stand up quickly, these receptors signal centers in the medulla oblongata, and nerve impulses go out that cause the heartbeat to increase and the arterioles to constrict to a greater degree. Now the blood pressure rises (fig. 12.17). On the other hand, if the blood pressure should rise too high, these same receptors signal the cardiac center, and subsequently, the arterioles dilate (due to lack of sympathetic stimulation) and the heartbeat slows. This lowers blood pressure.

Blood pressure is dependent on cardiac output and peripheral resistance, both of which are regulated by the autonomic nervous system.

Variations in Blood Pressure

Blood pressure is usually measured by using a sphygmomanometer. This instrument measures blood pressure in terms of millimeters of mercury. Normal resting blood pressure for a young adult is 120/80; the higher number is the systolic pressure, the pressure recorded in an artery when the left ventricle is contracting and the lower number is the diastolic pressure, the pressure recorded in an artery when the left ventricle is relaxing. It is estimated that about 20 percent of all Americans suffer from **hypertension** or high blood pressure. The reasons for the development of hypertension are various. One possible reason is associated with the secretion of renin by the kidneys (p. 268). The renin-angiotensin-aldostrone system leads to absorption of Na^+ and high blood pressure. The same effect can be brought about directly by an excess intake of salt in the diet.

Figure 12.18 Diagram illustrating how velocity and blood pressure are related to the total cross-sectional area of blood vessels. Capillaries have the greatest cross-sectional area and the least pressure and velocity. Skeletal muscle contraction, not blood pressure, accounts for the velocity of blood in the veins.

Figure 12.19 Skeletal muscle contraction moves blood in veins. *a.* Muscle contraction exerts pressure against vein and blood moves past valve. *b.* Once blood has passed the valve, it cannot slip back.

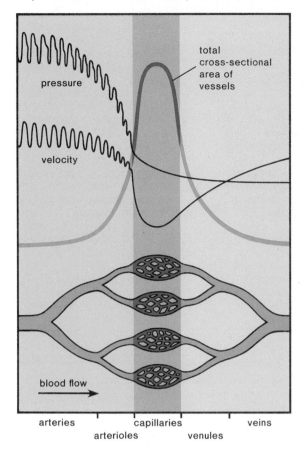

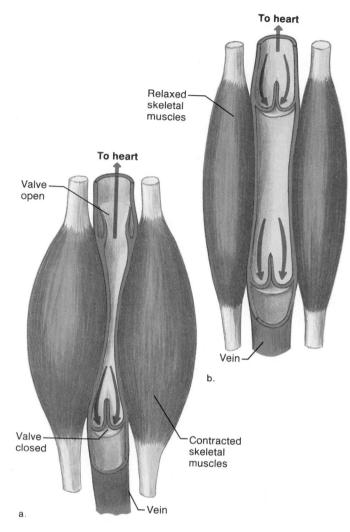

Blood pressure accounts for the movement of blood in the arteries and arterioles. Normal blood pressure (120/80) is the pressure that is measured within the brachial artery of the arm. Blood pressure in other arteries differs from this because blood pressure decreases with distance from the left ventricle. There is higher blood pressure in the arteries than the arterioles, and there is a sharp drop when the blood reaches the capillaries (fig. 12.18). The decrease may be correlated with the increase in the total cross-sectional area of the vessels; there are more arterioles than arteries, and many more capillaries than arterioles. Also, the blood moves much slower through the capillaries than it does in the aorta. This is important because the slow progress allows time for the exchange of molecules between the blood and the tissues.

Movement of blood in the veins is not due to blood pressure. Instead, it is due to skeletal muscle contraction. When skeletal muscles contract, they press against the weak walls of the veins, and this causes the blood to move past a *valve* (fig. 12.19). Once past the valve, the blood will not fall back. Blood pressure and flow gradually increase in the venous system due to a progressive reduction in the cross-sectional area as small venules join to form veins.

Blood pressure causes the flow of blood in the arteries and arterioles; skeletal muscle contraction causes the flow of blood in the venules and veins.

Summary

I. Heart. Humans have a four-chambered heart (two atria and two ventricles, in which the right side is separated from the left by a septum).

A. The right side of the heart pumps blood to the lungs, and the left side pumps blood to the tissues.

B. The heartbeat is divided into three phases: first, the atria contract and then, the ventricles contract. Finally, all chambers are in diastole.

The heart sounds are due to the closing of the heart valves.

The conduction system of the heart includes the SA node, the AV node, and the Purkinje fibers. The ECG allows one to determine if the conduction system and, therefore, the beat of the heart is regular.

The heart rate is regulated largely by the autonomic nervous system.

II. Vascular System

A. Arteries and arterioles carry blood away from the heart, veins and venules carry blood to the heart, and capillaries join arterioles to venules.

B. The pulmonary arteries take deoxygenated blood to the lungs, and the pulmonary veins return oxygenated blood to the heart.

The systemic circuit takes blood from the left ventricle of the heart to the right atrium of the heart. It serves the body proper.

All the arteries in the systemic system can eventually be traced from the aorta. All the veins in the systemic system can eventually be traced to the venae cavae.

The circulation to the brain includes the circle of Willis, a feature that protects the brain from reduced blood supply.

The circulation to the heart is dependent upon the proper functioning of the coronary arteries.

The hepatic portal system takes blood from the stomach and intestines to the liver.

III. Features of the Circulatory System

A. The pulse rate indicates the heartbeat rate.

B. Blood pressure is dependent on cardiac output and peripheral resistance, both of which are regulated by the autonomic nervous system.

Blood pressure steadily decreases from the aorta to the veins.

Blood pressure causes the flow of blood in the arteries and arterioles; skeletal muscle contraction causes the flow of blood in the venules and veins.

Study Questions

1. What types of blood vessels are there? Discuss their structure and function.
2. Trace the path of blood in the pulmonary circuit as it travels from and returns to the heart.
3. Describe the cardiac cycle (using the terms systole and diastole) and explain the heart sounds.
4. Describe an ECG and tell how its components are related to the cardiac cycle.
5. Trace the path of blood from the mesenteric arteries to the aorta, indicating which of the vessels are in the systemic circuit and which are in the pulmonary circuit.
6. What is blood pressure, and why is the average normal arterial blood pressure said to be 120/80?
7. In which type of vessel is blood pressure highest? Lowest? Velocity is lowest in which type vessel and why is it lowest? Why is this beneficial? What factors assist venous return of the blood?

Objective Questions

Fill in the blanks.

1. Arteries are blood vessels that take blood _____ from the heart.
2. When the left ventricle contracts blood enters the _____ .
3. The right side of the heart pumps blood to the _____ .
4. The _____ node is known as the pacemaker.
5. The blood vessels that serve the heart are the _____ arteries and veins.
6. Blood vessels to the brain end in a circular path known as the _____ .
7. The human body contains a portal system that takes blood from the _____ to the _____ .
8. The pressure of blood against the walls of a vessel is termed _____ .
9. Blood moves in arteries due to _____ and in veins due to _____
10. The major blood vessels in the arms are the _____ arteries and veins. Those in the legs are the _____ arteries and veins.

Medical Terminology Reinforcement Exercise

**Pronounce the following terms.
Analyze the meaning by word
parts as dissected:**

1. cryocardioplegia (kir-o-kar''de-o-ple'je-ah)—cryo/cardio/plegia

2. echocardiography (ek''o-kar''de-og'rah-fe)—echo/cardio/graphy

3. percutaneous transluminal coronary angioplasty (per''ku-ta'ne-us trans''loo'mĭ-nal kor'ŏ-na-re an'je-o-plas''te)—per/cutaneous trans/luminal coronary angio/plasty

4. vasoconstriction—(vas''o-kon-strik'shun)—vaso/con/stric/tion

5. valvuloplasty (val'vu-lo-plas''te)—valvulo/plasty

6. arteriosclerosis (ar-te''re-o-sklĕ-ro'sis)—arterio/scler/osis

7. tachycardia (tak''e-kar'de-ah) tachy/cardia

8. antihypertensive (an''tĭ-hi''per-ten'siv)—anti/hyper/tensive

9. arrhythmia (ah-rith'me-ah)—ar/rhythm/ia

10. thromboendarterectomy (throm''bo-end''ar-ter-ek'to-me) thrombo/end/arter/ectomy

13

Lymphatic System

Chapter Outline

Lymphatic System
  Lymphatic Vessels
  Bone Marrow
  Lymph Nodes
  Spleen
  Thymus
Immunity
  General Defense
  Specific Defense
  Immunotherapy

Learning Objectives

After you have studied this chapter, you should be able to:

1. Describe the structure and function of the lymphatic system.
2. Describe the manner in which excess tissue fluid is collected and returned to the cardiovascular system.
3. Describe the structure and function of lymph nodes.
4. Name and describe several disorders associated with the lymph nodes and the lymphatic system.
5. Describe the structure and function of the spleen and thymus.
6. Describe the general defense mechanisms of the body.
7. Contrast antibody-mediated immunity with cell-mediated immunity.
8. Give examples in which the body overdefends itself and examples in which the body underdefends itself.
9. Describe how it is possible to provide a patient with active and passive immunity.

Lymphatic System

The **lymphatic system** consists of lymphatic vessels (fig. 13.1) and the lymphoid organs. The lymphatic system is closely associated with the cardiovascular system because its vessels take up excess tissue fluid and transport it to the bloodstream. The lymphatic system has other functions also. You will recall that the lacteals in the villi of the small intestine were lymphatic vessels which absorbed fat molecules. The lymphoid organs also stores lymphocytes and macrophages—cells that are very important to the body's ability to defend itself from disease.

The lymphatic system (1) collects excess tissue fluid, (2) absorbs fat molecules in the intestines, and (3) plays a major role in the body's defense against disease.

Figure 13.1   Lymphatic vessels. The lymphatic vessels drains excess fluid from the tissues and returns it to the cardiovascular system. The thoracic duct begins at the chyle cistern, which receives lymph from the right and left lumbar lymphatic trunks and from the intestinal trunk. The enlargement shows the lymphatic vessels, called lacteals, which are present in the intestinal villi.

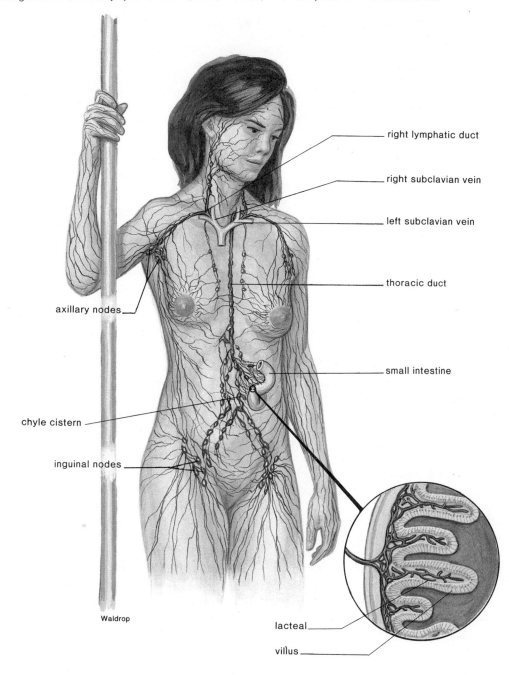

right lymphatic duct

right subclavian vein

left subclavian vein

thoracic duct

axillary nodes

small intestine

chyle cistern

inguinal nodes

Waldrop

lacteal

villus

Figure 13.2 Lymphatic capillaries lie close to cardiovascular capillaries. Arrows indicate that tissue fluid is formed as water leaves blood capillaries and that thereafter some of it is taken up by lymphatic capillaries.

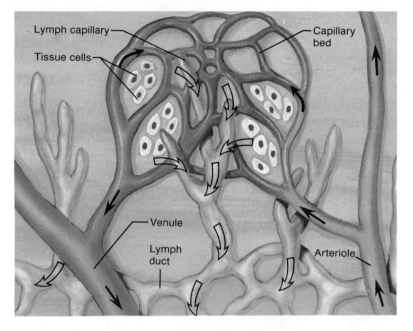

Lymphatic Vessels

The lymphatic system is quite extensive and every region of the body is richly supplied with lymph capillaries. The lymphatic vessels have a construction similar to cardiovascular veins, including the presence of valves. Also, movement of lymph within these vessels is dependent upon skeletal muscle contraction. When the muscles contract, the lymph is squeezed past a valve which closes preventing the lymph from flowing backwards.

The lymphatic system begins with lymphatic capillaries that lie near blood capillaries. They take up fluid that has diffused from the capillaries and has not been reabsorbed by them (fig. 13.2). Once tissue fluid enters the lymphatic vessels, it is called **lymph**. The lymphatic capillaries merge to form lymphatic vessels that enter a particular lymphatic trunk. Each lymphatic trunk serves a relatively large region of the body. For example, the left lumbar trunk drains lymph from the lymphatic vessels of the left legs, lower left abdominal wall, and left pelvic organs. Finally, each trunk enters one of two thoracic ducts.

The left thoracic duct is much larger than the right thoracic duct. It originates in the abdomen as the cisterna chyli, a dilated portion at the level of the second lumbar vertebrae, and travels with the aorta up through the diaphragm. In the thorax, it passes in front of the vertebral column and enters the left subclavian vein near the junction of the left jugular vein. The left thoracic duct receives lymph from the lower extremities, the abdomen, the left arm, and the left side of the head and neck.

The right thoracic duct receives lymph from the right arm and the right side of the head and neck. It begins in the right thorax and enters the right subclavian vein near the junction of the right jugular vein.

The lymphatic system is a one-way system. Lymph flows from a capillary to a lymphatic vessel to a lymphatic trunk and, finally, to a lymphatic duct, which enters a subclavian vein.

Bone Marrow

Bone marrow is a lymphoid organ (fig. 13.3) because red marrow produces lymphocytes and other blood cells necessary to immunity. In the adult, red bone marrow is present only in the bones of the skull, the sternum, the ribs, the clavicle, the spinal column, and the ends of the femur and the humerus. Red bone marrow consists of a network of connective tissue fibers called reticular fibers that are produced by cells called reticular cells. These and the developing blood cells are packed about thin-walled venous sinuses. Differentiated blood cells enter the bloodstream at these sinuses.

Red bone marrow produces lymphocytes and other cells necessary to the development of immunity.

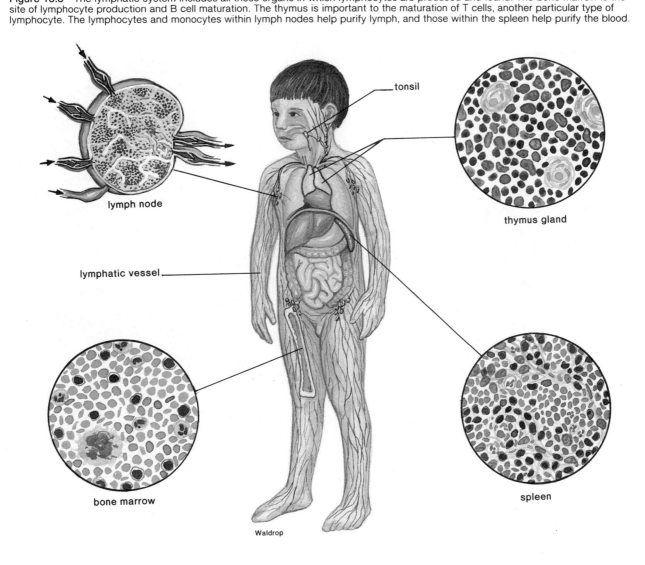

Figure 13.3 The lymphatic system includes all those organs in which lymphocytes are produced and found. The bone marrow is the site of lymphocyte production and B cell maturation. The thymus is important to the maturation of T cells, another particular type of lymphocyte. The lymphocytes and monocytes within lymph nodes help purify lymph, and those within the spleen help purify the blood.

tonsil

lymph node

thymus gland

lymphatic vessel

bone marrow

spleen

Waldrop

Lymph Nodes

At certain points along lymph veins, there occur small (about 2.5 cm) ovoid, or round, structures called **lymph nodes** (fig. 13.4). A lymph node has a fibrous capsule of connective tissue that dips down into the node and divides it into nodules. Each nodule contains a sinus (open space) filled with many lymphocytes and macrophages.

Lymph enters a lymph node by way of afferent lymphatic vessels and exits by way of efferent lymphatic vessels. In between, the lymph passes through the sinuses of the node where it is purified of infectious organisms and any other debris.

While nodules usually occur within lymph nodes, they can also occur singly or in groups. The *tonsils* are composed of partly encapsulated lymph nodules. There are also nodules within the intestinal wall called *Peyer's patches*.

The lymph nodes occur in groups in certain regions of the body. For example, there are the inguinal nodes of the lower extremities; lumbar nodes of the pelvic region; axillary nodes of the upper extremities; and cervical nodes of the neck.

Lymph nodes are divided into sinus-containing lobules where the lymph is cleansed by phagocytes.

Clinical Conditions

Lymphadenitis   When an infection sets in, there may be swelling and tenderness of the lymph nodes nearby. This is called **lymphadenitis.** If the infection is not contained, then lymphangitis may result. Red streaks can be seen through the skin and this indicates that the infection may spread even to the blood.

Figure 13.4 Anatomy of a lymph node. Lymph flows through a node in the direction indicated by the arrows and is facilitated by the placement of valves. As the lymph flows through the sinuses of the node, it is purified and filtered.

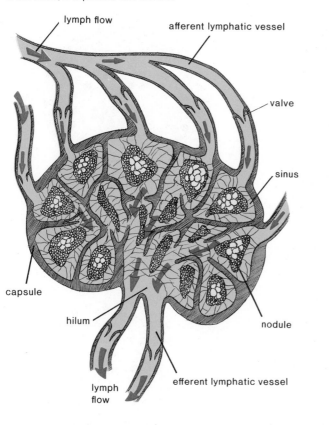

lymph flow

afferent lymphatic vessel

valve

sinus

capsule

hilum

nodule

lymph flow

efferent lymphatic vessel

Figure 13.5 The spleen has white pulp and red pulp. White pulp contains lymphocytes and macrophages. The red pulp contains red blood cells in addition to lymphocytes and macrophages. Blood is cleansed as it passes through the spleen.

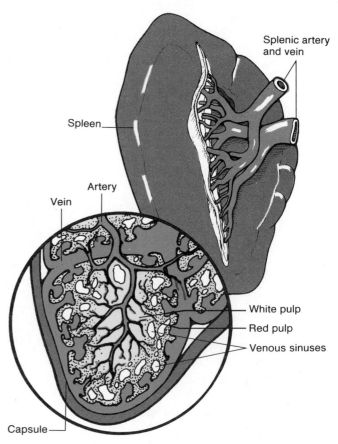

Splenic artery and vein

Spleen

Vein

Artery

White pulp

Red pulp

Venous sinuses

Capsule

Lymphoma Cancer within lymphoid tissue is called a **lymphoma.** In *Hodgkin's disease,* swollen lymph nodes in the neck lead to an involvement of the spleen, liver, and bone marrow. Prognosis is good, however, if Hodgkin's disease is caught early.

Metastasis **Metastasis** occurs when cancer cells break loose and relocate in some other area of the body. Cancer cells, for example, from a breast tumor may enter a lymphatic capillary along with tissue fluid and then travel to a lymph node where another cancerous growth begins. This is why nearby lymph nodes are often removed as a part of breast cancer surgery.

Edema If there is a blockage of a lymphatic vessel or if lymph nodes have been removed as in some cancer operations, then the excess tissue fluid is not removed and localized swelling occurs. This is called **edema.** Edema can also be due to a low osmotic pressure of the blood, in which more tissue fluid than usual forms, overwhelming the ability of the lymphatic vessels to take it up.

Spleen

The **spleen** (fig. 13.5) is located in the upper-left quadrant of the peritoneal cavity just beneath the diaphragm. The spleen is constructed as is a lymph node. Outer connective tissue divides the organ into lobules, which contain sinuses; but in the spleen, the sinuses are filled with blood instead of lymph. Especially since the blood vessels of the spleen can expand, it serves as a blood reservoir that makes blood available in times of low pressure or when the body needs extra oxygen carried in the blood.

A spleen nodule contains red pulp and white pulp. Red pulp contains red blood cells and lymphocytes and macrophages. The white pulp contains only lymphocytes and macrophages. Both types of pulp help purify the blood that passes through the spleen.

The spleen is divided into sinus-containing lobules where the blood is cleansed by lymphocytes and macrophages.

Figure 13.6 The complement system is a number of proteins always present in the plasma. When activated, some of these form pores in bacterial membranes, allowing fluids and salts to enter until the cell eventually bursts.

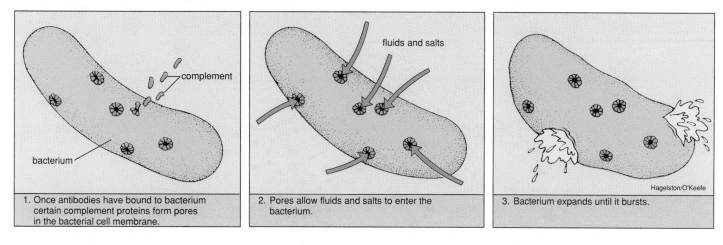

Hagelston/O'Keefe

1. Once antibodies have bound to bacterium certain complement proteins form pores in the bacterial cell membrane.

2. Pores allow fluids and salts to enter the bacterium.

3. Bacterium expands until it bursts.

Thymus

The **thymus** is located along the trachea behind the sternum in the upper thoracic cavity. This gland varies in size, but it is larger in children than adults. The thymus is also divided into lobules by connective tissue. Here, there are lymphocytes from bone marrow, which are destined to become T cells. The function of T cells is discussed in the next section.

The thymus is believed to secrete a hormone called thymosin. *Thymosin* stimulates the immune system to combat disease-causing agents.

The thymus is divided into lobules where lymphocytes called T (for thymus) cells are produced.

Immunity

The body is prepared to protect itself from foreign substances and cells, including infectious microbes. The first line of defense is immediately available because it consists of mechanisms that are nonspecific. The second line of defense takes a little longer because it is highly specific and contains mechanisms that are tailored to a particular threat.

General Defense

The environment contains many organisms that are able to invade and infect the body. The general defense mechanisms are useful against all of them.

Barriers to Entry

The secretions of the sebaceous glands in the skin contain chemicals that weaken or kill bacteria. The acid gastric secretions in the stomach inhibit the growth of many types of bacteria. The intestines and organs such as the vagina contain a mix of bacteria that normally reside there, and these prevent potential pathogens from also taking up residence. The respiratory tract is lined by cilia that sweep mucus and any trapped particles up into the throat where it may be swallowed.

Protective Chemicals

The blood and tissues contain chemicals that may prevent the multiplication of bacteria and viruses.

The **complement system** is a series of proteins produced by the liver that are present in the plasma. When these proteins are activated, an orderly cascade of reactions occur. Every protein molecule in the series activates many others in a predetermined sequence. In the end, certain proteins form pores in bacterial cell membranes. This allows fluids and salt to enter until the cell bursts (fig. 13.6).

Whenever a virus enters a tissue cell, the infected cell will produce and secrete **interferon.** Interferon binds to receptors on noninfected cells and this causes them to prepare for possible attack by producing substances that interfere with viral replication.

Inflammatory Response

Whenever the skin is broken due to a minor injury, a series of events occur that is known as the **inflammatory response,** because there is swelling and reddening at the site of the injury. Figure 13.7 illustrates how blood cells, some of which are produced by the lymphatic system, participate in the reaction:

1. Injured cells release bradykinin, a chemical that initiates nerve impulses, resulting in the sensation of pain. It also stimulates mast cells (derived from basophils) to release histamine. Histamine makes the capillary more permeable.

Figure 13.7 Inflammatory reaction. (*above*) When a capillary is ruptured due to injury, substances are released into the tissues. Among these are certain precursors that quickly become bradykinin, a chemical that initiates nerve impulses, resulting in the sensation of pain, and stimulates mast cells to release histamine. Histamine causes a capillary to become more permeable. (*below*) Now neutrophils (and monocytes) squeeze through the capillary wall and phagocytize bacteria that may have already been attacked by antibodies released by lymphocytes.

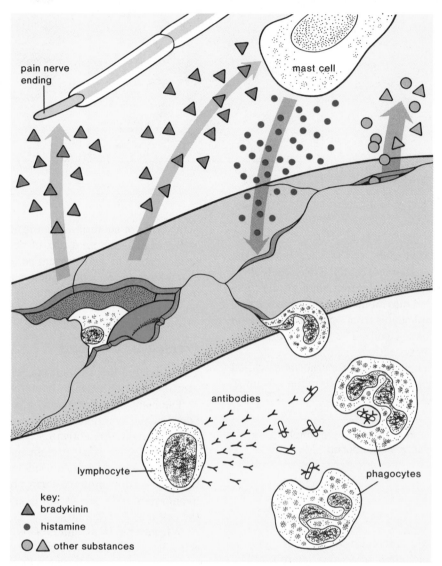

2. Neutrophils and macrophages leave the bloodstream and phagocytize foreign cells such as bacteria and viruses. Macrophages are large phagocytic cells derived from monocytes that are able to devour a hundred invaders and still survive. As the infection is being overcome, some neutrophils die and these—along with dead tissue, cells, and bacteria, and living white blood cells—form pus, which is a thick yellowish fluid. The presence of pus indicates that the body is trying to overcome the infection.

The first line of defense against disease is nonspecific. It consists of barriers to entry, protective chemicals, and the inflammatory response.

Specific Defense

Lymphocytes and monocytes allow the body to respond to a specific bacterial or viral invasion.

B and T Lymphocytes

There are two types of lymphocytes derived from bone marrow stem cells (fig. 13.8). Those called the **B cells** have not passed through the thymus, while those called the **T cells** have passed through the thymus. Both types of cells possess receptors within their cell membranes which are capable of recognizing and combining with an antigen. Antigens are substances that are foreign to the body. For example, microorganisms are antigenic in nature.

Figure 13.8   Bone marrow stem cells produce B and T cells. T cells but not B cells pass through the thymus. When a B cell recognizes an antigen, it divides to give antibody secreting plasma cells and memory cells. In this way B cells are responsible for antibody-mediated immunity. When helper T (T$_H$) cells recognize an antigen, because it has been presented to them by a macrophage (M), they produce lymphokines, which stimulate other immune cells, including B cells and killer T (T$_K$) cells. When T$_K$ cells recognize an antigen, because it has been presented to them by a macrophage (M), they attack any other cell bearing the antigen. T$_K$ cells are responsible for cell-mediated immunity.

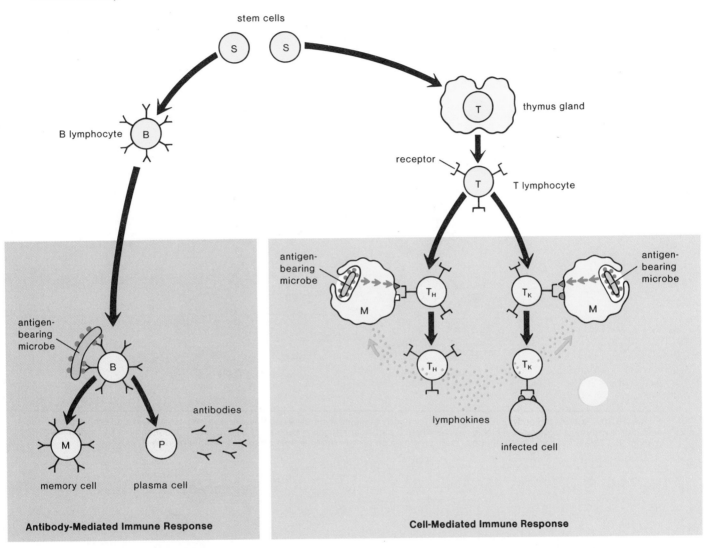

Antibody-Mediated Immunity

B cells are responsible for **antibody-mediated immunity.** When an antigen combines with a B cell receptor, the B cell enlarges and divides to give plasma cells and memory cells (fig. 13.9). The **plasma cells** produce antibodies (immunoglobulins) capable of attacking the antigen. The **memory cells** remain in the bloodstream for a long time and can quickly become plasma cells if the same bacterium or virus should enter the body again. In other words, the person is now immune. **Immunity** is present when the person already possesses antibodies capable of bringing an infection under control. When people are immune, they do not get sick even though they have been exposed to a particular bacterium or virus.

The immune reaction is very specific because only one type of antibody combines with a particular antigen (fig. 13.10). The body contains an enormous number of different types of B cells and only the one that produces the right kind of antibody responds to a particular microbial invasion by producing antibodies. After a microorganism is attacked by an antibody, it is often engulfed by a macrophage.

B cells are responsible for antibody-mediated immunity. Each one produces a specific type of antibody to counter a particular infection.

Each antibody contains variable regions that bind to an antigen in a lock-and-key manner and a constant region. Antibodies can be classified according to their

Figure 13.9 According to the clonal selection theory, the antigen selects the lymphocyte, which divides, producing, by the fifth day, a clone of many mature plasma cells, which actively secrete antibodies and memory cells that retain the ability to secrete these antibodies at a future time.

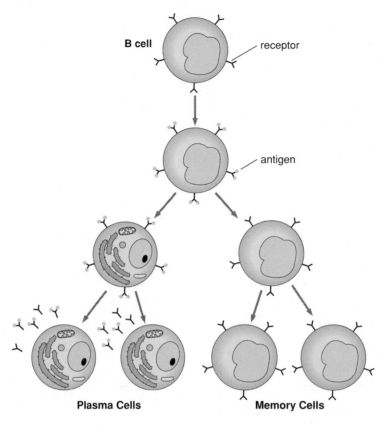

B cell

receptor

antigen

Plasma Cells

Memory Cells

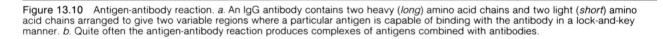

Figure 13.10 Antigen-antibody reaction. *a.* An IgG antibody contains two heavy (*long*) amino acid chains and two light (*short*) amino acid chains arranged to give two variable regions where a particular antigen is capable of binding with the antibody in a lock-and-key manner. *b.* Quite often the antigen-antibody reaction produces complexes of antigens combined with antibodies.

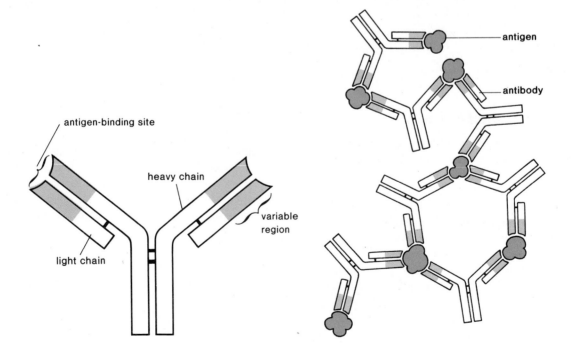

antigen-binding site

heavy chain

variable region

light chain

antigen

antibody

Table 13.1 The Immunoglobulins

Immunoglobulin	Examples of Functions
IgG	Main form of antibodies in circulation: production increased after immunization
IgA	Main antibody type in external secretions, such as saliva and mother's milk
IgE	Responsible for allergic symptoms in immediate hypersensitivity reactions
IgM	Function as antigen receptors on lymphocyte surface prior to immunization; secreted during primary response
IgD	Function as antigen receptors on lymphocyte surface prior to immunization; other functions unknown

From Stuart Ira Fox, *Human Physiology*, 3d ed. Copyright © 1990 Wm. C. Brown Publishers, Dubuque, Iowa. All Rights Reserved. Reprinted by permission.

constant regions. All antibodies that have the same basic kinds of constant regions are placed in a particular class. Table 13.1 lists the different classes of antibodies and their specific functions. Most antibodies belong to class Ig (immunoglobulin) G.

Cell-Mediated Immunity

T cells are responsible for **cell-mediated immunity** (fig. 13.8). T cells are activated by macrophages which "present" the antigen to them (fig. 13.11). After being activated, some T cells, called *helper T cells,* stimulate B cells. They also release chemicals called **lymphokines** that stimulate the immune system in general. The AIDS virus attacks helper T cells and that is why persons with AIDS have an impaired immune system.

Other types of T cells, upon being activated, divide and become "killer cells." *Killer cells* exhibit cytotoxicity, meaning that direct contact between a killer cell and a target cell (bearing a specific antigen) causes death of the target cell (fig. 13.12). Killer T cells specialize in providing resistance against antigens that appear on the cells of the body. They attack any tissue cell that bears an antigen, including the cells infected with viruses, a transplanted organ, a donated skin graft, or cancer. Cancer cells are abnormal cells that most likely display altered antigens. As long as T cells are capable of recognizing newly developed cancer cells, a cancerous growth cannot begin. Therefore, when cancer is present, it most likely signifies a failure of the immune system.

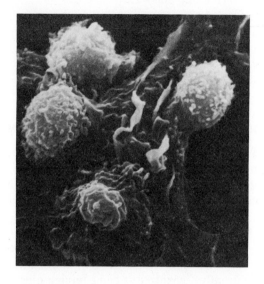

Figure 13.11 Scanning electron micrograph of three T cells attached to a macrophage. This step is believed to be necessary to stimulate T cells to become "angry killers."

Figure 13.12 *a.* Scanning electron micrograph showing lymphocytes attacking a larger cancer cell. *b.* Impending death of cancer cell is indicated by the blebs, or deep folds, that have appeared on its surface membrane.

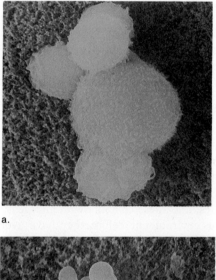

a.

b.

When a T cell attacks a foreign cell, the T cell releases chemicals that perforate the cell's membrane, and the cell shrivels up and dies.

T cells are responsible for cell-mediated immunity. They directly attack cells that bear antigens. For example, they protect us from cancer, but also attack transplanted organs.

Immunological Side Effects and Illnesses

The immune system protects us from disease because it can tell self from nonself. Sometimes, however, the immune system is underprotective, such as when an individual develops cancer, or is overprotective, such as when an individual has allergies.

Allergies **Allergies** are caused by an overactive immune system that forms antibodies to substances that are not usually recognized as being foreign substances. Unfortunately, allergies may be accompanied by cold-like symptoms or, even at times, severe systemic reactions such as shock, a sudden drop in blood pressure.

Of the five varieties of antibodies (table 13.1), it is the IgE type that causes allergies. IgE antibodies are found in the bloodstream, but they, unlike other types of antibodies, also reside in the membrane of mast cells found in the tissues. When an allergen, an antigen that provokes an allergic reaction, attaches to an IgE antibody on mast cells, these cells release histamine and other substances (fig. 13.13) that cause secretion of mucus and constriction of airways, resulting in the characteristic wheezing and labored breathing of someone with asthma.

Allergy injections sometimes prevent the occurrence of allergic symptoms. Injections of the allergen cause the body to build up high quantities of IgG type antibodies, and these combine with allergens received from the environment before they have a chance to reach the E type antibodies located in the membranes of mast cells.

Autoimmune Diseases **Certain human illnesses are** believed to be due to the production of antibodies that act against an individual's own tissues. In myasthenia gravis, autoantibodies attack the neuromuscular junctions so that the muscles do not obey nervous stimuli. Muscular weakness results. In MS (multiple sclerosis), antibodies attack the myelin sheath of nerve fibers, causing various neuromuscular disorders. A person with SLE (systemic lupus erythematosus) forms various antibodies to different constituents of the body, including the DNA of the cell nucleus. The disease sometimes results in death, usually due to kidney damage. In

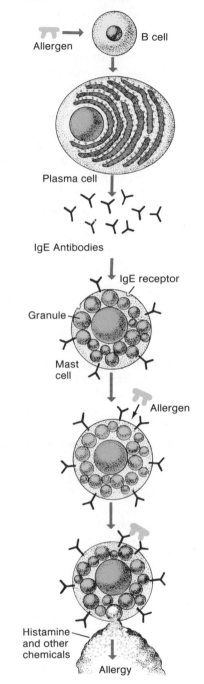

Figure 13.13 Allergic response. An allergen causes a plasma cell to secrete IgE antibodies. Some of these bind to mast cells. Then when the individual is exposed again, the allergen attaches to these IgE antibodies and the mast cell releases histamine and other chemicals that cause the allergic response.

Allergen → B cell

Plasma cell

IgE Antibodies

IgE receptor

Granule

Mast cell

Allergen

Histamine and other chemicals

Allergy

rheumatoid arthritis, it is the joints that are affected. When an autoimmune disease occurs, a viral infection of tissues has often set off an immune reaction to the body's own tissues in an attempt to attack the virus. There is evidence to suggest that type I diabetes is also the result of this sequence of events.

Immunotherapy

The immune system can be manipulated to help people avoid or recover from diseases.

Active Immunity

Active immunity, which provides long-lasting immunity, can develop after an infection. In many instances today, however, it is not necessary to suffer an illness to become immune because it is possible to be medically immunized against a disease. One possible recommended immunization schedule for children is given in figure 13.14. Immunization requires the use of vaccines, which are traditionally bacteria and viruses (antigens) that have been treated so that they are no longer virulent (able to cause disease). New methods of producing vaccines that utilize only a portion of a disease-causing agent have been developed recently. Such vaccines are safer because they cannot cause the disease they are designed to prevent.

After a vaccine is injected, it is possible to determine the amount of antibody present in a sample of serum—this is called the *antibody titer*. After the first exposure to an antigen, a primary response occurs. There is a period of several days during which no antibodies are present; then there is a slow rise in the titer, which is followed by a gradual decline (fig. 13.15). After a second exposure, a secondary response may occur. If so, the titer rises rapidly to a level much greater than before. The second exposure is often called the "*booster shot*" since it boosts the antibody titer to a high level. The antibody titer may now be high enough to prevent disease symptoms even if the individual is exposed to the disease. The individual is now immune to that particular disease.

A good secondary response can be related to the number of plasma and memory cells in the serum. Upon the second exposure, these cells are already present, and antibodies can be rapidly produced.

Vaccines can be used to make people actively immune.

Passive Immunity

Passive immunity occurs when an individual is given antibodies to combat a disease. Since these antibodies are not produced by the individual's B cells, passive immunity is short-lived. For example, newborn infants possess passive immunity because antibodies have crossed the placenta from their mother's blood. These antibodies soon disappear, however, so that within a few months infants become more susceptible to infections. Breast feeding prolongs the passive immunity an infant receives from its mother because there are antibodies in the mother's milk.

Figure 13.14 Suggested immunization schedule for infants and young children. Children who are not immunized are subject to childhood diseases that can cause serious health consequences.

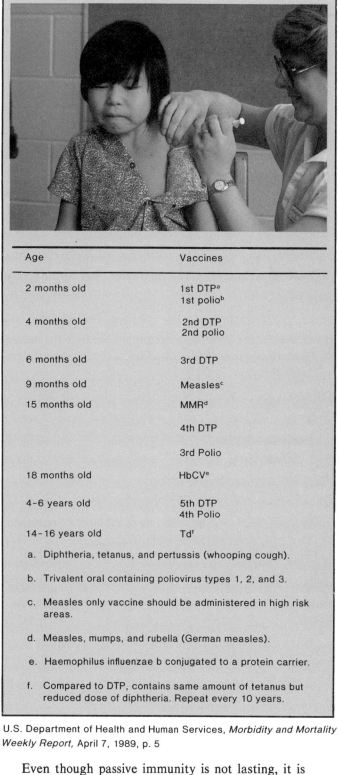

Age	Vaccines
2 months old	1st DTP[a] 1st polio[b]
4 months old	2nd DTP 2nd polio
6 months old	3rd DTP
9 months old	Measles[c]
15 months old	MMR[d]
	4th DTP
	3rd Polio
18 months old	HbCV[e]
4-6 years old	5th DTP 4th Polio
14-16 years old	Td[f]

a. Diphtheria, tetanus, and pertussis (whooping cough).

b. Trivalent oral containing poliovirus types 1, 2, and 3.

c. Measles only vaccine should be administered in high risk areas.

d. Measles, mumps, and rubella (German measles).

e. Haemophilus influenzae b conjugated to a protein carrier.

f. Compared to DTP, contains same amount of tetanus but reduced dose of diphtheria. Repeat every 10 years.

U.S. Department of Health and Human Services, *Morbidity and Mortality Weekly Report,* April 7, 1989, p. 5

Even though passive immunity is not lasting, it is sometimes used to prevent illness in a patient who has been unexpectedly exposed to an infectious disease. Usually, the person receives an injection of a serum containing antibodies. This may have been taken from

Figure 13.15 Immunization responses. The primary response after the first exposure to a vaccine is minimal, but the secondary response after the second exposure may show a dramatic rise in the amount of antibody present in serum.

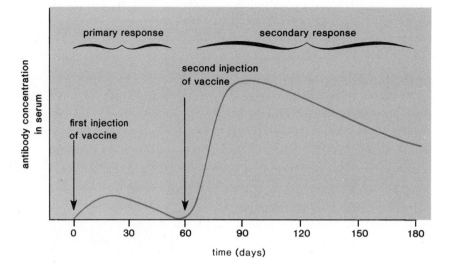

donors who have recovered from the illness. In other instances, horses were immunized, and serum taken from them to provide the needed antibodies. Horses were used to produce antibodies against diphtheria, botulism, and tetanus. Occasionally, a patient who received these antibodies became ill because the serum contained proteins that the individual's immune system recognized as foreign. This was called serum sickness.

A new method of producing antibodies has also been developed. Lymphocytes are removed from the body and exposed in vitro (in laboratory glassware) to a particular antigen. The stimulated lymphocytes divide and produce a clone of cells capable of producing only one type of antibody. These antibodies, called **monoclonal antibodies,** do not cause serum sickness.

Passive immunity is short-lived because the antibodies are administered to and not made by the individual.

Summary

I. Lymphatic System.
   A. The lymphatic system
      (1) collects excess tissue fluid,
      (2) absorbs fat molecules in the intestines, and (3) plays a major role in the body's defense against disease.
   B. The lymphatic capillaries take up excess tissue fluid. Lymphatic trunks drain lymph from a major region of the body and empty into one of two ducts, which enter the subclavian vein. The right thoracic duct serves the right arm, chest, and side of the head. The left thoracic duct serves the rest of the body.
   C. Bone marrow produces lymphacytes and other cells necessary to immunity.
   D. Lymph nodes cleanse the lymph. Afferent lymphatic vessels bring the lymph to the node and efferent lymphatic vessels take it away. In between, lymph passes through sinuses that are filled with phagocytes.
   E. Illnesses associated with the lymphatic system and lymph nodes are lymphadenitis (inflammation), lymphoma (cancer), metastasis (spread of cancer), and edema (swelling).
   F. The spleen contains blood sinuses filled with lymphocytes, macrophages, and red blood cells. It cleanses the blood.
   G. The thymus contains lobules where T cells, a special type of lymphocyte, mature.

II. Immunity
   A. General defense.
      Barriers to entry. The skin and mucous membranes of the body serve as mechanical barriers to possible entry by bacteria and viruses.
      Protective chemicals. Also present in the body and tissues are chemicals (complement and interferon) that may prevent the multiplication of microorganisms.
      Inflammation. The inflammatory reaction is a "call to arms"—it marshals phagocytic white blood cells and antibody-producing lymphocytes to a site of invasion by microorganisms.

B. Specific defense.

B lymphocytes are responsible for antibody-mediated immunity. An antibody combines with its antigen in a lock-and-key manner. The antigen-antibody reaction can lead to complexes that contain several antibodies and antigens.

T lymphocytes are responsible for cell-mediated immunity. Some T cells kill cells, such as cancer cells, directly and all T cells release lymphokines, chemicals that stimulate other immune cells to do their respective jobs.

There are immunological side effects and illnesses. Allergic symptoms are caused by the release of histamine and other substances from mast cells.

Autoimmune diseases seem to be preceded by a viral infection that fools the immune system into attacking the body's organ tissues.

C. Immunotherapy.

Vaccines can be used to make people actively immune.

Passive immunity is short-lived because the antibodies are administered to, and not made by, the individual.

Monoclonal antibodies are produced in pure batches—they are specific against just one antigen.

Study Questions

1. What is the lymphatic system, and what are its functions?
2. Trace the path of lymph from a lymphatic capillary to a lymphatic duct.
3. What causes lymph to flow in lymphatic vessels?
4. Name the lymphoid organs. Why is bone marrow a lymphoid organ?
5. What is the structure and function of a lymph node? conditions associated with lymph nodes and the lymphatic system.
7. What is the structure and function of the spleen?
8. What is the structure and function of the thymus?
9. Explain antibody-mediated and cell-mediated immunity.
10. Discuss allergies, tissue rejection, and autoimmune diseases.
11. How is active immunity achieved? passive immunity?

Objective Questions

Fill in the blanks.

1. Lymphatic vessels contain _____ , which close, preventing lymph from flowing backwards.
2. _____ and _____ are two types of white blood cells produced and stored in lymphoid organs.
3. Lymph nodes cleanse the _____ , while the spleen cleanses the _____ .
4. _____ occurs after cancer cells enter the lymph capillaries.
5. T lymphocytes have passed through the _____ .
6. A stimulated B cell divides and differentiates into antibody-secreting _____ cells and _____ cells that are ready to produce the same type antibody at a later time.
7. B cells are responsible for _____-mediated immunity.
8. Killer T cells are responsible for _____ -mediated immunity.
9. Allergic reactions are associated with the release of _____ from mast cells.
10. Immunization with _____ brings about active immunity.

Medical Terminology Reinforcement Exercise

Pronounce and analyze the meaning of the following dissected terms:

1. metastasis (mĕ-tas'tah-sis) meta/stasis
2. allergist (al'er-jist) allerg/ist
3. immunosuppressant (im''u-no-sŭ-pres'ant) immuno/suppress/ant
4. immunotherapy (ĭ-mu''no-ther'ah-pe) immuno/therapy
5. macrophage (mak'ro-fāj) macro/phage
6. splenorrhagia (sple''no-ra'je-ah) spleno/rrhagia
7. thymusectomy (thi''mus-ek'to-me) thymus/ectomy
8. lymphadenopathy (lim-fad''ĕ-nop'ah-the) lymph/adeno/pathy
9. lymphangiography (lim-fan''-je-og'rah-fe) lymph/angio/graphy
10. lymphedema (lim''fe-de'mah) lymph/edema

Respiration

Learning Objectives

After you have studied this chapter, you should be able to:

1. State and define the four processes involved in respiration.
2. State the path of airflow, and describe, in general, the structure and function of all organs mentioned.
3. Describe the mechanism by which breathing occurs, including inspiration and expiration.
4. Describe the events that occur during external and internal respiration.
5. Describe the various infections of the respiratory tract.
6. Describe the effects of smoking on the respiratory tract and on overall health.

Breathing

Breathing is only one aspect of respiration. Altogether, respiration refers to the complete process of getting oxygen to body cells for cellular respiration and to the reverse process of ridding the body of the carbon dioxide given off by cells (fig. 14.1). Respiration includes the following:

1. **Breathing:** entrance and exit of air into and out of the lungs
2. **External respiration:** the exchange of gases (O_2 + CO_2) between air and blood
3. **Internal respiration:** the exchange of gases between blood and tissue fluid
4. **Cellular respiration:** the production of ATP, the energy molecule, in cells.

In this chapter, we will begin with a consideration of breathing—the act of taking air in, or **inspiration** (inhalation), and the act of forcing air out, or **expiration** (exhalation). The normal breathing rate is about fourteen to twenty times per minute. Expired air contains less oxygen and more carbon dioxide than inspired air, indicating that the body takes in oxygen and gives off carbon dioxide.

Passage of Air

During inspiration and expiration, air is conducted toward or away from the lungs by a series of cavities, tubes, and openings, as illustrated in figure 14.2 and listed in table 14.1.

As air moves in along the air passages, it is filtered, warmed, and moistened. The filtering process is accomplished by coarse hairs and cilia in the region of the nostrils, and by cilia alone in the rest of the nose and windpipe. In the nose, the hairs and cilia act as a screening device. In the trachea, cilia beat upward, carrying mucus, dust, and occasional bits of food that "went the wrong way" into the pharynx where the accumulation may be swallowed or expectorated. The inspired air is warmed by heat given off by the blood vessels lying close to the surface of the lining of the air passages, and it is moistened by the wet surface of these passages.

On the other hand, as air moves out during expiration, it becomes progressively cooler and loses its moisture. As the gas cools, it deposits its moisture on the lining of the windpipe and nose, and the nose may even drip as a result of this condensation; but the air still retains so much moisture that upon expiration on a cold day, it condenses and forms a small cloud.

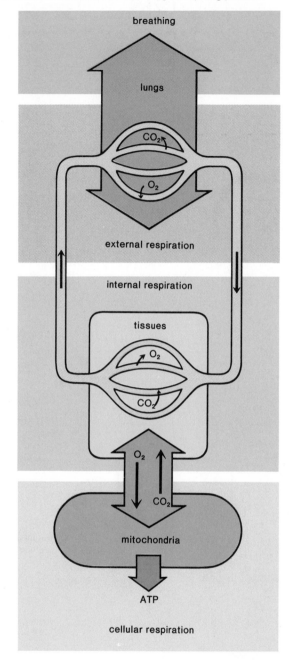

Figure 14.1 Respiration is divided into four components: breathing brings gas into the lungs; external respiration is the exchange of gases in the lungs; internal respiration is the exchange of gases in the tissues; and cellular respiration is the production of ATP in cells—an oxygen-requiring process.

Air is warmed, filtered, and moistened as it moves from the nose toward the lungs.

Each portion of the air passage also has its own unique structure and function, as described in the sections that follow.

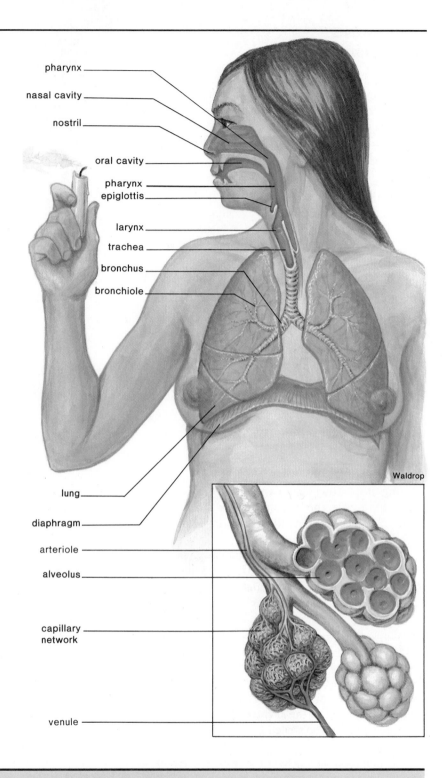

Figure 14.2 Diagram of human respiratory tract, with the internal structure of one lung revealed and an enlargement of a section of this lung. Gas exchange occurs in the alveoli, which are surrounded by a capillary network.

pharynx
nasal cavity
nostril
oral cavity
pharynx
epiglottis
larynx
trachea
bronchus
bronchiole

Waldrop

lung
diaphragm
arteriole
alveolus
capillary network
venule

Table 14.1 Path of Air

Structure	Function	Structure	Function
Nasal cavities	Filter, warm, and moisten	Trachea (windpipe)	Passage of air to thoracic cavity
Nasopharynx	Passage of air from nose to throat	Bronchi	Passage of air to each lung
Pharynx (throat)	Connection to surrounding regions	Bronchioles	Passage of air to each alveolus
Glottis	Passage of air	Alveoli	Air sacs for gas exchange
Larynx (voice box)	Sound production		

Figure 14.3 Anatomy of the larynx. *a.* Anterior view. *b.* Posterior view. There are three large cartilages in the larynx: the largest is the thyroid, the epiglottis is a leaflike covering for the glottis, and the cricoid forms a complete ring. There are also two pairs of smaller cartilages: the arytenoids and the corniculates. The arytenoid cartilages are attached to the vocal cords.

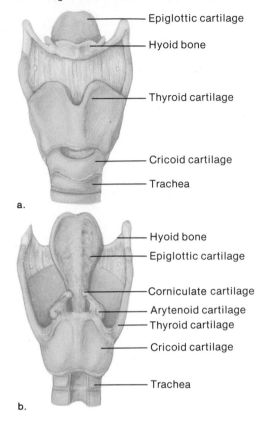

Epiglottic cartilage
Hyoid bone
Thyroid cartilage
Cricoid cartilage
Trachea

a.

Hyoid bone
Epiglottic cartilage
Corniculate cartilage
Arytenoid cartilage
Thyroid cartilage
Cricoid cartilage
Trachea

b.

Figure 14.4 Sections of larynx showing the vocal cords. *a.* Frontal section. *b.* Sagittal section. Two mucous membranes that extend across the larynx on either side of the glottis are called the true vocal cords. The false vocal cords, folds located just above the vocal cords, are not involved in voice production.

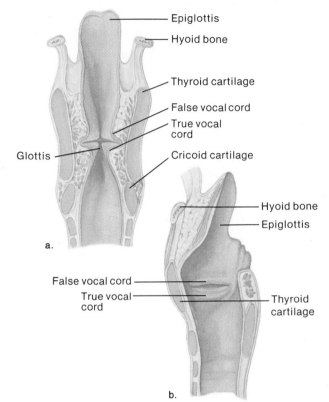

Epiglottis
Hyoid bone
Thyroid cartilage
False vocal cord
True vocal cord
Glottis
Cricoid cartilage

a.

Hyoid bone
Epiglottis
False vocal cord
True vocal cord
Thyroid cartilage

b.

Nose

The **nose** contains two nasal cavities, narrow canals with convoluted lateral walls that are separated from one another by a median septum. Up in the narrow recesses of the nasal cavities are special ciliated cells (fig. 8.3) that act as odor receptors. Nerves lead from these cells and go to the brain where the impulses are interpreted as smell.

The nasal cavities have a number of openings. The tears produced by lacrimal glands drain into the nasal cavities by way of lacrimal ducts. For this reason, crying produces a runny nose. The nasal cavities open into the cranial sinuses, air-filled spaces in the skull, and they empty into the nasopharynx, a chamber just beyond the soft palate. The *eustachian tubes* lead into the nasopharynx from the middle ears (fig. 8.12). This is why sore throats may lead to ear infections.

Pharynx

The air and food channels meet in the *pharynx* (fig. 10.4). Then the passage of air continues in the trachea (windpipe), which lies anterior to the esophagus

and normally opens only during the process of swallowing food. Just inferior to the pharynx lies the larynx, or voice box.

The nasal cavities contain the sense receptors for smell and open into the pharynx. The pharynx is the back of the throat where the air passage and food passage cross.

Larynx

The **larynx** may be imagined as a triangular box whose apex, the thyroid cartilage (Adam's apple), is located at the front of the neck (figs. 14.3 and 14.4). At the top of the larynx is a variable-sized opening called the *glottis.* When food is being swallowed, the glottis is covered by a flap of tissue called the *epiglottis* so that no food passes into the larynx. If, by chance, food or some other substance does gain entrance to the larynx, reflex coughing usually occurs to expel the substance. If this reflex is not sufficient, it may be necessary to resort to the *Heimlich maneuver* (fig. 14.5).

At the edges of the glottis, embedded in mucous membrane, are elastic ligaments called the **vocal cords** (fig. 14.4). These cords, which stretch from the back

Figure 14.5 Heimlich maneuver. More than eight Americans choke to death each day on food lodged in their trachea. A simple process termed the **abdominal thrust (Heimlich) maneuver** can save the life of a person who is choking. The abdominal thrust maneuver is performed as follows: If the victim is standing or sitting: (1) Stand behind the victim or the victim's chair, and wrap your arms around his or her waist. (2) Grasp your fist with your other hand, and place the fist against the victim's abdomen, slightly above the navel and below the rib cage. (3) Press your fist into the victim's abdomen with a quick upward thrust. (4) Repeat several times if necessary. If the victim is lying down: (1) Position the victim on his or her back. (2) Face the victim, and kneel on his or her hips. (3) With one of your hands on top of the other, place the heel of your bottom hand on the abdomen, slightly above the navel and below the rib cage. (4) Press into the victim's abdomen with a quick upward thrust. (5) Repeat several times if necessary. If you are alone and choking, use anything that applies force just below your diaphragm. Press into a table or a sink, or use your own fist.

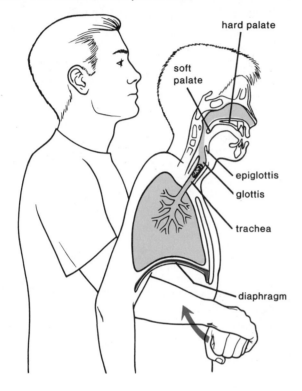

Figure 14.6 Site of tracheostomy to allow air to enter the trachea beneath the location of an obstruction.

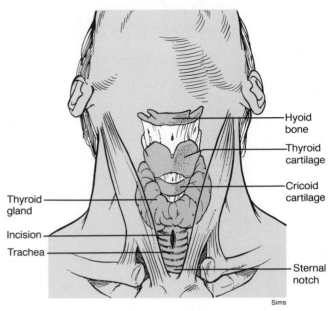

Sims

a more prominent thyroid cartilage and a deeper voice. The voice "breaks" in the young male due to his inability to control the longer vocal cords.

The larynx is the voice box because it contains the vocal cords at the sides of the glottis, an opening sometimes covered by the epiglottis.

Trachea

The **trachea** is a tube held open by cartilaginous rings. Ciliated mucous membrane (fig. 3.3) lines the trachea, and normally these cilia keep the windpipe free of debris. Smoking is known to destroy the cilia, and consequently the soot in cigarette smoke collects in the lungs. Smoking will be discussed more fully at the end of this chapter.

If the trachea is blocked because of illness or accidental swallowing of a foreign object, it is possible to insert a tube by way of an incision made in the trachea; this tube acts as an artificial air intake and exhaust duct. The operation is called a *tracheostomy* (fig. 14.6).

Bronchi

The trachea divides into two **bronchi** that enter the right and left lungs and branch into a great number of smaller passages called the **bronchioles.** The two bronchi resemble the trachea in structure, but as the bronchial

to the front of the larynx just at the sides of the glottis, vibrate when air is expelled past them through the glottis. Vibration of the vocal cords produces sound. The high or low pitch of the voice depends upon the length, thickness, and degree of elasticity of the vocal cords, and the tension at which they are held. The loudness, or intensity, of the voice depends upon the amplitude of the vibrations, or the degree to which vocal cords vibrate.

At the time of puberty, the growth of the larynx and the vocal cords is much more rapid and accentuated in the male than in the female, causing the male to have

Figure 14.7   Cast of lungs showing a large number of airways. The bronchi branch into the bronchioles that branch and rebranch until they terminate in the alveoli. Each lobe of the lung is in a different color in this cast.

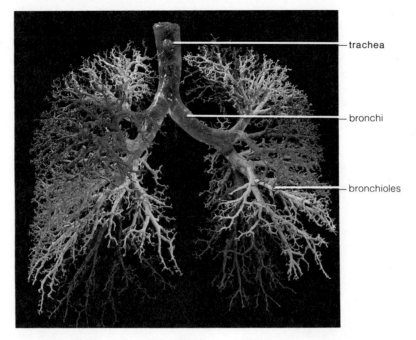

trachea

bronchi

bronchioles

tubes divide and subdivide, their walls become thinner and the small rings of cartilage do not occur (fig. 14.7). Each bronchiole terminates in an elongated space that is enclosed by a multitude of air pockets, or sacs, called **alveoli** which make up the lungs.

Lungs

Within the lungs, each alveolar sac is only one layer of squamous epithelium surrounded by blood capillaries. Gas exchange occurs between the air in the alveoli and the blood in the capillaries (fig. 14.2).

A film of lipoprotein called surfactant that lines the alveoli of mammalian lungs lowers the surface tension and prevents them from closing up. Some newborn babies, especially premature infants, lack this film, resulting in lung collapse. This condition, called *infant respiratory distress syndrome,* often results in death.

There are approximately 300,000,000 alveoli with a total cross-sectional area of 50 to 70 m². This is about forty times the surface area of the skin. Because of their many air spaces, the lungs are very light; normally, a piece of lung tissue dropped in a glass of water will float.

Air moves from the trachea and the two bronchi, held open by cartilaginous rings, into the lungs. The lungs are composed of air sacs, called alveoli.

Externally, the lungs are cone-shaped organs that lie on both sides of the heart in the thoracic cavity. Each lung has a narrow and rounded apex, which approaches the neck. The base of each is broad and concave to fit upon the convex surface of the diaphragm. The other surfaces of the lungs follow the contours of the ribs and the organs present in the thoracic cavity.

Figure 14.8 shows the relationship of the pulmonary vessels to the trachea and bronchial tubes. The branches of the pulmonary artery accompany the bronchial tubes and form a mass of capillaries around the alveoli. The four pulmonary veins collect blood from these capillaries and empty into the left atrium of the heart.

Mechanism of Breathing

In order to understand ventilation, the manner in which air is drawn into and expelled out of the lungs, it is necessary to remember first that when one is breathing, there is a continuous column of air from the pharynx to the alveoli of the lungs; that is, the air passages are open.

Secondly, we may note that the lungs lie within the sealed-off thoracic cavity. The **ribs,** which are hinged to the vertebral column at the back and to the sternum (breastbone) at the front, along with the muscles that lie between them, make up the top and sides of the thoracic cavity. The **diaphragm,** a dome-shaped horizontal

Figure 14.8 Posterior view of heart and lungs shows the relationship of the pulmonary vessels to the trachea and bronchial tubes. Trace the path of air to the left lung, and trace the path of blood from the heart to the left lung and return.

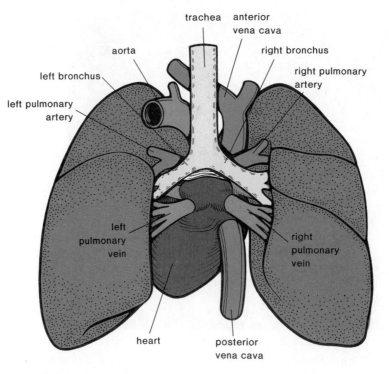

muscle, forms the floor of the thoracic cavity. The lungs themselves are enclosed by the **pleural membranes** (fig. 14.9), one of these, the *parietal,* adheres closely to the walls of the thoracic cavity and diaphragm, while the other, the *visceral,* is fused to the lungs. The two pleural layers lie very close to one another, being separated only by a thin film of fluid. Normally, the intrapleural pressure is less than atmospheric pressure. The importance of this reduced pressure is demonstrated when by design or accident air enters the intrapleural space: the lung is no longer able to expand to allow inspiration, and instead it actually collapses. (Air in the intrapleural space is called a *pneumothorax.*)

The lungs are completely enclosed and, by way of the pleural membranes, adhere to the walls of the thoracic cavity.

Inspiration

It can be shown that carbon dioxide and H^+ (hydrogen ions) are the primary stimuli that cause us to breathe. When the concentration of CO_2 (and subsequently H^+) reach a certain level in the blood, the *breathing center* in the medulla oblongata—the stem portion of the brain—is stimulated. This center is not affected by low oxygen levels, but there are chemoreceptors in the *carotid bodies* (located in the carotid arteries) and in the

aortic bodies (located in the aorta) that do respond to low blood oxygen in addition to carbon dioxide and H^+ concentration.

When the breathing center is stimulated, a nerve impulse goes out by way of nerves to the diaphragm and rib cage (fig. 14.10). In its relaxed state, the *diaphragm* is dome-shaped, but upon stimulation, it contracts and lowers. When the rib muscles contract, the *rib cage* moves upward and outward. Both of these contractions serve to increase the size of the thoracic cavity. As the thoracic cavity increases in size, the lungs expand. When the lungs expand, air pressure within the enlarged alveoli lowers and is immediately rebalanced by air rushing in through the nose or mouth.

Inspiration (fig. 14.9a and b) is the active phase of breathing. It is during this time that the diaphragm contracts, the rib muscles contract, and the lungs are pulled open, with the result that air comes rushing in. Note that the air comes in because the lungs have already opened up; the air does not force the lungs open. This is why it is sometimes said that *humans breathe by negative pressure.* It is the creation of a partial vacuum that sucks air into the lungs.

Stimulated by nerve impulses, the rib cage lifts up and out, and the diaphragm lowers to expand the thoracic cavity and lungs, allowing inspiration to occur.

Figure 14.9 Inspiration versus expiration. *a.* When the rib cage lifts up and outward and the diaphragm lowers, the lungs expand so that (*b*) air is drawn in. This sequence of events is only possible because the pressure within the intrapleural space, containing a thin film of water, is less than atmosphere pressure. *c.* When the rib cage lowers and the diaphragm rises (*d*), the lungs recoil so that air is forced out.

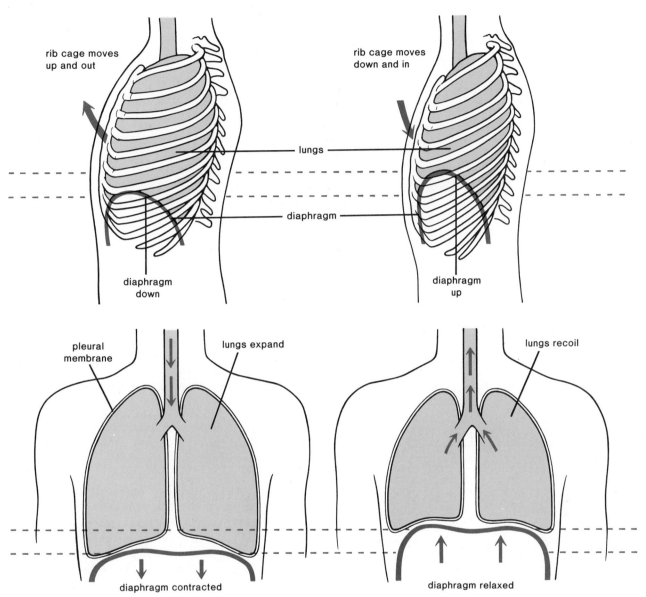

Expiration

When the lungs are expanded, the stretching of the alveoli stimulates special receptors in the alveolar walls, and these receptors initiate nerve impulses from the inflated lungs to the breathing center. When the impulses arrive at the medulla oblongata, the center is inhibited and stops sending signals to the diaphragm and the rib cage. The *diaphragm* relaxes and resumes its dome shape (fig. 14.9c and d). The abdominal organs press up against the diaphragm. The *rib cage* moves down and inward. The elastic lungs recoil and air is pushed outward. Table 14.2 summarizes the events causing inspiration and expiration. It is clear that while inspira-

tion is an active phase of breathing, normally expiration is passive since the breathing muscles automatically relax following contraction. But it is possible in deeper and more rapid breathing for both phases to be active, because there is another set of rib muscles whose contraction can forcibly cause the thoracic wall to move downward and inward. Also, when the abdominal wall muscles are contracted, there is an increase in pressure that helps expel air.

When nervous stimulation ceases, the rib cage lowers and the diaphragm rises, allowing the lungs to recoil and expiration to occur.

Figure 14.10 Nervous control of breathing. During inspiration, the respiratory center stimulates the rib (intercostal) muscles and the diaphragm to contract by way of the efferent (phrenic) nerve. Nerve impulses from the expanded lungs by way of the afferent (vagus) nerve then inhibit the respiratory center. Lack of stimulation causes the rib muscles and diaphragm to relax and expiration follows.

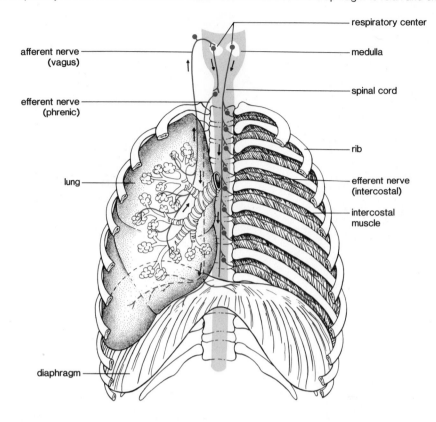

Table 14.2	Breathing Process
Inspiration	**Expiration**
Medulla sends stimulatory message to diaphragm and rib muscles.	Stretch receptors in lungs send inhibitory message to medulla.
Diaphragm contracts and flattens.	Diaphragm relaxes and resumes dome position.
Rib cage moves up and out.	Rib cage moves down and in.
Lungs expand.	Lungs recoil.
Negative pressure in lungs.	Positive pressure in lungs.
Air is pulled in.	Air is forced out.

Lung Capacities

When we breathe, the amount of air moved in and out with each breath is called the **tidal volume** (fig. 14.11). Normally, the tidal volume is about 500 ml, but we can increase the amount inhaled and exhaled by deep breathing. The total volume of air that can be moved in and out of the lungs during a single breath is called the **vital capacity.** First, we can increase inspiration to

as much as 3,500 ml. The increase beyond tidal volume is called the *inspiratory reserve volume.* Similarly, we can increase expiration beyond tidal volume by contracting the thoracic muscles. This is called the *expiratory reserve volume* and measures approximately 1,000 ml of air. Vital capacity is the sum of tidal, inspiratory reserve, and expiratory reserve volumes.

It can be noted in figure 14.11 that even after very deep breathing, some air (about 1,000 ml) remains in the lungs. This is called the **residual volume.** This air is no longer useful for gas exchange purposes. In some lung diseases, such as emphysema and asthma, the residual volume builds up because the individual has difficulty emptying the lungs. This means that the lungs tend to be filled with useless air and, as you can see from examining figure 14.11, the vital capacity is reduced.

Dead Space

Some of the inspired air never reaches the lungs; instead, it fills the conducting airways. These passages are not used for gas exchange and, therefore, they are said to contain **dead space.** To ventilate the lungs, then, it is better to breathe more slowly and deeply to insure that a greater percentage of the tidal volume will reach the lungs.

Figure 14.11 Vital capacity. *a.* This individual is using a spirometer, which measures the amount of air that can be maximally inhaled and exhaled. When he inspires, a pen moves up, and when he expires, a pen moves down. *b.* The resulting pattern such as the one shown here is called a spirograph.

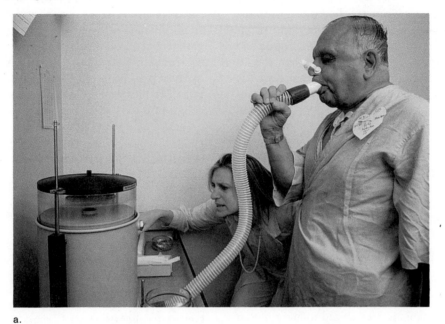

a.

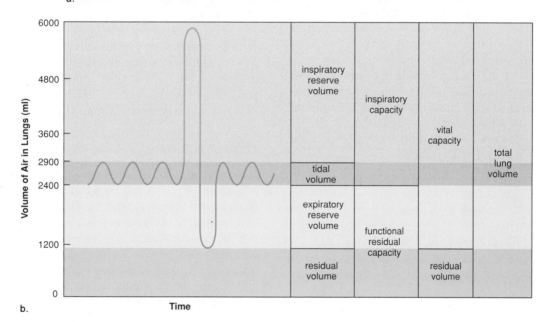

b.

If we breathe through a tube, we increase the amount of dead space and increase the amount of air that never reaches the lungs. Any device that increases the amount of dead space beyond maximal inhaling capacity means death to the individual because the air inhaled would never reach the alveoli.

The manner in which we breathe has physiological consequences.

External and Internal Respiration

External Respiration

The term *external respiration* refers to the exchange of gases between the air in the alveoli and the blood within the pulmonary capillaries (fig. 14.2). The wall of an alveolus consists of a thin, single layer of cells, and the wall of a blood capillary also consists of such a layer. Since neither of the walls offers resistance to the passage of gases, *diffusion* is believed to govern the exchange of oxygen and carbon dioxide between alveolar air and the blood. Active cellular absorption and secretion do not appear to play a role. Rather, the direction in which the gases move is determined by the pressure or tension gradients between blood and inspired air.

Atmospheric air contains little carbon dioxide, but blood flowing into the lung capillaries is almost saturated with this gas. Therefore, *carbon dioxide diffuses out of the blood into the alveoli.* The pressure pattern is the reverse for oxygen. Blood coming into the pulmonary capillaries is deoxygenated, and alveolar air is oxygenated; therefore, *oxygen diffuses into the capillary.* Breathing at high altitudes is less effective than at low altitudes because the air pressure is lower, making the concentration of oxygen (and other gases) lower than normal; therefore less oxygen diffuses into the blood. Breathing problems do not occur in airplanes because the cabin is pressurized to maintain an appropriate pressure. Emergency oxygen is available in case the pressure should, for one reason or another, be reduced.

As blood enters the pulmonary capillaries (fig. 14.12), most of the carbon dioxide is being carried as the **bicarbonate ion, HCO_3^-.** As soon as some carbon dioxide starts to diffuse out of the blood into the alveoli, the bicarbonate ion reacts to release more carbon dioxide. The enzyme *carbonic anhydrase,* present in red blood cells, speeds up this reaction.

As oxygen enters the capillaries, it combines with hemoglobin:

$$Hb + O_2 \rightarrow HbO_2$$

External respiration, the exchange of oxygen for carbon dioxide between air within the alveoli and the blood in pulmonary capillaries, is dependent on the process of diffusion.

Internal Respiration

As blood enters the systemic capillaries (fig. 14.12), oxygen leaves hemoglobin and diffuses out into tissue fluid:

$$HbO_2 \rightarrow Hb + O_2$$

Diffusion of oxygen out of the blood and into the tissues occurs because the oxygen concentration in tissue fluid is low due to the fact that the cells are continuously taking it up and using it. On the other hand,

carbon dioxide concentration is high because carbon dioxide is continuously being produced by the cells. Therefore, *carbon dioxide will diffuse into the blood.*

Carbon dioxide enters the red blood cells where a small amount of carbon dioxide is carried by hemoglobin, but most reacts with water and becomes the bicarbonate ion (HCO_3^-). The enzyme carbonic anhydrase present in red blood cells speeds up the reaction. The bicarbonate ion diffuses out of the red blood cells and is carried in the plasma.

Internal respiration, the exchange of oxygen for carbon dioxide between blood in the tissue capillaries and tissue fluid, is dependent on the process of diffusion.

Respiratory Infections and Lung Disorders

Respiratory Infections

Germs frequently spread from one individual to another by way of the respiratory tract. Droplets from one single sneeze may be loaded with billions of bacteria or viruses. The mucous membranes are protected by the production of mucus and by the constant beating of the cilia; but if the number of infective agents is large and/or the resistance of the individual is reduced, an upper respiratory infection may result.

Common Cold

A cold is a viral infection that usually begins as a scratchy sore throat, followed by a watery mucous discharge from the nasal cavities. There is rarely a fever, and symptoms are usually mild, requiring little or no medication. Although colds have a short duration, the immunity they provide is also brief. Since there are estimated to be over 150 cold-causing viruses, it is very difficult to gain immunity to all of them.

Vaccines for these infections are not in wide use, and because they are viral in nature, antibiotics are not helpful. Since viruses take over the machinery of the cell when they reproduce, it is difficult to develop drugs that will affect the virus without affecting the cell itself.

Influenza

"Flu" is a viral infection of the respiratory tract, which is accompanied by aches and pains in the joints. There is usually a fever and the illness lasts for a longer length of time than a cold. Immunity is possible, but only the vaccine developed for the particular virus prevalent that

Figure 14.12 Diagram illustrating external and internal respiration. During external respiration in the lungs, CO_2 leaves the blood and O_2 enters the blood. During internal respiration in the tissues, O_2 leaves the blood and CO_2 enters the blood.

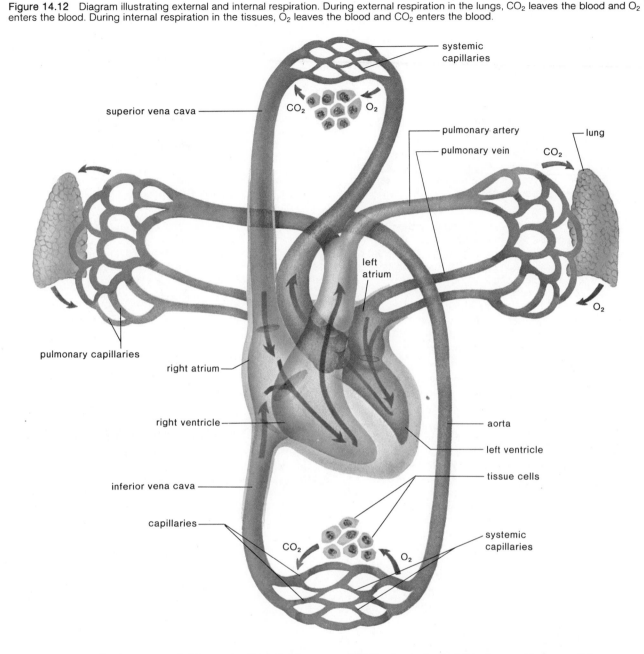

season can usually be successful in protecting the individual during a current flu epidemic. Since flu viruses constantly mutate, there can be no buildup in immunity and a new viral illness rapidly spreads from person to person and from place to place. Pandemics, in which a newly mutated flu virus spreads about the world, have occurred regularly, about every ten years.

Bronchitis

Viral infections can spread from the nasal cavities to the sinuses (sinusitis), to the middle ears (otitis media), to the larynx (laryngitis), and to the bronchi (bron-

chitis). Acute bronchitis is usually caused by a secondary bacterial infection of the bronchi, resulting in a heavy mucous discharge with much coughing. Acute bronchitis usually responds to antibiotic therapy.

Chronic bronchitis is not necessarily due to infection. It is often caused by a constant irritation of the lining of the bronchi, which as a result undergoes degenerative changes, with the loss of cilia preventing the normal cleansing action. There is frequent coughing, and the individual is more susceptible to upper respiratory infections. Chronic bronchitis most often affects cigarette smokers.

More about . . .

Risks of Smoking and Benefits of Quitting

Risks of Smoking

Based on available statistics, the American Cancer Society informs us that smoking carries a high risk. Among the risks of smoking are the following:

Shortened Life Expectancy A twenty-five-year-old who smokes two packs of cigarettes a day has a life expectancy 8.3 years shorter than a nonsmoker. The greater the number of packs smoked, the shorter the life expectancy.

Lung Cancer The first event appears to be a thickening of the cells that line the bronchi. Then there is a loss of cilia so that it is impossible to prevent dust and dirt from settling in the lungs. Following this, cells with atypical nuclei appear in the thickened lining. A disordered collection of cells with atypical nuclei may be considered to be cancer in situ (at one location). A final step occurs when some cells break loose and penetrate the other tissues, a process called metastasis. Tumors now develop (b).

Cancer of the Larynx, Mouth, Esophagus, Bladder, and Pancreas The chances of developing these cancers are from 2 to 17 times higher in cigarette smokers than in nonsmokers.

Emphysema Cigarette smokers have 4 to 25 times greater risk of developing emphysema. Damage is seen in the lungs of even young smokers. Smoking causes the lining of the bronchioles to thicken. . . . If a large part of the lungs is involved, the lungs are permanently inflated and the chest balloons out due to this trapped air. The victim is breathless and has a cough. Since the surface area for gas exchange is reduced, not enough oxygen reaches the heart and brain. The heart works furiously to force more blood through the lungs, which may lead to a heart condition. Lack of oxygen for the brain may make the person feel depressed, sluggish, and irritable.

Coronary Heart Disease Cigarette smoking is the major factor in 120,000 additional U.S. deaths from coronary heart disease each year.

Reproductive Effects Smoking mothers have more stillbirths and low-birthweight babies who are more vulnerable to disease and death. Children of smoking mothers are smaller and underdeveloped physically and socially even seven years after birth.

Benefits of Quitting

In the same manner, the American Cancer Society informs smokers of the benefits of quitting. These benefits include the following:

Risk of Premature Death is Reduced Do not smoke for 10 to 15 years, and the risk of death due to any one of the cancers mentioned approaches that of the nonsmoker.

Health of Respiratory System Improves The cough and excess sputum disappear during the first few weeks after quitting. As long as cancer has not yet developed, all the ill effects mentioned can reverse themselves and the lungs can become healthy again. In patients with emphysema, the rate of alveoli destruction is reduced and lung function may improve.

Strep Throat

This is a very severe throat infection caused by the bacterium *Streptococcus pyogenes*. Swallowing may be difficult, and there is a fever. Unlike a viral infection, strep throat should be treated with antibiotics. If not treated, it may lead to complications, such as rheumatic fever, from which the heart valves may be permanently affected.

Upper respiratory infections due to a viral infection are not treatable by antibiotics, but bacterial ones are treatable by antibiotics.

Lung Disorders

Pneumonia and tuberculosis, two infections of the lungs, formerly caused a large percentage of deaths in the United States. Now they are controlled by antibiotics. Two other illnesses discussed in the following paragraphs, emphysema and lung cancer, are not due to infections; in most instances, they are due to cigarette smoking.

Pneumonia

Most forms of pneumonia are caused by bacteria or viruses that infect the lungs. The demise of AIDS patients is usually due to a particularly rare form of pneumonia caused by a protozoan *Pneumocystis carinii*. Sometimes pneumonia is localized in specific lobes of the lungs, and these become inoperative as they fill with mucus and pus. The more lobes involved, the more serious the infection.

Tuberculosis

Tuberculosis is caused by the tubercle bacillus. It is possible to tell if a person has ever been exposed to tuberculosis by use of a skin test in which a highly diluted

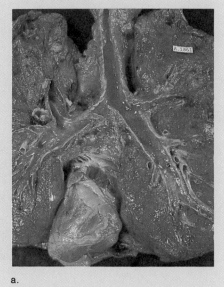

a.

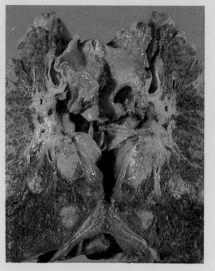

b.

extract of the bacilli is injected into the skin of the patient. A person who has never been in contact with the bacillus will show no reaction, but one who has developed immunity to the organism will show an area of inflammation that peaks in about forty-eight hours. If these bacilli do invade the lung tissue, the cells build a protective capsule about the foreigners to isolate them from the rest of the body. This tiny capsule is called a *tubercle.* If the resistance of the body is high, the imprisoned organisms may die, but if the resistance is low, the organisms may eventually be liberated. If a chest X ray detects the presence of tubercles, the individual is put on appropriate drug therapy to ensure the localization of the disease and the eventual destruction of any live bacterial organisms.

Emphysema

Emphysema refers to the destruction of lung tissue, with accompanying ballooning or inflation of the lungs due to trapped air. The trouble stems from the destruction

and collapsing of the bronchioles. When this occurs, the alveoli are cut off from renewed oxygen supply and the air within them is trapped. The trapped air very often causes rupturing of the alveolar walls (fig. 14.13) together with a fibrous thickening of the walls of the small blood vessels in the vicinity (see the section on emphysema in the reading above).

Chronic bronchitis and emphysema, two conditions most often caused by smoking, together are called chronic obstructive pulmonary disease (COPD).

Pulmonary Fibrosis

Inhaling particles such as silica (sand), coal dust, and asbestos (fig. 14.14) can lead to pulmonary fibrosis, in which fibrous connective tissue builds up in the lungs. Breathing capacity can be seriously impaired, and the development of cancer is not rare. Since asbestos has been so widely used, as a fire-proofing and insulating agent, unwarranted exposure has occurred.

Figure 14.13 Scanning electron micrograph of the lungs of a person with emphysema. There are large cavities in the lungs due to the breakdown of alveoli.

Figure 14.13 Scanning electron micrograph of the lungs of a person with emphysema. There are large cavities in the lungs due to the breakdown of alveoli.

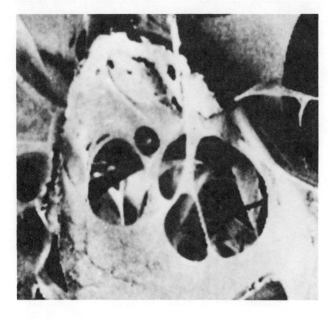

Figure 14.14 Macrophage impaled on an asbestos fiber. When asbestos fibers get caught in the lungs, fibrous tissue develops, and eventually there may be cancer.

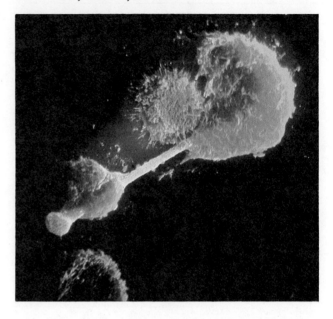

Lung Cancer

Lung cancer used to be more prevalent in men than women, but recently lung cancer has surpassed breast cancer as a cause of death in women. This can be linked to an increase in the number of women who smoke. Autopsies on smokers have revealed the progressive steps by which lung cancer commonly develops in the bronchi (see the section on lung cancer in the reading on p. 256).

The tumor may grow until the bronchus is blocked, cutting off the supply of air to that lung. The lung then collapses, and the secretions trapped in the lung spaces become infected, with a resulting pneumonia or the formation of a lung abscess. The only treatment that offers a possibility of cure, before secondary growths have had time to form, is to remove the lung completely. This operation is called *pneumonectomy*.

The incidence of lung cancer is much higher in individuals who smoke compared to those who do not smoke.

Summary

I. Breathing
   A. Passage of air. The nasal cavities contain the sense receptors for smell and open into the pharynx. The pharynx is the back of the throat where the air passage and food passage meet.

   The larynx is the voice box because it contains the vocal cords at the sides of the glottis, an opening sometimes covered by the epiglottis.

   Air moves from the trachea and the two bronchi, held open by cartilaginous rings, into the lungs. The lungs are composed of air sacs, called alveoli.

   B. Mechanism of breathing. The lungs are completely enclosed and, by way of the pleural membranes, adhere to the walls of the thoracic cavity.

   Stimulated by nerve impulses, the rib cage lifts up and out, and the diaphragm lowers to expand the thoracic cavity and lungs, allowing inspiration to occur.

   When nervous stimulation ceases, the rib cage lowers and the diaphragm rises allowing the lungs to recoil and expiration to occur.

   C. Lung capacities. Tidal volume is the amount of air inhaled and exhaled with each breath. Vital capacity is the total volume of air that can be moved in and out of the lungs during a single breath.

   After breathing out as hard as we can, there is still some air in the lungs. This is called the residual volume.

   The space within the airways is called dead space because no gas exchange takes place in the airways.

II. External and Internal Respiration
   A. External respiration, the exchange of oxygen for carbon

dioxide between air within the alveoli and the blood in pulmonary capillaries, is dependent on the process of diffusion.

B. Internal respiration, the exchange of oxygen for carbon dioxide between blood in the tissue capillaries and tissue fluid, is dependent on the process of diffusion.

III. Respiratory Infections and Lung Disorders
A. Colds, influenza, bronchitis, and strep throat are all respiratory infections.
   Viral infections are not treatable by antibiotics, but bacterial ones are treatable by antibiotics.
B. Lung disorders. Pneumonia and tuberculosis are caused by bacteria; emphysema, pulmonary fibrosis, and lung

cancer are often associated with an environmental factor.
   Chronic bronchitis and emphysema, two conditions most often caused by smoking, together are called chronic obstructive pulmonary disease (COPD).
   The incidence of lung cancer is much higher in individuals who smoke compared to those who do not smoke.

Study Questions

1. What are the four parts of respiration? In which of these is oxygen actually used up and carbon dioxide produced?
2. List the parts of the respiratory tract. What are the special functions of the nasal cavity, larynx, and alveoli?
3. What are the steps in inspiration and expiration? How is breathing controlled?

4. Why can't we breathe through a very long tube?
5. What physical process is believed to explain gas exchange?
6. How is carbon dioxide carried in the blood? How is oxygen carried?
7. Explain gas exchange for both internal and external respiration.

8. Name some infections of the respiratory tract.
9. What are emphysema and pulmonary fibrosis, and how do they affect one's health?
10. What is the most common cause of lung cancer? Why is the incidence of lung cancer rising among women?

Objective Questions

Fill in the blanks.

1. In tracing the path of air, the _____ immediately follows the pharynx.
2. The lungs contain air sacs called _____ .
3. The breathing rate is primarily regulated by the amount of _____ in the blood.

4. Air enters the lungs after they have _____ .
5. Carbon dioxide is carried in the blood as the _____ ion.
6. During external respiration, oxygen _____ the blood.
7. Gas exchange is dependent on the physical process of _____ .

8. During internal respiration, carbon dioxide _____ the blood.
9. The most likely cause of emphysema and chronic bronchitis is _____ _____ .
10. Most cases of lung cancer actually begin in the _____ .

Medical Terminology Reinforcement Exercise

Pronounce and analyze the meaning of the following words as dissected:

1. eupnea (ūp-ne'ah) eu/pnea
2. nasopharyngitis (na''zo-fah''in-ji'tis) naso/pharyng/itis
3. tracheostomy (tra''ke-os'to-me) trache/ostomy
4. pneumonomelanosis (nu-mo''no-mel''ah-no'sis) pneumono/melan/osis
5. pleuropericarditis (ploor''o-per''i-kar''di'tis) pleuro/peri/card/itis
6. bronchoscopy (brong-kos'ko-pe) broncho/scopy

7. dyspnea (disp'ne-ah) dys/pnea
8. laryngospasm (lah-rin'go-spazm) laryngo/spasm
9. hemothorax (he''mo-tho'raks) hemo/thorax
10. otorhinolaryngology (o''to-ri''no-lar''in-gol'o-je) oto/rhino/laryngo/logy

Circle the plural form of the following anatomical parts and pronounce both the singular and plural forms.

11. bronchus (brong'kus) bronchi (brong'ki)
12. thoraces (tho'races) thorax (tho'raks)
13. pleurae (ploor'e) pleura (ploor'ah)
14. alveoli (al-ve'o-li) alveolus (al-ve'o-lus)
15. metastasis (me-tas'tah-sis) metastases (me-tas'tah-sez)

15

Excretory System

Chapter Outline

Learning Objectives

After you have studied this chapter, you should be able to:

1. Name the organs of excretion and tell what wastes they excrete.
2. Trace the path of urine, and describe, in general, the structure and function of each organ mentioned.
3. Describe the gross anatomy of the kidney.
4. State the parts of a kidney nephron, and relate these to the gross anatomy of the kidney.
5. Describe the three steps in urine formation and relate these to parts of a nephron.
6. State, in general, the contents of urine.
7. Describe how water excretion is regulated and how the pH of the blood is adjusted by the kidneys.
8. State the significance of having sugar in the urine.
9. Describe the symptoms of renal failure.
10. Describe how the kidney machine works.

Urinary System

There are various organs of excretion (fig. 15.1) and only the kidneys are part of the urinary system. The sweat glands in the skin excrete perspiration—a solution of water, salt, and some urea. As mentioned previously, the liver excretes bile pigments and the lungs excrete carbon dioxide. Certain salts, such as those of iron and calcium, are excreted directly into the cavity of the intestine by the epithelial cells lining it. These salts leave the body in the feces.

At this point, it might be helpful to remember that the term defecation, and not excretion, is used to refer to the elimination of feces from the body. Substances that are excreted are those that are waste products of metabolism. Undigested food and bacteria, which make up feces, have never been a part of the functioning of the body, but salts that are passed into the gut are excretory substances because they were once metabolites in the body.

The urinary system includes the kidneys and other structures which are illustrated in figure 15.2 and listed in table 15.1. The structures are listed in order according to the path of urine.

Figure 15.1   The organs of excretion include not only the kidneys but also the skin, lungs, liver, and large intestine. The lungs excrete carbon dioxide; the liver excretes hemoglobin breakdown products; and the intestine excretes certain salts. Excretion, ridding the body of metabolic wastes, should not be confused with defecation, ridding the body of nondigestible remains.

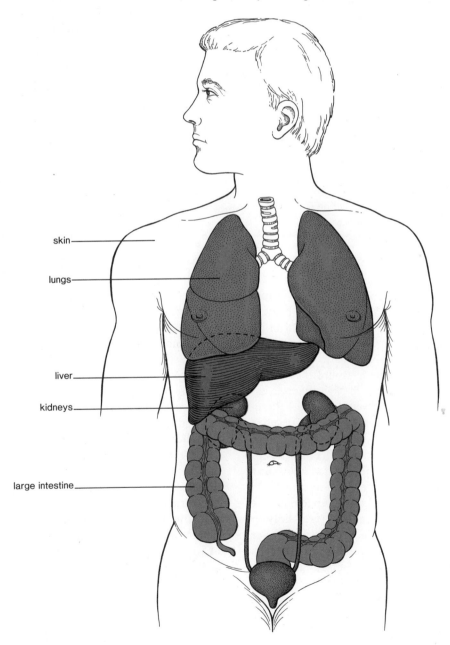

skin

lungs

liver

kidneys

large intestine

Figure 15.2 The urinary system. Urine is only found within the kidneys, ureters, bladder, and urethra.

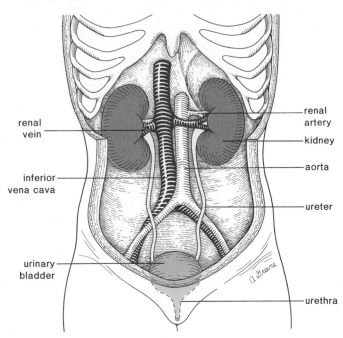

Figure 15.3 Longitudinal section of a male urethra leaving the bladder. Note the position of the prostate gland, which can enlarge to obstruct the flow of urine.

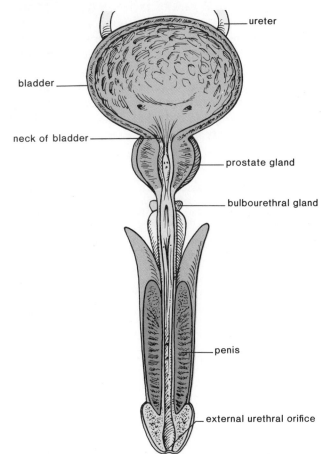

Table 15.1	Urinary System
Organ	**Function**
Kidneys	Produce urine
Ureters	Transport urine
Bladder	Stores urine
Urethra	Eliminates urine

Path of Urine

Urine is made by the kidneys, which are bean-shaped, reddish-brown organs, about the size of a fist. They are located one on either side of the vertebral column just below the diaphragm in the upper lumbar region. The kidneys lie in depressions against the deep muscles of the back beneath the peritoneum where they receive some protection from the lower rib cage. Each is covered by a tough capsule of fibrous connective tissue, called the renal capsule, overlaid by adipose tissue.

The **ureters** are muscular tubes that convey the urine from the kidneys toward the bladder by peristaltic contractions. Urine enters the bladder in jets that occur at the rate of one to five per minute.

The **urinary bladder,** which can hold up to 600 ml of urine, is a hollow muscular organ that gradually expands as urine enters. In the male, the bladder lies ventral to the rectum, seminal vesicles, and vas deferens. In the female, it is ventral to the uterus and upper vagina.

The **urethra,** which extends from the urinary bladder to an external opening, differs in length in females and males. In females, the urethra lies ventral to the vagina and is only about 2.5 cm long. The short length of the female urethra invites bacterial invasion and explains why females are more prone to bladder infections. In males, the urethra averages 15 cm when the penis is relaxed. As the urethra leaves the bladder, it is encircled by the prostate gland (fig. 15.3). In older men, enlargement of the prostate gland may prevent urination, a condition that can usually be surgically corrected.

There is no connection between the genital (reproductive) and urinary systems in females, but there is a connection in males. During urination in males, the urethra carries urine, and during sexual orgasm, the urethra transports semen. This double function does not alter the path of urine, and it is important to realize that urine is found only in those structures listed in table 15.1.

Figure 15.4 An anterior view of the kidneys, adrenal glands, and associated vessels. The right kidney is longitudinally dissected to show the internal structures.

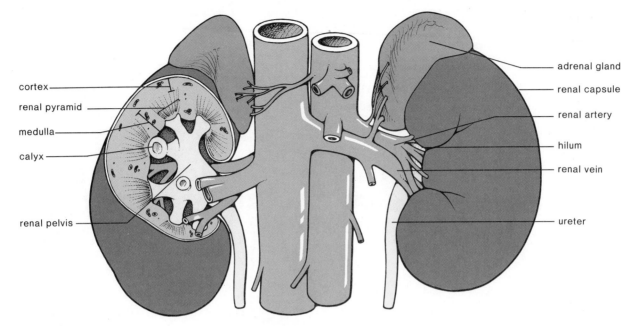

cortex
renal pyramid
medulla
calyx
renal pelvis

adrenal gland
renal capsule
renal artery
hilum
renal vein
ureter

Urination

When the bladder fills with urine, stretch receptors send nerve impulses to the spinal cord; nerve impulses leaving the cord then cause the bladder to contract and the sphincters to relax so that urination may take place. In older children and adults, it is possible for the brain to control this reflex, delaying urination until a suitable time.

Only the urinary system consisting of the kidneys, bladder, ureters, and urethra ever hold urine.

Kidneys

On the concave side of each kidney there is a depression, the hilum, where the renal blood vessels and the ureters enter. When a kidney is sliced lengthwise, it is possible to make out three regions (1) an outer granulated layer called the **cortex,** which dips down in between; (2) a radially striated, or lined, layer called the **medulla;** and (3) an inner space, or cavity, called the renal **pelvis** which is continuous with the ureter (fig. 15.4).

Upon closer examination, it can be seen that the medulla contains conical-shaped masses of tissue called *renal pyramids.* At the tip of each pyramid, there is a tube, called a *calyx* that empties into the pelvis.

Nephrons

Microscopically, the kidney is composed of over one million **nephrons,** sometimes called renal tubules. Each nephron is made up of several parts (fig. 15.5a). The blind end of the nephron is pushed in on itself to form a cuplike structure called **Bowman's capsule,** within which there is a capillary tuft called the *glomerulus.* Next, there is the **proximal** (meaning near the Bowman's capsule) **convoluted tubule,** which makes a U-turn to form the portion of the tubule called the loop of Henle. This leads to the **distal** (far from Bowman's capsule) **convoluted tubule,** which enters a **collecting duct.** Figure 15.5b indicates the position of a single nephron within the kidney. Bowman's capsules and convoluted tubules lie within the cortex and account for the granular appearance of the cortex. Loops of Henle and collecting ducts lie within the triangular-shaped *pyramids* of the medulla. Since these are longitudinal structures, they account for the striped appearance of the pyramids.

Urine Formation

Urine contains the substances listed in table 15.2. The kidneys are the chief excretory organs for nitrogenous wastes. *Ammonia* arises when amino acids are broken down in cells. Much of this ammonia is carried in the blood to the liver where it is converted to *urea,* a less

Figure 15.5 *a.* Diagram of nephron (kidney tubule) gross anatomy. You may trace the path of blood about the nephron by following the arrows. Note that the dotted line indicates which portions of the nephron are in the cortex and which portions are in the medulla of the kidney. *b.* Each kidney receives a renal artery that divides into arterioles within the kidney. Venules leaving the kidney join to form the renal vein. This drawing shows how one nephron is placed in the kidney so that some parts are in the cortex and other parts are in the medulla.

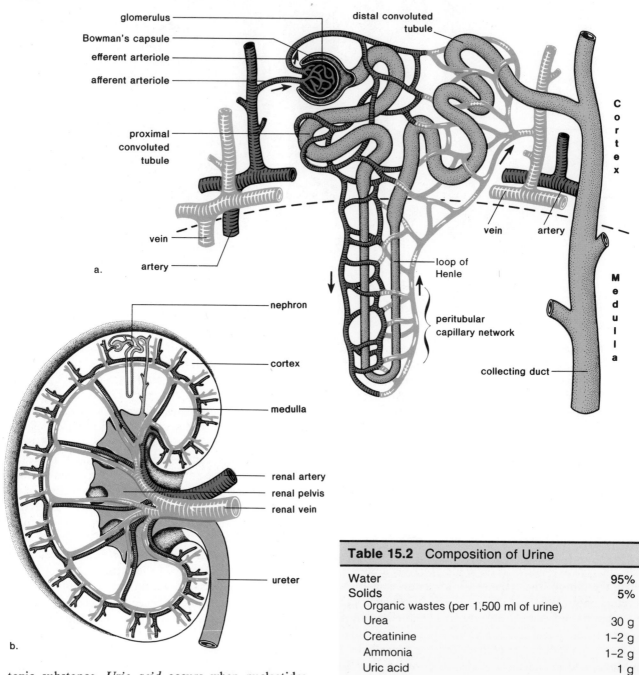

Table 15.2	Composition of Urine	
Water		95%
Solids		5%
Organic wastes (per 1,500 ml of urine)		
Urea		30 g
Creatinine		1–2 g
Ammonia		1–2 g
Uric acid		1 g
Ions (Electrolytes)		25 g
Positive	*Negative*	
Sodium	Chlorides	
Potassium	Sulfates	
Magnesium	Phosphates	
Calcium		

toxic substance. *Uric acid* occurs when nucleotides (p. 323) are broken down in cells. If uric acid is present in excess, it will precipitate out of the plasma. Crystals of uric acid sometimes collect in the joints, producing a painful ailment called gout. *Creatinine* is an end product of muscle metabolism. It results when creatine phosphate (p. 99) breaks down.

Salts are excreted because their proper concentration in the blood is so important to keeping the proper osmotic pressure and electrolyte balance of the blood. The balance of potassium (K^+) and sodium (Na^+) is important to nerve conduction, and the level of calcium (Ca^{++}) in the blood affects muscle contraction, for example.

Figure 15.6   Diagram of nephron showing steps in urine formation: filtration, reabsorption, and tubular excretion. Note also that water enters the tissues at the loop of Henle and collecting duct.

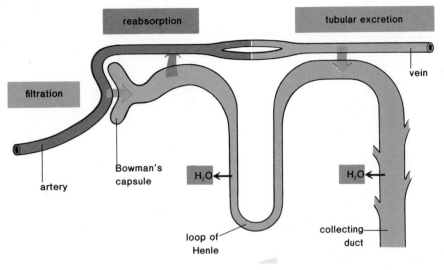

Steps in Urine Formation

Each nephron has its own blood supply (fig. 15.6), including two capillary regions: the **glomerulus** is a capillary tuft inside Bowman's capsule and the **peritubular capillary network** surrounds the rest of the nephron. Urine formation requires the movement of molecules between these capillaries and the nephron.

1. *Pressure filtration* occurs at Bowman's capsule. During pressure filtration, water, nutrient molecules, and waste molecules move from the glomerulus to the inside of Bowman's capsule. The blood has been *filtered* because large molecules, such as protein molecules, remain within the blood while small molecules, such as glucose and urea, leave the blood to enter the tubule.
2. *Selective reabsorption* occurs primarily at the proximal convoluted tubule. During selective reabsorption, nutrient and salt (Na^+Cl^-) molecules are actively reabsorbed from the proximal convoluted tubule into the peritubular capillary and water follows passively.
3. *Tubular excretion* occurs primarily at the distal convoluted tubule. Tubule excretion occurs when large waste molecules, such as creatine, are actively secreted into the distal convoluted tubule. This step in urine formation plays a minor role in comparison to the first two steps.

Concentrated Urine   Humans excrete a urine that contains only waste molecules dissolved in a minimum amount of water. This concentrated urine results because water is reabsorbed not only at the proximal convoluted tubule but along the entire length of the nephron, particularly at the loop of Henle and the col-

Table 15.3 Nephron		
Name of Part	**Location in Kidney**	**Function**
Bowman's capsule	Cortex	Forms filtrate
Proximal convoluted tubule	Cortex	Selective reabsorption
Loop of Henle	Medulla	Extrusion of sodium and reabsorption of water
Distal convoluted tubule	Cortex	Tubular excretion
Collecting duct	Medulla	Reabsorption of water

lecting duct. Table 15.3 lists the parts of a nephron, their location within the kidney, and their contribution to urine formation.

Wastes, nutrients, and water are all filtered into a nephron, but nutrients and water are reabsorbed so that humans excrete a concentrated solution of wastes.

Special Features of the Nephron

The cells along the length of the nephron (fig. 15.7) are specialized to carry on their respective functions.

The juxtaglomerular apparatus (fig. 15.8) occurs at a region of contact between the afferent arteriole and the distal convoluted tubule. The cells in this region are involved in regulating sodium (Na^+) reabsorption and in maintaining blood volume, as we shall see.

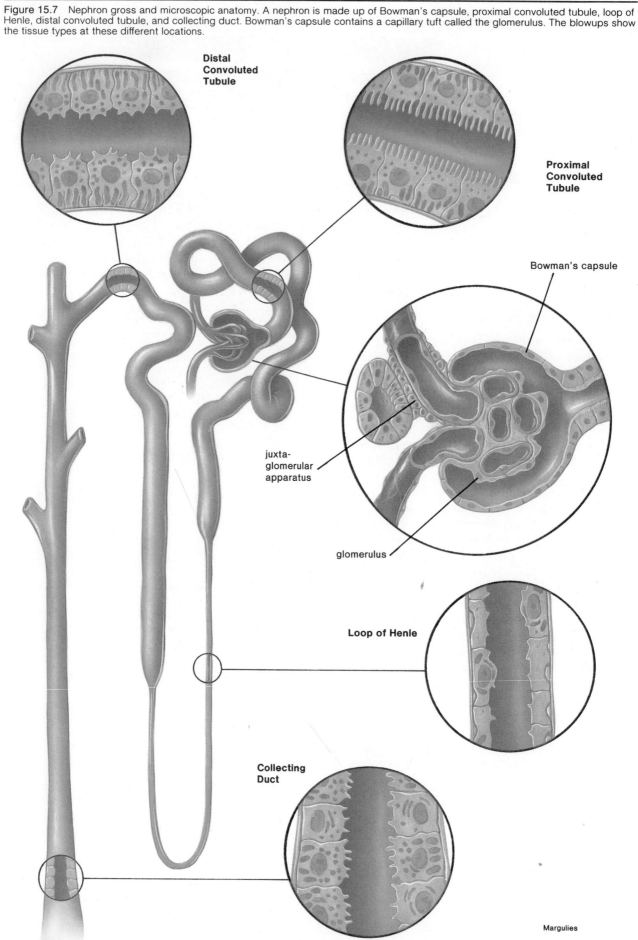

Figure 15.7 Nephron gross and microscopic anatomy. A nephron is made up of Bowman's capsule, proximal convoluted tubule, loop of Henle, distal convoluted tubule, and collecting duct. Bowman's capsule contains a capillary tuft called the glomerulus. The blowups show the tissue types at these different locations.

Distal
Convoluted
Tubule

Proximal
Convoluted
Tubule

Bowman's capsule

juxta-
glomerular
apparatus

glomerulus

Loop of Henle

Collecting
Duct

Margulies

Figure 15.8 Drawing of glomerulus and adjacent distal convoluted tubule. The cells of the juxtaglomerular apparatus (*circled*) are sensitive to blood pressure and release renin if this pressure falls below normal.

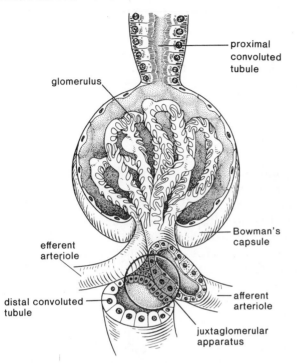

Figure 15.9 Electron micrograph of cells that line the lumen (*inside*) of a proximal convoluted tubule, where selective reabsorption takes place. The cells have a brush border composed of microvilli that greatly increase the surface area exposed to the lumen. Each cell has many mitochondria that supply the energy needed for active transport. A red blood cell (*lower right*) is seen in the peritubular capillary (Mv = microvilli; M = mitochondria).

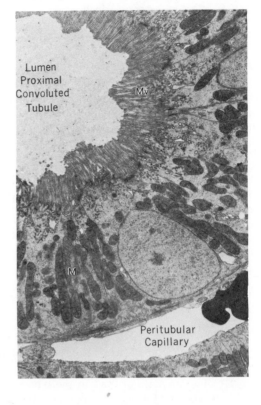

Each region of the nephron is anatomically suited to its task in urine formation.

The inner layer of Bowman's capsule is made up of podocytes that have long cytoplasmic processes. When these cling to the capillary walls of the glomerulus, they leave pores through which filtration can take place.

The cells lining the proximal convoluted tubule have numerous microvilli, about one micron in length, that increase the surface area for reabsorption (fig. 15.9). In addition, the cells contain numerous mitochondria, which produce the energy necessary for active transport. Reabsorption occurs until the threshold level of a substance is obtained. Thereafter, the substance will appear in the urine. For example, the threshold level of glucose is about 180 mg glucose per 100 ml of blood. After this amount is reabsorbed, any excess present in the filtrate will appear in the urine. In diabetes mellitus, the filtrate contains excess glucose because the liver fails to store glucose as glycogen or the body cells fail to take it up.

The loop of Henle is made up of a descending (going down) and ascending (going up) limb. The ascending limb extrudes sodium so that the tissues of the medulla become hypertonic to the fluid in the descending limb and the collecting duct. It is these features that allow the loop of Henle to perform its function of concentrating the urine.

The cells of the distal convoluted tubule also have numerous mitochondria, but they lack a brush border. This is consistent with their role in actively moving molecules from the blood into the tubule rather than in the opposite direction.

Regulatory Functions of the Kidney

Blood Volume

Maintenance of blood volume and electrolyte balance is under the control of hormones. ADH (antidiuretic hormone) is a hormone secreted by the posterior pituitary that primarily maintains blood volume. ADH increases the permeability of the collecting duct so that more water can be reabsorbed. In order to understand the function of this hormone, consider its name. Diuresis means increased amount of urine, and antidiuresis means decreased amount of urine. When ADH is present, more water is reabsorbed, and a decreased amount of urine results. This hormone is secreted according to whether blood volume needs to be increased or decreased. When water is reabsorbed at the collecting duct, blood volume increases, and when water is not reabsorbed, blood volume decreases. In practical terms (table 15.4), if an individual does not drink much

Table 15.4 Antidiuretic Hormone

Increase in ADH	Increased reabsorption of water	Less urine
Decrease in ADH	Decreased reabsorption of water	More urine

water on a certain day, the posterior lobe of the pituitary releases ADH; more water is reabsorbed; blood volume is maintained at a normal level; and, consequently, there is less urine. On the other hand, if an individual drinks a large amount of water and does not perspire much, the posterior lobe of the pituitary does not release ADH; more water is excreted; blood volume is maintained at a normal level; and a greater amount of urine is formed.

Drinking alcohol causes diuresis because it inhibits the secretion of ADH. The dehydration that follows is believed to contribute to the symptoms of a "hangover." Drugs called diuretics are often prescribed for high blood pressure. The drugs cause increased urinary excretion and, therefore, reduce blood volume and blood pressure. Concomitantly, any *edema* present is also reduced.

Aldosterone secreted by the adrenal cortex is a hormone that primarily maintains Na^+ and K^+ balance. It causes the distal convoluted tubule to reabsorb Na^+ and excrete K^+. The increase of Na^+ in the blood causes water to be reabsorbed, leading to an increase in blood volume and blood pressure.

Blood pressure is constantly monitored by the afferent arteriole cells within the juxtaglomerular apparatus (fig. 15.8). The afferent arteriole cells in the region secrete renin when the blood pressure is insufficient to promote efficient filtration in the glomerulus. Renin is an enzyme that converts a large plasma protein angiotensinogen into angiotensin I, a molecule that is converted to angiotensin II in the lungs. Angiotensin II stimulates the adrenal cortex to release aldosterone. The renin-angiotensin-aldosterone system seems to be always active in some people who have hypertension.

Electrolyte Balance

The kidneys regulate the electrolyte balance in the blood by controlling the excretion and reabsorption of various ions. Sodium, Na^+, is an important ion in plasma that must be regulated, but the kidneys also excrete or reabsorb other ions such as HCO_3^-, K^+, and Mg^{++} as needed.

This action of the kidneys also controls the acid/base balance of the body. Body fluids are usually slightly basic.

The kidneys contribute to homeostasis not only by excreting nitrogen wastes, but also by controlling the volume, electrolyte balance, and pH of the blood.

Clinical Conditions

Urinalysis, or examination of the urine, indicates if there are any abnormal substances in the urine. The most likely of these are discussed in the following paragraphs.

Diabetes Glucose in the urine, which usually means that the individual has diabetes mellitus, a condition in which either the liver fails to store glucose as glycogen or the cells fail to take glucose up. In both cases there is an abnormally high blood glucose level. This makes the filtrate level of glucose high and, because the proximal convoluted tubule only absorbs an amount appropriate to the normal blood glucose level, glucose appears in the urine.

Renal Disease The urinary tract is subject to attack by a number of different bacteria. If the infection is localized in the urethra, it is called **urethritis.** If it invades the bladder, it is called **cystitis.** And, finally, if the kidneys are affected, it is called **nephritis.**

In renal disease, the glomerular membrane becomes more permeable than usual. Therefore albumin, white cells, or even red cells may appear in the urine. One of the first indications of renal disease may be kidney edema.

Kidney Edema When plasma proteins are excreted by way of the urine, the osmotic pressure of the blood is reduced. Water collects in the tissues because water uptake at the venous end of the capillaries (fig. 11.7) is dependent on the osmotic pressure of the blood. When the amount of tissue fluid increases beyond the capability of the lymph vessels to absorb it, edema, particularly in the abdomen, occurs.

As the blood pressure lowers due to a loss in blood volume, the kidneys reabsorb more salt and water, but this, in the end, only serves to increase the edema. The only permanent solution is to cure the underlying cause of the edema.

Uremia Uremia, the presence of urea in the blood, is concomitant with kidney failure. Death from kidney failure, however, is not due to the buildup of nitrogenous wastes, rather it is due to an imbalance of the plasma ions. Studies have shown that if urea is high but the ions are stabilized at their normal levels, the patient usually recovers from the symptoms of uremia. An ion

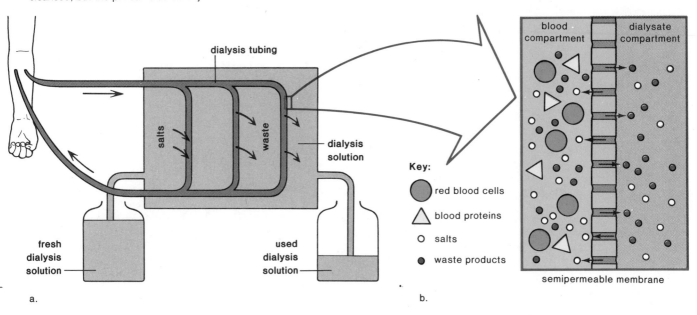

Figure 15.10 Diagram of an artificial kidney. *a.* As the patient's blood circulates through dialysis tubing, it is exposed to a solution. *b.* Wastes exit from the blood into the solution because of a preestablished concentration gradient. In this way, the blood is not only cleansed, but the pH can also be adjusted.

a.

Key:

🔴 red blood cells

🔺 blood proteins

○ salts

● waste products

b.

blood compartment   dialysate compartment

semipermeable membrane

imbalance, however, particularly the accumulation of potassium in the blood, interferes with the heartbeat and leads to heart failure.

Kidney Transplant Patients with renal failure can sometimes undergo a kidney transplant operation during which they receive a functioning kidney from a donor. Because the body can function quite well with only one kidney, both the donor and recipient can expect to lead normal lives. As with all organ transplants, there is the possibility of organ rejection, so that receiving a kidney from a close relative has the highest chance of success. Identical twins have the same transplant antigens, which allows them to readily donate organs to each other.

Kidney transplants and hemodialysis are available procedures for persons who have suffered renal failure.

Hemodialysis If a satisfactory donor cannot be found for a kidney transplant, which is frequently the case, the patient may undergo hemodialysis treatments, utilizing either a kidney machine or Continuous Ambulatory Peritoneal Dialysis, CAPD.

Dialysis is defined as the diffusion of dissolved molecules through a semipermeable membrane. These molecules will, of course, move across a membrane from the area of greater concentration to one of lesser concentration. While attached to a kidney machine (fig. 15.10), the patient's blood is passed through a semi-

permeable membranous tube that is in contact with a balanced salt solution, or dialysate. Substances more concentrated in the blood diffuse into the dialysate. Conversely, substances more concentrated in the dialysate diffuse into the blood. Accordingly, the artificial kidney can be utilized either to extract substances from the blood, including waste products or toxic chemicals and drugs, or to add substances to the blood, for example, bicarbonate ions in acidosis. In the course of a six-hour hemodialysis, from 50 to 250 g of urea can be removed from a patient, which greatly exceeds the urea clearance of normal kidneys. Therefore, a patient need undergo treatment only about twice a week.

In CAPD (fig. 15.11), a fresh amount of dialysate is introduced into the abdominal cavity from a bag attached to a permanently implanted plastic tube. Waste and water molecules pass into the fluid before it is collected four to eight hours later. The individual can go about his or her normal activities during CAPD unlike hemodialysis.

Kidney Stones Kidney stones are formed when fairly insoluble substances, such as uric acid and calcium salts, precipitate out of the urine instead of remaining in solution. The stones, which usually form in the renal pelvis, although the bladder can be another site of formation, either pass naturally or they may be surgically removed. A new treatment is to "smash" them with ultrasonic waves after which the fragments will pass naturally.

Figure 15.11 CAPD (Continuous Ambulatory Peritoneal Dialysis). *a.* Dialysate fluid is introduced into the abdominal cavity by way of a plastic bag. *b.* After bag is securely placed at the waist, patient can move freely about. *c.* After four to eight hours, the old fluid is removed before the procedure is repeated.

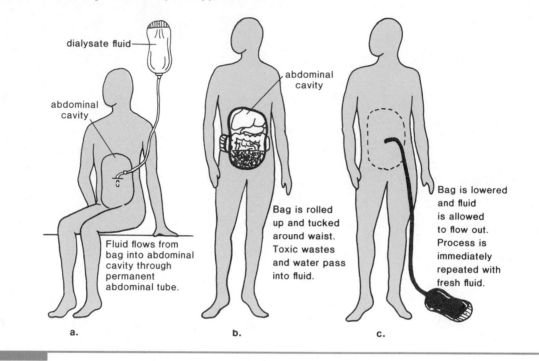

dialysate fluid

abdominal cavity

Fluid flows from bag into abdominal cavity through permanent abdominal tube.

a.

abdominal cavity

Bag is rolled up and tucked around waist. Toxic wastes and water pass into fluid.

b.

Bag is lowered and fluid is allowed to flow out. Process is immediately repeated with fresh fluid.

c.

Summary

There are various organs that excrete metabolic wastes, but the kidneys are the primary organs.

I. Urinary System
  A. Path of urine. The path of urine is the kidneys, ureters, urinary bladder, and urethra.
  B. Kidneys. Macroscopically, the kidneys are divided into the cortex, medulla, and pelvis. Microscopically, they contain the nephrons.

II. Urine Formation
  A. Steps in urine formation. The steps in urine formation are pressure filtration, selective reabsorption, and tubular excretion.
  B. Regulatory functions of the kidney. The kidneys regulate the blood volume, adjust the pH, and maintain the electrolyte balance.

III. Illnesses
  Renal disease, kidney edema, and uremia are illnesses of extreme concern that may require hemodialysis or a kidney transplant.

Study Questions

1. Name several excretory organs and the substances they excrete.
2. Give the path of urine.
3. Describe the macroscopic anatomy of the kidney.
4. Name the parts of the nephron and describe the anatomy of each part.
5. What is the composition of urine?
6. Name three nitrogenous end products and explain how each is formed in the body.
7. Describe how urine is made by telling what happens at each part of the tubule.
8. Explain these terms: pressure filtration, selective reabsorption, and tubular excretion.
9. How does the nephron regulate the pH of the blood?
10. Explain how the artificial kidney machine and CAPD work.

Objective Questions

Fill in the blanks.

1. The primary nitrogenous end product of humans is _____ .
2. The intestines are an organ of excretion because they rid the body of _____ .
3. Urine leaves the bladder in the _____ .
4. The capillary tuft inside Bowman's capsule is called the _____ .

5. _____ is a substance that is found in the filtrate, is reabsorbed, and still is in urine.
6. _____ is a substance that is found in filtrate, is not reabsorbed, and is concentrated in urine.
7. Tubular excretion takes place at the _____ , a portion of the nephron.

8. Reabsorption of water from the collecting duct is regulated by the hormone _____ .
9. In addition to excreting nitrogenous wastes, the kidneys adjust the _____ and _____ of blood.
10. Persons who have nonfunctioning kidneys are often on _____ machines.

Medical Terminology Reinforcement Exercise

Pronounce and analyze the meaning of the following dissected terms:

1. hematuria (hem″ah-tu-re′ah) hemat/uria
2. oliguria (ol″ĭ-gu′re-ah) olig/uria
3. polyuria (pol″e-u′re-ah) poly/uria
4. extracorporal shock wave lithotripsy (ESWL) (eks″trah-kor-op′re-al lith″o-trip′se) extra/corporal shock wave litho/tripsy
5. antidiuretic (an″tĭ-di″u-ret′ik) anti/di/uret/ic
6. urethratresia (u-re″thrah-tre′ze-ah) urethr/atresia
7. cystopyelonephritis (sis″to-pi″e-lo-ne-frit′tis) cysto/pyelo/nephr/itis

8. nocturia (nok-tu′re-ah) noct/uria
9. glomerulonephritis (glo-mer″u-lo-nĕ-fri′tis) glomerulo/nephr/itis
10. ureterovesicostomy (u-re″ter-o-ves″ĭ-kos′to-me) uretero/vesico/stomy

V

Reproduction and Development

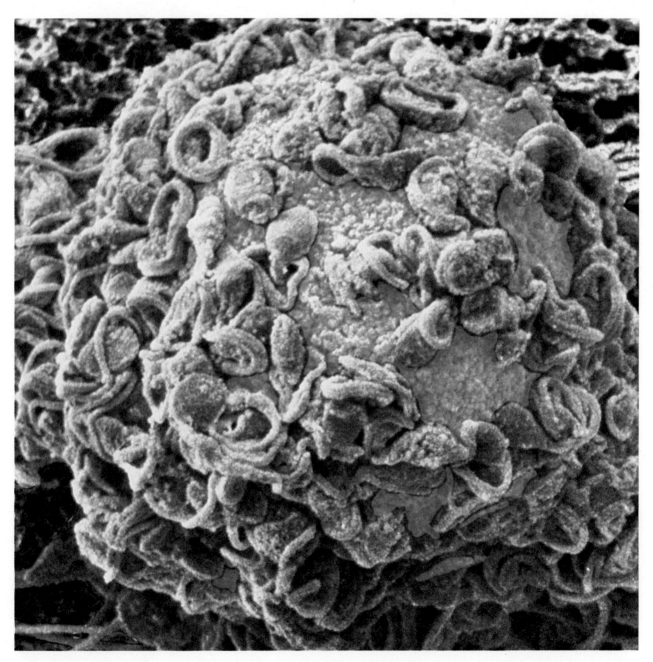

The human egg shown here is covered by sperm, yet only one of these will penetrate the egg and contribute chromosomes to the new individual. As this sperm enters the zona pellucida, a chemical change occurs that prevents other sperm from penetrating this thin transparent layer outside the egg.

16

Reproductive System

Chapter Outline

Male Reproductive System
  Testes
  Genital Tract
  Penis
  Hormonal Regulation in the Male
Female Reproductive System
  Ovaries
  Genital Tract
  External Genitalia
  Hormonal Regulation in the Female
  (Simplified)
Control of Reproduction
  Birth Control
  Infertility

Learning Objectives

After you have studied this chapter, you should be able to:

1. Name and state the function of the reproductive structures in the male.
2. State the path that sperm take from their site of production to the site of fertilization.
3. Describe the macroscopic and microscopic anatomy of the testes.
4. Name the glands and their products that contribute to the composition of semen.
5. Describe the anatomy of the penis and the events preceding and during ejaculation.
6. Discuss hormonal regulation in the male in the same depth as the text.
7. Name at least six actions of testosterone, including both primary and secondary sex characteristics.
8. Name and state the functions of the reproductive structures in the female.
9. Describe the macroscopic and microscopic anatomy of the ovaries.
10. Label a diagram of the external female genitalia.
11. Contrast male orgasm with female orgasm.
12. Describe the ovarian and uterine cycles.
13. Discuss hormonal regulation in the female, including feedback control, in the same depth as the text.
14. Name at least six actions of estrogen and progesterone, including both primary and secondary sex characteristics.
15. Categorize birth control measures according to the system used in the text.
16. State the basis for pregnancy tests and describe the physical signs of pregnancy.

Male Reproductive System

Human beings practice sexual reproduction, and there are two separate sexes, males and females. Even though the anatomy differs between the sexes, there are certain similarities. For example, each sex produces sex cells, or **gametes,** within primary reproductive organs that are usually termed the **gonads.** After puberty, males produce sperm within the testes and females produce eggs within the ovaries. The gonads not only produce the gametes, they also secrete hormones, which are necessary for the growth and maintenance of the gonads and the accessory reproductive organs. The accessory reproductive organs include the ducts through which the sperm and egg pass after leaving the gonads and certain other structures that we will be discussing. The sex hormones account for the development of secondary sex characteristics by which we usually distinguish males and females without the need of examining the genitals. The secondary sex characteristics become obvious at the time of puberty.

Figure 16.1 shows the reproductive system of the male and table 16.1 lists the anatomical parts of this system.

Testes

The **testes** lie outside the abdominal cavity of the male within the **scrotum,** a pouch that consists of a layer of smooth muscle covered by skin. The smooth muscle layer forms two cavities, one for each testis. Normally,

Table 16.1 Male Reproductive System

Organ	Function
Testes	Produce sperm and sex hormones
Epididymis	Maturation and some storage of sperm
Vas deferens	Conducts and stores sperm
Seminal vesicles	Contribute to seminal fluid
Prostate gland	Contributes to seminal fluid
Urethra	Conducts sperm
Cowper's glands	Contribute to seminal fluid
Penis	Organ of copulation

Figure 16.1 Side view of male reproductive system. Trace the path of the genital tract from a testis to the exterior. The seminal vesicles, Cowper's gland, and prostate gland produce seminal fluid and do not contain sperm. Notice that the penis in this drawing is not circumcised since the foreskin is present.

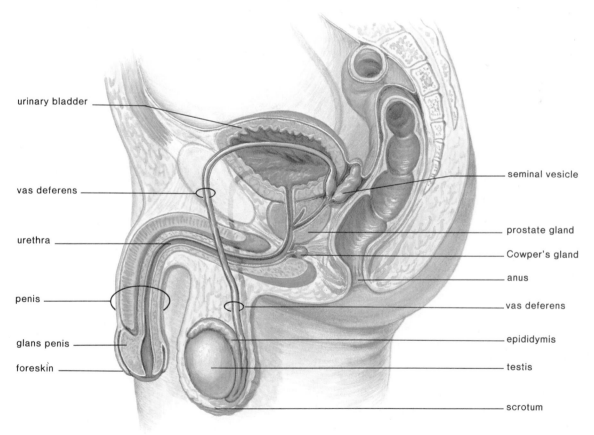

Figure 16.2 Testis anatomy and physiology. *a.* Longitudinal section showing lobules containing seminiferous tubules. *b.* Cross section of a tubule showing germ cells in the various stages of spermatogenesis. *c.* Diagram of spermatogenesis. Primary spermatocytes have 46 chromosomes, but the other cells including the sperm have 23 chromosomes. Four spermatids are shown because there are two cell divisions called meiosis I and meiosis II.

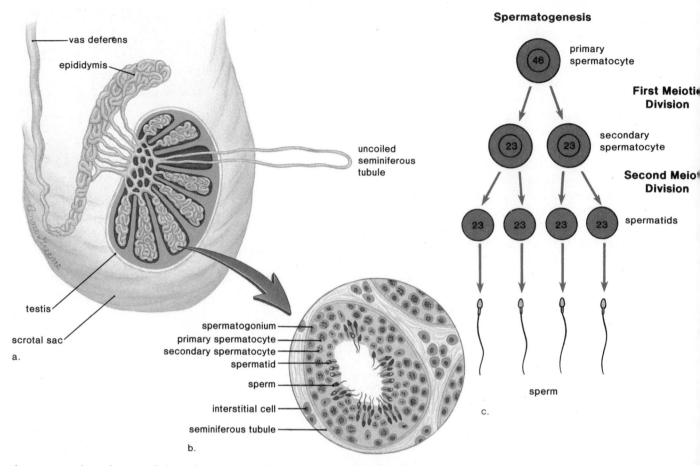

the scrotum is rather pendulous, but on occasion, certain stimuli such as fright or cold temperature can cause contraction of the muscle so that the scrotum is held closer to the body.

The testes begin their development inside the abdominal cavity, but descend into the scrotal sacs during the last two months of fetal development. If, by chance, the testes do not descend and the male is not treated or operated on to place the testes in the scrotum, sterility—the inability to produce offspring—usually follows. This is because the internal temperature of the body is too high to produce viable sperm.

Seminiferous Tubules

Fibrous connective tissue forms the wall of each testis and divides it into lobules (fig. 16.2). Each lobule contains one to three tightly coiled **seminiferous tubules,** which have a combined length of approximately 250 m. A microscopic cross section through a tubule shows it is packed with cells undergoing spermatogenesis. These cells are derived from undifferentiated germ cells, called spermatogonia (singular spermatogonium), that lie just inside the outer wall and divide mitotically, always producing new spermatogonia. Some spermatogonia move

away from the outer wall to increase in size and become primary spermatocytes that undergo meiosis, the type of cell division described in figure 16.2c. Although primary spermatocytes have forty-six chromosomes, they divide (called meiosis I) to give secondary spermatocytes, each with twenty-three chromosomes. Secondary spermatocytes divide (called meiosis II) to give spermatids, that also have twenty-three chromosomes, but are single stranded. Spermatids then differentiate into spermatozoa, or mature sperm.

Sperm The mature sperm, or spermatozoan (fig. 16.3), has three distinct parts: a head, a mid-piece, and a tail. The *tail* flagella (fig. 2.10) and the *mid-piece* contain energy-producing mitochondria. The *head* contains the twenty-three chromosomes within a nucleus. The tip of the nucleus is covered by a cap called the **acrosome,** which is believed to contain enzymes needed for fertilization. The human egg is surrounded by several layers of cells and a mucoprotein substance. The acrosome enzymes are believed to aid the sperm in reaching the surface of the egg and allowing a single sperm to penetrate the egg. It is hypothesized that each acrosome contains such a minute amount of enzyme

Figure 16.3 Microscopic anatomy of sperm.

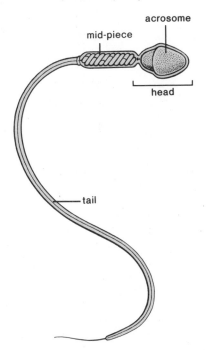

that it requires the action of many sperm to allow just one to actually penetrate the egg. This may explain why so many sperm are required for the process of fertilization. A normal human male usually produces several hundred million sperm per day, an adequate number for fertilization. Sperm are continually produced throughout a male's reproductive life.

In males, spermatogenesis occurs within the seminiferous tubules of the testes. Sperm have a head, capped by an acrosome, where twenty-three chromosomes reside in the nucleus; a mitochondria-containing mid-piece; and a tail that contains microtubules.

Interstitial Cells

The male sex hormones, the androgens, are secreted by cells that lie between the seminiferous tubules and are, therefore, called **interstitial cells.** The most important of the androgens is testosterone, whose functions are discussed on page 279.

Genital Tract

Sperm are produced in the testes, but they mature in the **epididymis** (fig. 16.1), a tightly coiled tubule about 6 m in length that lies just outside each testis. During the two-to-four-day maturation period, the sperm develop their characteristic swimming ability. Also, it is possible that during this time, defective sperm are removed from the epididymis. Each epididymis joins with a **vas deferens,** which ascends through a canal called the *inguinal canal* and enters the abdomen where it

curves around the bladder and empties into the urethra. Sperm are stored in the last part of the epididymis and the first part of the vas deferens. They pass from each vas deferens into the urethra only when ejaculation is imminent.

Seminal Fluid

At the time of ejaculation, sperm leave the penis in a fluid called **seminal fluid.** This fluid is produced by three types of glands—the seminal vesicles, the prostate gland, and Cowper's glands. The **seminal vesicles** lie at the base of the bladder, and each has a duct that joins with a vas deferens. The **prostate gland** is a single doughnut-shaped gland that surrounds the upper portion of the urethra just below the bladder. In older men, the prostate may enlarge and cut off the urethra, making urination painful and difficult. This condition may be treated medically or surgically. **Cowper's glands** are pea-sized organs that lie inferior to the prostate on either side of the urethra.

Each component of seminal fluid seems to have a particular function. Sperm are more viable in a basic solution; seminal fluid, which is white and milky in appearance, has a slightly basic pH (about 7.5). Swimming sperm require energy, and seminal fluid contains the sugar fructose, which presumably serves as an energy source. Seminal fluid also contains prostaglandins, chemicals that cause the uterus to contract. Some investigators now believe that uterine contraction is necessary to propel the sperm and that the sperm swim only when they are in the vicinity of the egg.

Penis

The **penis** has a long shaft and an enlarged tip called the glans penis. At birth, the penis is covered by a layer of skin called the **foreskin** or prepuce. Gradually, over a period of five to ten years, the foreskin becomes separated from the penis and may be retracted. During this time, there is a natural shedding of cells between the foreskin and penis. These cells along with an oil secretion that begins at puberty is called smegma. In the child, no special cleansing method is needed to wash away smegma, but in the adult, the foreskin can be retracted to do so. **Circumcision** is the surgical removal of the foreskin soon after birth.

The penis is the copulatory organ of males. When the male is sexually aroused, the penis becomes erect and ready for intercourse (fig. 16.4). **Erection** is achieved because blood sinuses within the erectile tissue of the penis become filled with blood. Parasympathetic impulses dilate the arteries of the penis, while the veins are passively compressed so that blood flows into the erectile tissue under pressure. If the penis fails to become erect, the condition is called **impotency.** Although it was formerly believed that almost all cases

Figure 16.4 Erection contrasted to a flaccid penis and relaxed scrotum. The erectile tissue fills with blood when the penis becomes erect. The penis has three columns of erectile tissue—only one is shown in this diagram. Also note that the penis lacks a foreskin, due to circumcision.

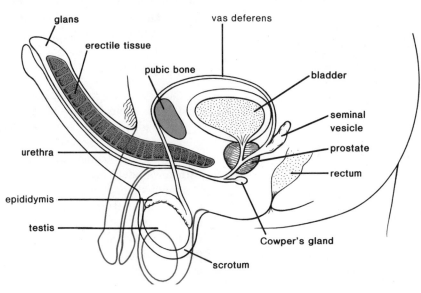

of impotency were due to psychological reasons, it has recently been reported that some cases may be due to hormonal imbalances. Treatment consists of finding the precise imbalance and restoring the proper level of testosterone.

Ejaculation

As sexual stimulation becomes intense, sperm enter the urethra from each vas deferens and the glands secrete seminal fluid. Sperm and seminal fluid together are called **semen.** Once semen is in the urethra, rhythmical muscle contractions cause it to be expelled from the penis in spurts. During ejaculation, a sphincter closes off the bladder so that no urine enters the urethra. (Notice that the urethra carries either urine or semen at different times.)

The contractions that expel semen from the penis are a part of male **orgasm,** the physiological and psychological sensations that occur at the climax of sexual stimulation. The psychological sensation of pleasure is centered in the brain, but the physiological reactions involve the genital (reproductive) organs and associated muscles as well as the entire body. Marked muscle tension is followed by contraction and relaxation.

Following ejaculation and/or loss of sexual arousal, the penis returns to its normal flaccid state. After ejaculation, a male typically experiences a period of time, called the refractory period, during which stimulation does not bring about an erection.

There may be in excess of 400 million sperm in 3.5 ml of semen expelled during ejaculation. The sperm count can be much lower than this, however, and fertilization will still take place.

Sperm mature in the epididymis and are also stored in the vas deferens before entering the urethra just prior to ejaculation. The accessory glands (seminal vesicles, prostate gland, and Cowper's gland) produce seminal fluid. Semen, which contains sperm and seminal fluid, leaves the penis during ejaculation.

Hormonal Regulation in the Male

The hypothalamus has ultimate control of the testes' sexual functions because it secretes a gonadotropic-releasing hormone (GnRH) that stimulates the anterior pituitary to produce the gonadotropic hormones. Two gonadotropic hormones, **FSH** (follicle-stimulating hormone) and **LH** (luteinizing hormone), are named for their function in females, but exist in both sexes, stimulating the appropriate gonads in each. FSH promotes spermatogenesis in the seminiferous tubules and LH promotes the production of testosterone in the interstitial cells. Sometimes LH in males is given the name interstitial cell-stimulating hormone (ICSH).

The hormones mentioned are involved in a negative feedback process (fig. 16.5) that maintains the production of testosterone at a fairly constant rate. For example, when the amount of testosterone in the blood rises to a certain level, it causes the hypothalamus to decrease its secretion of releasing hormone, which causes the anterior pituitary to decrease its secretion of LH. As the level of testosterone begins to fall, the hypothalamus increases secretion of the releasing hormone and the anterior pituitary increases its secretion of LH, and stimulation of the interstitial cells reoccurs. It should be emphasized that only minor fluctuations

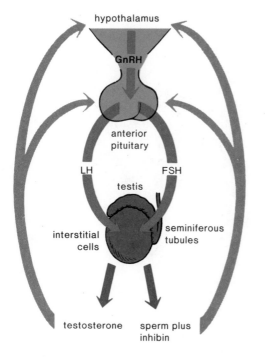

Figure 16.5 Hypothalamic-pituitary-gonad system as it functions in the male. GnRH is a hypothalamic-releasing hormone that stimulates the anterior pituitary to secrete LH and FSH. These gonadotropic hormones act on the testes. LH promotes the production of testosterone, and FSH promotes spermatogenesis. Negative feedback controls the level of all hormones involved.

of testosterone level occur in the male and that the feedback mechanism in this case acts to maintain testosterone at a normal level. It has long been suspected that the seminiferous tubules produce a hormone that blocks FSH secretion. This substance, termed inhibin, has recently been isolated.

Testosterone

The male sex hormone, **testosterone,** has many functions. It is essential for the normal development and functioning of the primary sex organs, which we have just discussed. It is also necessary for the maturation of sperm, probably after diffusion from the interstitial cells into the seminiferous tubules.

Greatly increased testosterone secretion at the time of puberty stimulates the growth of the penis and testes. Testosterone also brings about and maintains the secondary sex characteristics in males that develop at the time of puberty. Testosterone causes growth of a beard, axillary (underarm) hair, and pubic hair. It prompts the larynx and vocal cords to enlarge, causing the voice to change. It is responsible for the greater muscle strength of males, and this is the reason why some athletes take supplemental amounts of *anabolic steroids,* which are either testosterone or related chemicals. The contraindications of taking anabolic steroids are dis-

cussed in the reading on page 175. Testosterone also causes oil and some sweat glands in the skin to secrete; therefore, it is largely responsible for acne and body odor. Another side effect of testosterone activity is baldness. Genes for baldness are probably inherited by both sexes, but baldness is seen more often in males because of the presence of testosterone.

Testosterone is believed to be largely responsible for the sex drive and may even contribute to the supposed aggressiveness of males.

In males, FSH promotes spermatogenesis and LH promotes testosterone production. Testosterone stimulates growth of the male genitals during puberty, and is necessary for the maturation of sperm and development of the secondary sex characteristics.

Female Reproductive System

Table 16.2 lists the anatomical parts of this system, and figure 16.6 shows the reproductive system of the female.

Table 16.2 Female Reproductive System	
Organ	**Function**
Ovaries	Produce egg and sex hormones
Uterine tubes	Conduct egg
Uterus (womb)	Location of developing fetus
Cervix	Contains opening to uterus
Vagina	Organ of copulation and birth canal

Ovaries

The ovaries lie in shallow depressions, one on each side of the upper pelvic cavity. A longitudinal section through an ovary shows that it is made up of an outer cortex and an inner medulla. The cortex contains ovarian **follicles** at various stages of maturation. A female is born with a large number of follicles (400,000) in both ovaries, each containing a potential egg. In contrast to the male, the female produces no new primary oocytes after she is born. Only a small number of follicles (about 400) ever mature because a female produces only one egg per month during her reproductive years. Since follicles are present at birth, they age as the woman ages. This is one possible reason why older women are more likely to produce children with genetic defects.

As the follicle undergoes maturation, it develops from a primary to a secondary to a **Graafian follicle.**

Figure 16.6 Parasagittal view of female reproductive system. The ovaries produce one egg a month; fertilization occurs in the oviduct and development occurs in the uterus. The vagina is the birth canal and organ of copulation.

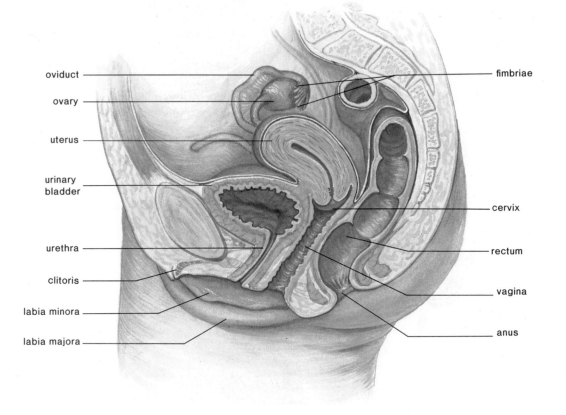

In a primary follicle a primary oocyte divides meiotically (called meiosis I) into two cells, each having twenty-three chromosomes (fig. 16.7). One of these cells, termed the secondary oocyte, receives almost all the cytoplasm, nutrients, and enzymes. The other is a polar body that disintegrates. A secondary follicle contains the secondary oocyte pushed to one side of a fluid-filled cavity. In a Graafian follicle, the fluid-filled cavity increases to the point that the follicle wall balloons out on the surface of the ovary and bursts, releasing the secondary oocyte, surrounded by the zona pellucida and a few cells; this is called **ovulation.** Once a follicle has lost its secondary oocyte (sometimes called an egg), it develops into a **corpus luteum,** a glandlike structure. Actually another division, called meiosis II, is required before there is a mature egg. Meiosis II does not occur until the secondary oocyte is fertilized by a sperm. If fertilization does not occur, the secondary oocyte disintegrates. If fertilization and pregnancy do not occur, the corpus luteum begins to degenerate after about ten days. If fertilization and pregnancy do occur, the corpus luteum persists for three to six months.

The follicle and corpus luteum secrete the female sex hormones estrogen and progesterone, which is discussed on page 284.

In females, oogenesis occurs within the ovaries, where one follicle reaches maturity each month. The follicle balloons out of the ovary and bursts to release the egg. The ruptured follicle develops into a corpus luteum. The follicle and corpus luteum produce the female sex hormones estrogen and progesterone.

Genital Tract

The female genital tract includes the oviducts, uterus, and vagina.

Oviducts

The oviducts (uterine or fallopian tubes) extend from the uterus to the ovaries. The oviducts are not attached to the ovaries but instead have fingerlike projections, called **fimbria,** that sweep over the ovary at the time of ovulation. When the egg bursts (fig. 16.7) from the ovary during ovulation, it is usually swept up into an oviduct by the combined action of the fimbria and the beating of cilia that line the tubes.

Figure 16.7 Ovary anatomy and physiology. *a.* Diagram of oogenesis. Primary oocyte has 46 chromosomes, but the other cells including the egg have 23 chromosomes. One egg and at least two polar bodies result because there are two cell divisions called meiosis I and meiosis II. The second division does not occur unless there is fertilization. *b.* Longitudinal section of ovary showing the development and maturation of a follicle; ovulation; and formation and degeneration of corpus luteum.

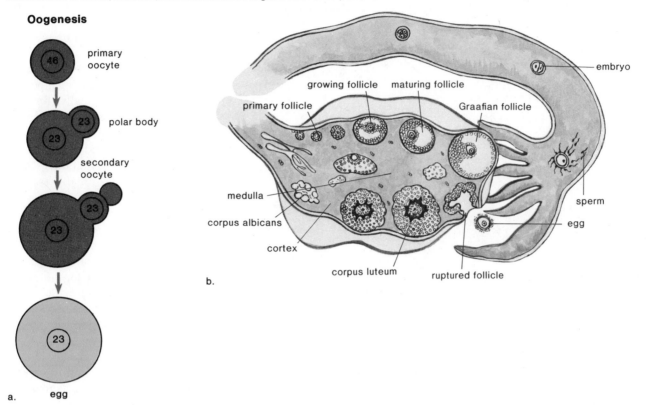

Oogenesis

Uterus

The **uterus** is a thick-walled, muscular organ about the size and shape of an inverted pear. Normally it lies above, and is tipped over, the urinary bladder. The oviducts join the uterus anteriorly, and posteriorly the cervix enters into the vagina nearly at a right angle. A small opening in the cervix leads to the vaginal canal. Development of the embryo normally takes place in the uterus. This organ, sometimes called the womb, is approximately 5 cm wide in its usual state but is capable of stretching to over 30 cm to accommodate the growing baby.

The wall of the uterus has three layers. The *perimetrium* is a layer of peritoneum that also covers the broad ligament, a band of connective tissue on both sides that anchors the uterus to the body wall. The *myometrium* is a thick layer of smooth muscle, and the **endometrium** is a layer of mucous membrane that lines the cavity of the uterus. In the nonpregnant female, the thickness of the endometrium varies according to a monthly reproductive cycle, called the uterine cycle (p. 283).

Because the egg must traverse a small space before entering an oviduct, it is possible for the egg to get lost and instead enter the abdominal cavity. Such eggs usually disintegrate but in some rare cases have been fertilized in the abdominal cavity and have implanted themselves in the wall of an abdominal organ. Very rarely, such embryos have come to term, the child being delivered by surgery.

Once in an oviduct, the egg is propelled slowly by cilia movement and tubular muscular contraction toward the uterus. Fertilization usually occurs in an oviduct because the egg only lives approximately six to twenty-four hours. The developing embryo normally arrives at the uterus after several days and then embeds, or implants, itself in the uterine lining, which has been prepared to receive it. Occasionally, the zygote becomes embedded in the wall of an oviduct, where it begins to develop. Tubular pregnancies cannot succeed because the tubes are not anatomically capable of allowing full development to occur. An *ectopic pregnancy* is one that begins outside the uterus.

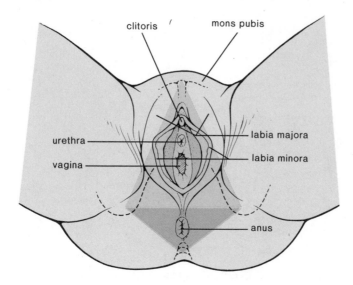

Figure 16.8 External genitalia of female. At birth, the opening of the vagina is partially occluded by a membrane called the hymen. Physical activities and sexual intercourse disrupt the hymen.

clitoris

mons pubis

urethra

vagina

labia majora

labia minora

anus

Cancer of the cervix is a common form of cancer in women. Early detection is possible by means of a **Pap test,** which requires that the doctor remove a few cells from the region of the cervix for microscopic examination. If the cells are cancerous, a hysterectomy may be recommended. A hysterectomy is the removal of the uterus. Removal of the ovaries in addition to the uterus, is termed an ovariohysterectomy. Since the vagina remains, the woman may still engage in sexual intercourse.

Vagina

The **vagina** is a tube that makes a 45-degree angle with the small of the back. The mucosal lining of the vagina lies in folds that extends as the fibromuscular wall stretches. This capacity to extend is especially important when the vagina serves as the birth canal, and it may also facilitate intercourse when the vagina receives the penis during copulation.

External Genitalia

The external genital organs of the female (fig. 16.8) are known collectively as the **vulva.** The vulva includes two large, hair-covered folds of skin called the **labia majora.** They extend backward from the *mons pubis,* a fatty prominence underlying the pubic hair. The **labia minora** are two small folds lying just inside the labia majora. They extend forward from the vaginal opening to encircle and form a foreskin for the *clitoris,* an organ that is homologous to the penis. Although quite small, the clitoris has a shaft of erectile tissue and is capped

by a pea-shaped glans. The clitoris also has sense receptors that allow it to function as a sexually sensitive organ.

The *vestibule,* a cleft between the labia minora, contains the openings of the urethra and the vagina. The vagina may be partially closed by a ring of tissue called the hymen. The hymen is ordinarily ruptured by initial sexual intercourse; however, it can also be disrupted by other types of physical activities. If the hymen persists after sexual intercourse, it can be surgically ruptured.

Notice that the urinary and reproductive systems in the female are entirely separate. For example, the urethra carries only urine and the vagina serves only as the birth canal and the organ for sexual intercourse.

Orgasm

Sexual response in the female may be more subtle than in the male, but there are certain corollaries. The clitoris is believed to be an especially sensitive organ for initiating sexual sensations. It is possible for the clitoris to become ever so slightly erect as its erectile tissues becomes engorged with blood. But vasocongestion is more obvious in the labia minora, which expand and deepen in color. Erectile tissue within the vaginal wall also expands with blood, and the added pressure in these blood vessels causes small droplets of fluid to squeeze through the vessel walls and lubricate the vagina.

Release from muscle tension occurs in females, especially in the region of the vulva and vagina, but also throughout the entire body. Increased uterine motility may assist the transport of sperm toward the uterine tubes. Since female orgasm is not signaled by ejaculation, there is a wide range in normality regarding sexual response.

The egg must cross a small space to enter an oviduct, which conducts it toward the uterus. The vagina, the copulatory organ in females, opens into the vestibule where the urethra also opens. The vestibule is bounded by the labia minora, which come together at the clitoris, a highly sensitive organ. Outside the labia minora are the labia majora. There is no ejaculation in the female, and, therefore, orgasm is harder to detect and varies widely in normality.

Hormonal Regulation in the Female (Simplified)

Hormonal regulation in the female is quite complex, so this presentation has been simplified for easy understanding. The following glands and hormones are involved in hormonal regulation:

Table 16.3 Ovarian and Uterine Cycles (Simplified)

Ovarian Cycle Phases	Events	Uterine Cycle Phases	Events
Follicular phase (days 1–13)	FSH secretion by pituitary	Menstruation phase (days 1–5)	Endometrium breaks down
	Follicle maturation and secretion of estrogen	Proliferation phase (days 6–13)	Endometrium rebuilds
Ovulation Day 14[a]			
Luteal phase (days 15–28)	LH secretion by pituitary	Secretory phase (days 15–28)	Endometrium thickens and glands are secretory
	Corpus luteum formation and secretion of progesterone		

[a] Assuming a 28-day cycle.

Hypothalamus: secretes GnRH (gonadotropic-releasing hormone)

Anterior pituitary: secretes FSH (follicle-stimulating hormone) and LH (luteinizing hormone), the gonadotropic hormones

Ovaries: secrete estrogen and progesterone, the female sex hormones

Ovarian Cycle

The gonadotropic and sex hormones are not present in constant amounts in the female and instead are secreted at different rates during a monthly **ovarian cycle,** which lasts an average of twenty-eight days but may vary widely in specific individuals. For simplicity's sake, it is convenient to emphasize that during the first half of a twenty-eight-day cycle (days one to thirteen, table 16.3), FSH from the anterior pituitary is promoting the development of a follicle in the ovary and that this follicle is secreting estrogen. As the blood estrogen level rises, it exerts feedback control over the anterior pituitary secretion of FSH so that this follicular phase comes to an end (fig. 16.9). The end of the follicular phase is marked by ovulation on the fourteenth day of the twenty-eight-day cycle. Similarly, it may be emphasized that during the last half of the ovarian cycle (days fifteen to twenty-eight, table 16.3) anterior pituitary production of LH is promoting the development of a corpus luteum, which is secreting progesterone. As the blood progesterone level rises, it exerts negative feedback control over anterior pituitary secretion of LH so that the corpus luteum begins to degenerate. As the luteal phase comes to an end, menstruation occurs.

Uterine Cycle

The female sex hormones estrogen and progesterone have numerous functions, one of which is discussed here. The effect these hormones have on the endometrium of

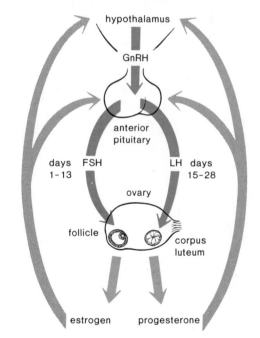

Figure 16.9 Hypothalamic-pituitary-gonad system (simplified) as it functions in the female. GnRH is a hypothalamic-releasing hormone that stimulates the anterior pituitary to secrete LH and FSH. These gonadotropic hormones act on the ovaries. FSH promotes the development of the follicle that later, under the influence of LH, becomes the corpus luteum. Negative feedback controls the level of all hormones involved.

the uterus causes the uterus to undergo a cyclical series of events known as the *uterine cycle* (table 16.3). Cycles that last twenty-eight days, are divided as follows:

During *days one to five,* there is a low level of female sex hormones in the body, causing the uterine lining to disintegrate and its blood vessels to rupture. A flow of blood, known as the *menses,* passes out of the vagina during a period of **menstruation,** also known as the menstrual period.

During *days six to thirteen,* increased production of estrogen by an ovarian follicle causes the endometrium to thicken and become vascular and glandular. This is called the proliferation phase of the uterine cycle.

Ovulation usually occurs on the fourteenth day of the twenty-eight-day cycle. (A secretion of LH, called the LH surge, is believed to cause ovulation to occur.)

During *days fifteen to twenty-eight,* increased production of progesterone by the corpus luteum causes the endometrium to double in thickness and the uterine glands to become mature, producing a thick mucoid secretion. This is called the secretory phase of the uterine cycle. The endometrium is now prepared to receive the developing embryo, but if pregnancy does not occur, the corpus luteum degenerates and the low level of sex hormones in the female body causes the uterine lining to break down. This is evident, due to the menstrual discharge that begins at this time. Even while menstruation is occurring, the anterior pituitary begins to increase its production of FSH and a new follicle begins maturation. Table 16.3 indicates how the ovarian cycle controls the menstrual cycle.

Hormonal regulation in the female results in an ovarian cycle. During the first half of the cycle, FSH causes maturation of the follicle, which secretes estrogen. After ovulation, LH converts the follicle into the corpus luteum, which produces progesterone. Estrogen and progesterone regulate the uterine cycle. Estrogen causes the endometrium to rebuild. Ovulation usually occurs on the fourteenth day of a twenty-eight-day cycle. As progesterone is produced by the corpus luteum, the endometrium thickens and becomes secretory. Then, a low level of hormones causes the endometrium to break down as menstruation occurs.

Pregnancy

If pregnancy occurs, menstruation does not occur. Instead, the developing embryo embeds itself in the endometrium lining several days following fertilization. Once this process, called **implantation,** is complete, a female is *pregnant.* During implantation, an outer membrane surrounding the zygote produces a gonadotropic hormone (**HCG,** or **h**uman **c**horionic **g**onadotropin) that prevents degeneration of the corpus luteum and, instead, causes it to secrete even larger quantities of progesterone. The corpus luteum may be maintained, as much as six months, even after the placenta is fully developed.

The **placenta** (fig. 17.6) originates from both maternal and fetal tissue, and is the region of exchange of molecules between fetal and maternal blood although there is no mixing of the two types of blood. After its formation, the placenta continues production of HCG and begins production of progesterone and estrogen. The latter hormones have two effects: they shut down the anterior pituitary so that no new follicles mature,

and they maintain the lining of the uterus so that the corpus luteum is not needed. There is no menstruation during the nine months of pregnancy.

Pregnancy Tests Pregnancy tests are based on the fact that HCG is present in the blood and urine of a pregnant woman.

Before the advent of monoclonal antibodies, only a blood test requiring the use of radioactive material in the hospital was available to detect pregnancy before the first missed menstrual period. Now there is a monoclonal antibody (p. 242) test for the detection of pregnancy ten days after conception. This test can be done on a urine sample, and the results are available within the hour.

The physical signs that might prompt a woman to have a pregnancy test are cessation of menstruation, increased frequency of urination, morning sickness, and increase in the size and fullness of the breasts, as well as darkening of the areolae (fig. 16.10).

Female Sex Hormones

The female sex hormones, estrogen and progesterone, have many effects on the body. In particular, estrogen secreted at the time of puberty stimulates the growth of the uterus and vagina. Estrogen is necessary for egg maturation and is largely responsible for the secondary sex characteristics in females. For example, it is responsible for the onset of the uterine cycle, as well as female body hair and fat distribution. In general, females have a more rounded appearance than males because of a greater accumulation of fat beneath the skin. Also, the pelvic girdle enlarges in females so that the pelvic cavity has a larger relative size compared to males; this means that females have wider hips. Both estrogen and progesterone are also required for breast development.

Breasts A female breast contains fifteen to twenty-five lobules (fig. 16.10), each with its own milk duct, which begins at the nipple and divides into numerous other ducts that end in blind sacs called *alveoli.* In a nonlactating (nonmilk-producing) breast, the ducts far outnumber the alveoli because alveoli are made up of cells that can produce milk.

Milk is not produced during pregnancy. *Lactogenic hormone* (prolactin) is needed for lactation (milk production) to begin, and the production of this hormone is suppressed because of the feedback inhibition estrogen and progesterone have on the pituitary during pregnancy. It takes a couple of days after delivery for milk production to begin and, in the meantime, the breasts produce a watery, yellowish white fluid called

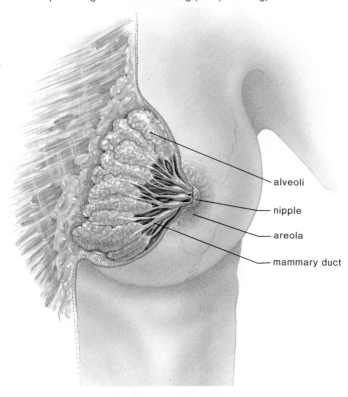

Figure 16.10 Anatomy of breast. The female breast contains lobules consisting of ducts and alveoli. The alveoli are lined by milk-producing cells in the lactating (milk-producing) breast.

- alveoli
- nipple
- areola
- mammary duct

colostrum, which differs from milk in that it contains more protein and less fat. Colostrum is a source of passive immunity for the baby. The production of milk requires continued breast feeding. When a breast is suckled, the nerve endings in the areola are stimulated and nerve impulses travel to the hypothalamus, which causes oxytocin to be released by the posterior pituitary. When this hormone arrives at the breasts, it causes contraction of the lobules so that milk flows into the ducts (called milk letdown).

Menopause

Menopause, the period in a woman's life during which the menstrual cycle ceases, is likely to occur between ages forty-five and fifty-five. The ovaries are no longer responsive to the gonadotropic hormones produced by the anterior pituitary, and the ovaries no longer secrete estrogen or progesterone. At the onset of menopause, the menstrual cycle becomes irregular, but as long as menstruation occurs it is still possible for a woman to conceive and become pregnant. Therefore, a woman is usually not considered to have completed menopause until there has been no menstruation for a year. The hormonal changes during menopause often produce

physical symptoms, such as "hot flashes" that are caused by circulatory irregularities, dizziness, headaches, insomnia, sleepiness, and depression. Again, there is a great variation among women, and any of these symptoms may be absent altogether.

Women sometimes report an increased sex drive following menopause, and it has been suggested that this may be due to androgen production by the adrenal cortex.

Estrogen and progesterone affect the female genitals, promote development of the egg, and maintain the secondary sex characteristics. Lactogenic hormone causes the breasts to begin to secrete milk after delivery, but another hormone, oxytocin, is responsible for milk letdown. When menopause occurs, FSH and LH are still produced, but the ovaries are no longer able to respond.

Control of Reproduction

Birth Control

Several means of birth control have been available for quite some time (table 16.4). The use of these contraceptive methods decreases the probability of pregnancy. A common way to discuss pregnancy rate is to indicate the number of pregnancies expected per 100 women per year. For example, it is expected that 80 out of 100 young women, or 80 percent, who are regularly engaging in unprotected intercourse will be pregnant within a year. Another way to discuss birth control methods is to indicate their effectiveness, in which case the emphasis is placed on the number of women who will not get pregnant. For example, with the least effective method given in table 16.4, we expect that 70 out of 100, or 70 percent, sexually active women will not get pregnant while 30 women will get pregnant within a year.

Future Means of Birth Control

There are four areas in which birth control investigations have been directed: morning-after medication, a long-lasting method, a medication that is specifically for males, and new barrier methods.

There is a new birth control pill, called Ru 486, on the market in France consisting of a synthetic steroid that prevents progesterone from acting on the uterine lining because it has a high affinity for progesterone receptors. In clinical tests, the uterine lining sloughed off within four days in 85 percent of women who were

Table 16.4 Common Birth Control Methods

Name	Procedure	Methodology	Effectiveness*	Action Needed	Risk†
Vasectomy	Vas deferentia are cut and tied	No sperm in semen	Almost 100%	None	Irreversible sterility
Tubal ligation	Oviducts are cut and tied	No eggs in oviduct	Almost 100%	None	Irreversible sterility
Pill	Must take medication daily	Shuts down pituitary	Almost 100%	None	Thromboembolism
IUD	Must be inserted into uterus by physician	Prevents implantation	More than 90%	None	Infection
Diaphragm	Plastic cup inserted into vagina to cover cervix	Blocks entrance of sperm into uterus	With jelly about 90%	Must be inserted each time before intercourse	—
Cervical cap	Rubber cup held by suction over cervix	Delivers spermicide near cervix	Almost 85%	Must be placed in cervix before intercourse	Cancer of cervix?
Condom	Sheath that fits over erect penis	Traps sperm	About 85%	Must be placed on penis at time of intercourse	—
Coitus interruptus (withdrawal)	Male withdraws penis before ejaculation	Prevents sperm from entering vagina	About 80%	Intercourse must be interrupted before ejaculation	—
Jellies, creams, foams	Contain spermicidal chemicals	Kill a large number of sperm	About 75%	Must be inserted before intercourse	—
Rhythm method	Determine day of ovulation by record keeping; testing by various methods	Avoid day of ovulation	About 70%	Limits sexual activity	—
Douche	Cleanses vagina and uterus after intercourse	Washes out sperm	Less than 70%	Must be done *immediately* after intercourse	—

*Effectiveness is the average percentage of women who did not become pregnant in a population of 100 sexually active women using the technique for one year.

†Only condoms offer protection against AIDS.

less than a month pregnant. To improve the success rate, the drug is administered in conjunction with a small dose of prostaglandin that causes contraction of the uterus to expel an embryo. The promoters of this treatment are using the term "contragestation" to describe its effects; however, it should be recognized that this medication, rather than preventing implantation, brings on an *abortion,* the loss of an implanted fetus. One day the medication might be used by many women who are experiencing delayed menstruation without knowing whether they are actually pregnant.

In this country, DES, a synthetic estrogen that affects the uterine lining making implantation difficult, is sometimes given following intercourse. Since large doses are required, causing nausea and vomiting, DES is usually given only for incest or rape.

Depo-Provera is an injectable contraceptive that is commercially available in many countries outside the United States. The injection contains crystals that gradually dissolve over a period of three months. The crystals contain a chemical related to progesterone and this chemical suppresses ovulation. The drug has not been approved for use in the United States because cancer developed in some test animals receiving the injections. More animal studies are now underway. An even more potent progesterone-like molecule for implantation under the skin is now close to being approved for sale in the United States. The *implant* consists of narrow tubes that slowly release the drug over a period of five years.

Various possibilities exist for a "*male pill.*" Scientists have made analogs of gonadotropic-releasing hormones that interfere with the action of this hormone and prevent it from stimulating the pituitary. The seminiferous tubules produce a hormone termed inhibin that inhibits FSH production by the pituitary (p. 279). Testosterone and/or related chemicals can be used to inhibit spermatogenesis in males, but there are usually feminizing side effects because the excess is changed to estrogen by the body.

There has been a revival of interest in barrier methods of birth control and a "female condom" is now being studied to determine its effectiveness against pregnancy and sexually transmitted diseases. The closed end of a large plastic tube is anchored by a plastic ring in the upper vagina and the open end of the tube is held in place by a thinner ring that rests just outside the vagina.

There are numerous well-known birth control methods and devices available to those who wish to prevent pregnancy. These differ as to their effectiveness. In addition, new methods are expected to be developed.

Infertility

Sometimes couples do not need to prevent pregnancy; conception or fertilization does not occur despite frequent intercourse. The American Medical Association estimates that 15 percent of all couples in this country are unable to have any children and are, therefore, properly termed sterile; another 10 percent have fewer children than they wish and are, therefore, termed *infertile.*

Infertility can be due to a number of factors. It is possible that fertilization takes place but the embryo dies before implantation takes place. One area of concern is that radiation, chemical mutagens, and the use of psychoactive drugs can contribute to sterility, possibly by causing chromosomal mutations that prevent development from proceeding normally. The lack of progesterone can also prevent implantation and, therefore, the proper administration of this hormone is sometimes helpful.

It is also possible that fertilization never takes place. There may be a congenital malformation of the reproductive tract or there may be an obstruction of the oviduct or vas deferens due to infection. Sometimes these physical defects can be corrected surgically. If no obstruction is apparent, it is possible to give females a substance rich in FSH and LH that is extracted from the urine of postmenopausal women. This treatment causes multiple ovulations and sometimes multiple pregnancies, however.

When reproduction does not occur in the usual manner, couples today are seeking alternative reproductive methods that may include the following:

Artificial Insemination by Donor (AID) Since the 1960s, there have been hundreds of thousands of births following artificial insemination in which sperm is placed in the vagina by a physician. Sometimes a woman is artificially inseminated with her husband's sperm. This is especially helpful if the husband has a low sperm count—the sperm can be collected over a period of time and concentrated so that the sperm count is sufficient to result in fertilization. Often, however, a woman is inseminated by sperm acquired from a donor who is a complete stranger to her.

In Vitro Fertilization (IVF) Over a hundred babies have been conceived using this method. First, a woman is given appropriate hormonal treatment. Then laparoscopy may be done. The laparoscope is a metal tube about the size of a pencil that is equipped with a tiny light and telescopic lens. In this instance, it is also fitted with a tube for retrieving eggs. After insertion through a small incision near the woman's naval, the physician guides the laparoscope to the ovaries where the eggs are sucked up into the tube. Alternately, it is possible to place a needle through the vaginal wall and guide it, by the use of ultrasound, to the ovaries where the needle is used to retrieve the eggs. This method is called transvaginal retrieval.

Concentrated sperm from the male is placed in a solution that approximates the conditions of the female genital tract. When the eggs are introduced, fertilization occurs. The resultant zygotes begin development and after about two to four days, embryos are inserted into the uterus of the woman, who is now in the secretory phase of her menstrual cycle. If implantation is successful, development is normal and continues to term.

Gamete Intrafallopian Transfer (GIFT) This method was devised as a means to overcome the low success rate (15 to 20 percent) of in vitro fertilization. The method is exactly the same as in vitro fertilization except the eggs and sperm are immediately placed in the oviducts after they have been brought together. This procedure would be helpful in couples whose eggs and sperm never make it to the oviducts; sometimes the egg gets lost between the ovary and the oviducts, and sometimes the sperm never reach the oviducts. GIFT has an advantage in that it is a one-step procedure for the woman—the eggs are removed and reintroduced all in the same time period. For this reason, it is less expensive, $1,500, compared with $3,000 and up for in vitro fertilization.

Surrogate Mothers Over a hundred babies have been born to women paid to have them by other individuals who have contributed sperm (or egg)

to the fertilization process. If all the alternative methods discussed above are considered, it is possible to imagine that a baby could have five parents: (1) sperm donor, (2) egg donor, (3) surrogate mother, (4) adoptive mother, and (5) adoptive father.

Some couples are infertile. There may be a hormonal imbalance or a blockage of the oviducts. When corrective medical procedures fail, it is possible today to consider an alternative method of reproduction.

Summary

I. Male Reproductive System
   A. Testes. The testes contain the interstitial cells and seminiferous tubules where spermatogenesis occurs. Sperm are the reproductive cells of males.
   B. Genital tract. Sperm mature in the epididymis and are also stored in the vas deferens before entering the urethra just prior to ejaculation. The accessory glands (seminal vesicles, prostate gland, and Cowper's glands) produce seminal fluid.
   C. Penis. Semen, which contains sperm and seminal fluid, leaves the penis during ejaculation.
   D. Hormonal regulation in the male. In males, FSH promotes spermatogenesis and LH promotes testosterone production. Testosterone stimulates growth of the male genitals during puberty, and is necessary for the maturation of sperm and development of the secondary sex characteristics.

II. Female Reproductive System
   A. Ovaries. Oogenesis occurs in the ovaries, where one follicle reaches maturity each month. This follicle balloons out of the ovary and bursts to release the egg. The ruptured follicle

develops into a corpus luteum. The follicle and corpus luteum produce the female sex hormones estrogen and progesterone.
   B. Genital tract. The egg must cross a small space to enter an oviduct, which conducts it toward the uterus. The vagina, the copulatory organ in females, opens into the vestibule where the urethra also opens.
   C. External genitalia. The vestibule is bounded by the labia minora, which come together at the clitoris, a highly sensitive organ. Outside the labia minora are the labia majora.
   D. Hormonal regulation in the female. Hormonal regulation in the female results in an ovarian cycle. During the first half of the cycle, FSH causes maturation of the follicle, which secretes estrogen. After ovulation, LH converts the follicle into the corpus luteum, which produces progesterone. Estrogen and progesterone regulate the uterine cycle. Ovulation occurs on the fourteenth day of a twenty-eight-day cycle. As progesterone is produced by the corpus luteum, the

endometrium thickens and becomes secretory. Then, a low level of hormones causes the endometrium to break down as menstruation occurs.
   Estrogen and progesterone affect the female genitals, promote development of the egg, and maintain the secondary sex characteristics. Lactogenic hormone causes the breasts to begin to secrete milk after delivery, but oxytocin is responsible for milk letdown. When menopause occurs, FSH and LH are still produced, but the ovaries are no longer able to respond.

III. Control of Reproduction
   A. Birth control. There are numerous well-known birth control methods and devices available to those who wish to prevent pregnancy. These differ as to their effectiveness. In addition, new methods are expected to be developed.
   B. Infertility. Some couples are infertile. There may be a hormonal imbalance or a blockage of the oviducts. When corrective medical procedures fail, it is possible today to consider an alternative method of reproduction.

Study Questions

1. Discuss the anatomy and physiology of the testes. Describe the structure of sperm.
2. Give the path of sperm.
3. What glands produce seminal fluid?
4. Discuss the anatomy and physiology of the penis. Describe ejaculation.
5. Discuss hormonal regulation in the male. Name three functions of testosterone.
6. Discuss the anatomy and physiology of the ovaries. Describe ovulation.
7. Give the path of the egg. Where do fertilization and implantation occur? Name two functions of the vagina.
8. Describe the external genitalia in females.
9. Compare male and female orgasm.
10. Discuss hormonal regulation in the female. Give the events of the uterine cycle, and relate them to the ovarian cycle. In what way is menstruation prevented if pregnancy occurs?
11. Name four functions of the female sex hormones. Describe the anatomy and physiology of the breast.
12. Discuss the various means of birth control and their relative effectiveness.

Objective Questions

Fill in the blanks.

1. If one were tracing the path of sperm, the structure that follows the epididymis is the _____ .
2. The prostate gland, Cowper's glands, and the _____ , all contribute to seminal fluid.
3. The primary male sex hormone is _____ .
4. An erection is caused by the entrance of _____ into the penis.
5. In the female reproductive system, the uterus lies between the oviducts and the _____ .
6. In the ovarian cycle, once each month a _____ produces an egg. In the uterine cycle, the _____ lining of the uterus is prepared to receive the embryo.
7. The female sex hormones are _____ and _____ .
8. Pregnancy in the female is detected by the presence of _____ in the blood or urine.
9. The most effective means of birth control are _____ in males and _____ in females.
10. In vitro fertilization occurs in _____ .

Medical Terminology Reinforcement Exercise

Pronounce and analyze the meaning of the following terms as dissected:

1. orchidopexy (or''kĭ-do-pek''se) orchido/pexy
2. transurethral resection of prostate (TURP) (trans''u-re'thral re-sek'shun pros'tāt)—trans/urethr/al re/sect/ion of the prostate
3. gonadotropic (gon''ah-do-trōp'ik) gonado/trop/ic
4. contraceptive (kon''trah-sep'tiv) contra/cept/ive
5. gynecomastia (jin''ē-ko-mas'te-ah) gyneco/mast/ia
6. hysterosalpingo-oophorectomy (his''ter-o-sal-ping''-go-o''of-o-rek'to-me) hystero/salpingo-/oophor/ectomy
7. endometriosis (en''do-me''tre-o'sis) endo/metri/osis
8. colporrhaphy (kol''po-ra'ah-fe) colpo/rrhaphy
9. menometrorrhagia (men''o-met''ro-ra'je-ah) meno/metro/rrhag/ia
10. multipara (mul-tip'ah-rah) multi/para
11. hyperemesis gravidarum (hi''per-em'ē-sis gravid'um) hyper/emesis gravi/darum
12. dysmenorrhea (dis''men-o-re'ah) dys/meno/rrhea

Development

Learning Objectives

After you have finished this chapter, you should be able to:

1. State the four processes of development.
2. Describe the weekly events of embryonic development and the monthly events of fetal development.
3. Describe the structure and function of the placenta.
4. List and discuss the stages of development after birth.

Embryonic Development

Human development is very often divided into *embryonic development* (first two months) and *fetal development* (third through ninth month). The embryonic period consists of early development, during which all the major organs form, and fetal development consists of a refinement of these structures. The fetus, not the embryo, is recognizable as a human being (fig. 17.1).

The following developmental processes occur during embryonic development:

Cleavage Immediately after fertilization, the zygote begins to divide so that at first there are 2, then 4, 8, 16, and 32 cells, and so forth. Since increase in size does not accompany these divisions, the embryo is at first no larger than the zygote was. Cell division during **cleavage** is mitotic, and each cell receives a full complement of chromosomes and genes.

Growth Later cell division is accompanied by an increase in size of the daughter cells, and **growth** in the true sense of the term takes place.

Morphogenesis **Morphogenesis** refers to the shaping of the embryo and is first evident when certain cells are seen to move, or migrate, in relation to other cells. By these movements, the embryo begins to assume various shapes.

Figure 17.1   Human development is divided into the embryonic period (first two months) and fetal development (third to ninth month). *a.* Embryo is not recognizably human. *b.* Fetus is recognizably human.

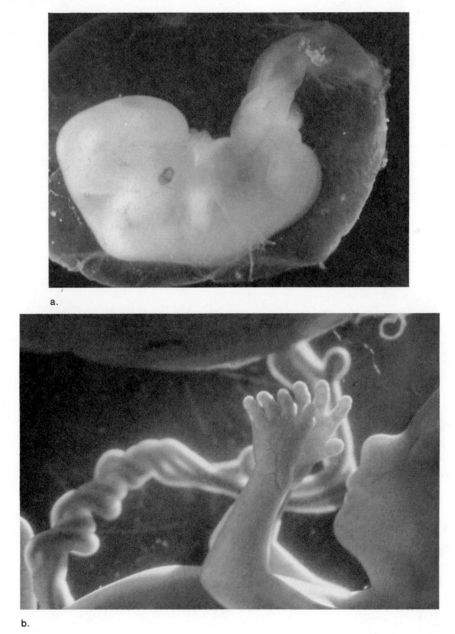

a.

b.

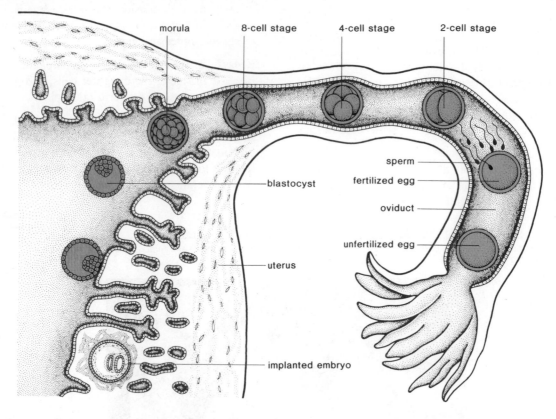

Figure 17.2 Fertilization and implantation of human embryo. After the egg is fertilized, it begins cleavage as it moves toward the uterus. At the time of implantation, the embryo is developing the germ layers.

Differentiation When cells take on a specific structure and function, **differentiation** occurs. The first system to become visibly differentiated is the nervous system.

We will now consider embryonic development week by week.

First Week

Immediately after fertilization within the oviduct, the human zygote begins to undergo cleavage as it travels down the oviduct to the uterus (fig. 17.2). By the time the embryo reaches the uterus on the third day, it is a hollow ball containing an inner cell mass, which will become the fetus. Sometimes during human development, the inner cell mass splits, and two embryos start developing rather than one. These two embryos are *identical twins* because they have inherited exactly the same chromosomes. *Fraternal twins,* which arise when two different eggs are fertilized by two different sperm, do not have identical chromosomes.

Second Week

At the end of the first week, the embryo begins the process of implanting itself in the wall of the uterus and an outer membrane, **the chorion,** secretes human chorionic gonadotropin (HCG), the hormone that is the basis for the pregnancy test and serves to maintain the corpus luteum past the time it would normally disintegrate. Because of this, the endometrium is maintained and menstruation does not occur. The embryo is now about the size of the period at the end of this sentence.

Third Week

During the third week of development, the inner cell mass flattens into the embryonic disk composed of two layers of cells (fig. 17.3a). Development is rapid and by the end of the third week, the nervous system, which develops from neural folds; the heart, which develops from the pericardial area; and the muscles, which develop from blocks of tissues called somites, are all visible (fig. 17.3b). The embryo is surrounded by amniotic fluid within a membrane, the **amnion.**

Fourth to Fifth Week (fig. 17.4)

4.8 mm (3/16 inch)
Heart pulsating and pumping blood
All somites (40) present
Eyes, ears, and nose forming
Digestive system forming
Limb buds begin to form
Body flexed; C-shaped
Presence of tail gives nonhuman appearance

Figure 17.3   *a.* Superior view of the embryo at sixteen days. The primitive node and streak are where morphogenesis is occurring and this will lead to tissue and organ formation. *b.* Human embryo at twenty-one days. The nervous system is forming as the neural folds close. The pericardial area contains the primitive heart, and the somites are the precursors of the muscles.

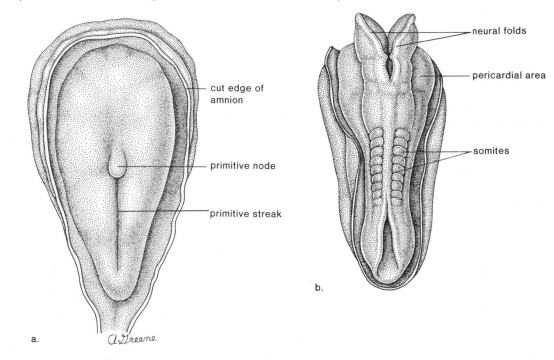

Figure 17.4   Human embryo at beginning of fifth week. *a.* Scanning electron micrograph. *b.* Drawing. The embryo is curled so that the head touches the heart, two organs whose development is further along than the rest of the body. The organs of the gastrointestinal tract are forming. The presence of the tail is an evolutionary remnant; its bones will regress and become those of the coccyx. The arms and legs will develop from the bulges that are called limb buds.

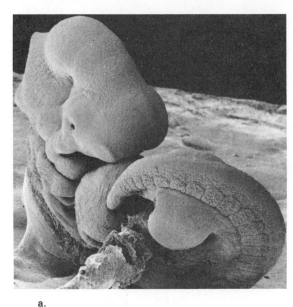

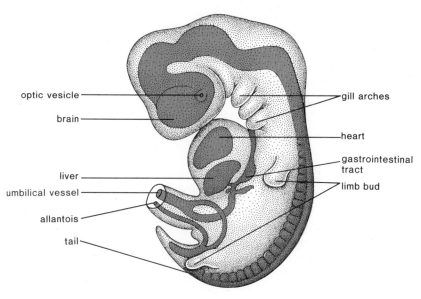

Sixth to Seventh Week

14.3 mm (9/16 inch)
Head becomes disproportionately large
Face and neck forming
Limb buds developing; digits forming
Cartilaginous skeleton forming
Tail regressing

Eighth Week (Two Months)

28.6 mm; .95 gm (1⅛ inches; about 1/30 ounce)
Nose flat, eyes far apart, eyelids fused
Limbs beginning to take shape; digits well formed
Ossification beginning
All internal organs have formed
Recognizable as human

Fetal Development

Third Month

7.62 cm; 28.4 gm (3 inches; 1 ounce)
Head prominent
Eyes formed, but lids still fused
External ears present; nose gains bridge
Tooth sockets and buds forming in jawbones
Nails forming
Ossification continuing
Heartbeat can be detected with special
instruments
Sex can be determined by inspection

Fourth Month

16.5 cm to 17.8 cm; 113.4 gm (6½ to 7 inches;
4 ounces)
Eyes, ears, nose, and mouth have typical human
appearance; eyebrows appear
Skin bright pink, transparent, and covered with
fine, downlike hair
Bony skeleton now visible, body catching up with
head size
Active muscles; movement may be felt as baby
moves in womb

Fifth Month (fig. 17.5)

25.4 cm to 30.5 cm; 226.8 gm to 453.6 gm (10 to
12 inches; ½ to 1 pound)
Eyelids still completely fused; some hair may be
present on head
Skin bright red and still covered by fine hair
Internal organs maturing; heartbeat can be heard
without special instruments

Sixth Month

27.9 cm to 35.6 cm; .57 kg to .68 kg (11 to 14
inches; 1¼ to 1½ pounds)
Eyelids finally separated and eyelashes formed
Skin quite wrinkled and somewhat red; covered
with heavy protective creamy coating
Nails now extend to end of digits

Seventh Month

35.6 cm to 43.2 cm; 1.36 kg (14 to 17 inches; 3
pounds)
Eyes are open
Skin still quite red and covered with wrinkles
Testes have descended into scrotal sacs
Premature baby at this stage has a slight chance

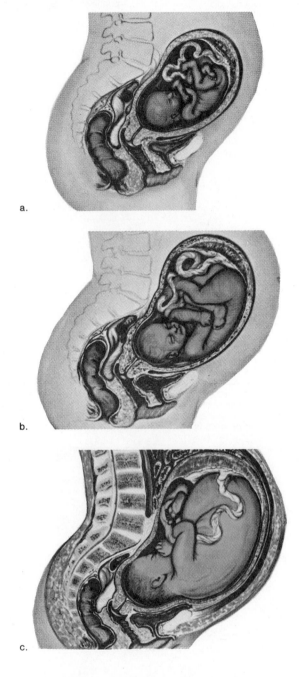

Figure 17.5 Later human development. *a.* Four-to-five month fetus. *b.* Six-to-seven month fetus. *c.* Eight-to-nine month fetus.

for survival in nurseries staffed by skilled
physicians and nurses

Eighth Month

41.9 cm to 45.7 cm; 2.27 kg (16½ to 18 inches; 5
pounds)
Subcutaneous fat deposition leads to weight gain
Bones of head soft and flexible
Growth and maturation of baby in last two
months extremely valuable for survival

Figure 17.6 Anatomy of the placenta. The placenta is composed of both fetal and maternal tissues. Fetal blood vessels penetrate the uterine lining and are surrounded by maternal blood. Exchange of molecules between fetal and maternal blood takes place across the walls of the fetal blood vessels.

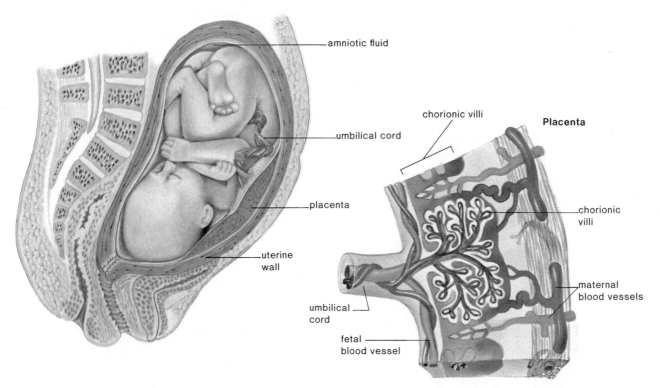

Ninth Month

Average baby weighs about 3.14 kg (7 pounds) if a girl and 3.4 kg (7½) if a boy
Length about 50.8 cm (20 inches)
Skin still coated with creamy coating
Fine downy body hair has largely disappeared
Fingernails may protrude beyond ends of fingers
Size of fontanels between bones of skull varies considerably from one child to another, but generally will close within twelve to eighteen months

Placenta

The **placenta** serves the needs of the growing embryo and fetus. It is the place where oxygen and nutrients enter embryonic (fetal) circulation and where carbon dioxide and nitrogenous wastes enter maternal circulation. *There is no mixing of maternal and embryonic (fetal) blood at the placenta,* however. The exchange takes place across embryonic blood vessels bathed by maternal blood (fig. 17.6).

The placenta has a fetal side and a maternal side. Projections from the chorion called the chorionic villi, extend into the myometrium of the uterus. The chorionic villi contain blood vessels that join to form

the umbilical arteries and veins. The **umbilical** cord stretches between the placenta and the fetus. Although it may seem that the umbilical cord travels from the placenta to the intestines, actually the umbilical cord is simply taking fetal blood to and from the placenta. The umbilical cord is the lifeline of the fetus because it contains those vessels which transport waste molecules to the placenta for disposal, and transport oxygen and nutrient molecules from the placenta to the rest of the fetal circulatory system.

The placenta begins to form soon after the embryo implants itself into the uterine lining, but it is not fully functioning until the embryo is about one month old. At first, the chorionic villi project from the entire chorion, but later the chorionic villi are concentrated on the side where the maternal blood vessels of the myometrium are found. Here the complex, interlocking nature of fetal and maternal blood vessels forms. In the newborn, the placenta is an oval disk with a diameter of 15–20 cm (8 inches) and a thickness of 2.5 cm (1 inch) and weighs 500–600 gm (over a pound). It forms a significant part of the afterbirth (p. 297).

Harmful substances can cross the placenta, and this is of particular concern because some birth defects are caused by mothers taking drugs that can affect embryonic development. Each organ or part seems to have a

Figure 17.7 Three stages of parturition. *a.* Position of fetus just before birth begins. *b.* Dilation of cervix. *c.* Birth of baby. *d.* Expulsion of afterbirth.

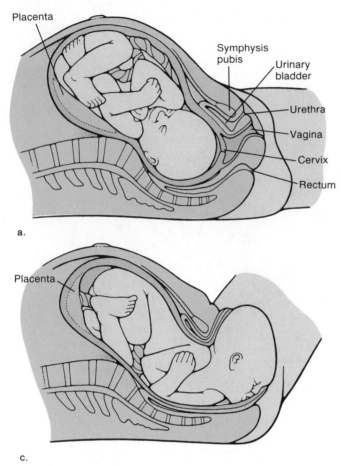

a.

b.

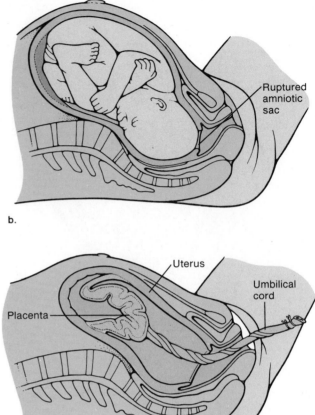

c.

d.

Schenk

sensitive period, during which a substance can alter its normal formation. Some bacteria (e.g., syphilis) and viruses (e.g., AIDS) can cross the placenta and, in this way, infected babies can be born. Also, drugs such as heroin and cocaine can cross the placenta and, in this way, addicted babies can be born.

Parturition

Development encompasses the time from *conception* (fertilization) to *parturition* (birth). In humans the *gestation period,* or length of pregnancy, is approximately nine months. It is customary to calculate the time of birth by adding 280 days to the start of the last menstruation because this date is usually known,

whereas the day of fertilization is usually unknown. Because the time of birth is influenced by so many variables, only about 5 percent of babies actually arrive on the forecasted date.

The uterus characteristically contracts throughout pregnancy. At first, light, often indiscernible contractions last about twenty to thirty seconds and occur every fifteen to twenty minutes, but near the end of pregnancy they become stronger and more frequent so that the woman may falsely think that she is in labor. The onset of true labor is marked by uterine contractions that occur regularly every fifteen to twenty minutes and last for forty seconds or more. **Parturition,** which includes labor and expulsion of the fetus, is usually considered to have three stages (fig. 17.7).

Stages

During the *first stage,* the cervix dilates; during the *second,* the baby is born; and during the *third,* the afterbirth is expelled.

The events that cause parturition are still not entirely known but there is now evidence suggesting the involvement of prostaglandins. It may be, too, that the prostaglandins cause the release of oxytocin from the maternal posterior pituitary. Both prostaglandins and oxytocin do cause the uterus to contract, and either can be given to induce parturition.

Stage 1

Prior to the first stage of parturition or concomitant with it, there may be a "bloody show" caused by the expulsion of a mucus plug from the cervical canal. This plug prevents bacteria and sperm from entering the uterus during pregnancy.

Uterine contractions during the first stage of labor occur in such a way that the cervical canal slowly disappears (fig. 17.7b) as the lower part of the uterus is pulled upward toward the baby's head. This process is called *effacement,* or "taking up the cervix." With further contractions, the baby's head acts as a wedge to assist cervical dilation. The baby's head usually has a diameter of about 10 cm (4 inches), and, therefore, the cervix has to dilate to this diameter in order to allow the head to pass through. If it has not occurred already, the amnion is apt to rupture now, releasing the amniotic fluid, which escapes out the vagina. The first stage of labor ends once the cervix is completely dilated.

Stage 2

During the second stage, the uterine contractions occur every one to two minutes and last about one minute each. They are accompanied by a desire to push or bear down. As the baby's head gradually descends into the vagina, the desire to push becomes greater. When the baby's head reaches the exterior, it turns so that the back of the head is uppermost (fig. 17.7c). Since the vagina may not expand enough to allow passage of the head without tearing, an *episiotomy* is often performed. This incision, which enlarges the opening, is stitched later and will heal more perfectly than a tear would. As soon as the head is delivered, the baby's shoulders rotate so that the baby faces either to the right or left. The physician may at this time hold the head and guide it downward while one shoulder and then the other emerges. The rest of the baby follows easily.

Once the baby is breathing normally, the umbilical cord is cut and tied, severing the child from the placenta. The stump of the cord shrivels and leaves a scar, which is the navel.

Stage 3

The placenta, or **afterbirth,** is delivered during the third stage of labor (fig. 17.7d). About fifteen minutes after delivery of the baby, uterine muscular contractions shrink the uterus and dislodge the placenta. The placenta is then expelled into the vagina. As soon as the placenta and other membranes are delivered, the third stage of labor is complete.

During the first stage of birth, the cervix dilates; during the second, the child is born; and during the third, the afterbirth is expelled.

Development after Birth

Development does not cease once birth has occurred; it continues throughout the stages of life: infancy, childhood, adolescence, and adulthood.

Infancy

Infancy lasts until about two years of age. The newborn has certain innate reflexes that help it establish a relationship with its caretakers. For example, there is the (*a*) *rooting reflex.* A baby will turn its head in the direction of a touch on its cheek and open its mouth. This helps the baby find the nipple. (*b*) In manifesting the *suckling reflex,* a baby will suck on any object that touches the mouth. This reflex assures that the infant will obtain nourishment from the breast. (*c*) Babies cry (*crying reflex*) when they are in discomfort or are hungry. This helps to assure that their needs will be met by caretakers. (*d*) The *smile reflex* is obvious at two to three months of age, when an infant develops the "social smile," a means of interacting with caretakers. Before this time, what appears to be a smile may simply be due to gas pains.

Table 17.1 Milestones in Sensorimotor Development

Age	Sensorimotor Achievements	Vocalization and Language
3 months	Grasps objects, smiles spontaneously, holds head steady when sitting, lifts up head when on stomach, follows object readily with his eyes.	Squeals, coos especially in response to social interaction, laughs.
4½ months	Reaches awkwardly for some objects that he sees, frequently looks at hands, readily brings objects in hand to mouth, holds head steady in most positions, sits with props, bears some weight on legs.	Eyes seem to search for speaker, some consonants are mixed in with cooing sounds, babbling begins.
6 months	Usually reaches for near objects that he sees and looks at objects that he grasps, sits without support leaning forward on his hands, bears his weight on legs but must be balanced by adult, brings feet to mouth when lying on back.	Simple babbling of sounds, like "mamama" and "bababa," turns to voice, laughs easily.
9 months	Grasps small objects with thumb and fingertips, shows first preference for one hand (usually the right), sits upright with good control, stands holding on, crawls, often imitates.	Repetition of sounds in babbling becomes common, some production of intonation patterns of parents' language, some imitation of sounds, understands "no."
1 year	Neat grasp of small objects, stands alone, walks with one hand held by adult, seats self on floor, mouths objects much less, drinks from cup but messily, often imitates simple behaviors, cooperates in dressing.	Produces a few words such as "mama" and "dada," understands a few simple words and commands, produces sentence-like intonation patterns called expressive jargon.
1½ years	Puts cubes in bucket, dumps contents from bottle, walks alone and falls only rarely, uses spoon with little spilling, undresses self, scribbles spontaneously.	Produces between five and fifty single words, produces complex intonation patterns, understands many words and simple sentences.
2 years	Turns single pages in a book, builds tower of blocks, shifts easily between sitting and standing, runs, throws and kicks balls, washes hands, puts on clothing.	Produces more than fifty words, produces a few short "sentences," understands much in concrete situations, shows much interest in language and communication.

Source: Frankenburg & Dodds 1967; Tlg, Ames & Baker 1981; Knoblock & Pasamanick 1974; Lenneberg 1967; Ramsay 1984 as appeared in *Human Development: From Conception Through Adolescence*. By Kurt W. Fischer and Arlyne Lazerson. Copyright © 1984 by Kurt W. Fischer and Arlyne Lazerson. Reprinted with permission by W. H. Freeman and Company.

Infants are not only responsive to touch; they also are capable of seeing, hearing, smelling, and tasting. Their sensorimotor development is outlined in table 17.1. Their ability to perform ever more difficult operations can be correlated with the development of the brain during this time period.

Childhood and Adolescence

Childhood lasts until puberty (around ten years in girls and twelve years in boys), and then adolescence continues until adulthood, a term that has a social as well as a biological definition. The growth of girls and boys during this time is shown in figure 17.8.

At the time of puberty, the sex organs mature, and the secondary sex characteristics begin to appear. The cause of puberty is related to the level of sex hormones in the body. It is now recognized that the hypothalamic-pituitary-gonad system functions long before puberty, but the level of hormones is low because the hypothalamus is supersensitive to feedback control. At the start of puberty, the hypothalamus becomes less sensitive to feedback control and begins to increase its production of releasing hormones, causing the pituitary and the gonads to mature and increase their production of hormones. The sensitivity of the hypothalamus continues to decrease until the gonadotropic and sex hormones reach the adult level.

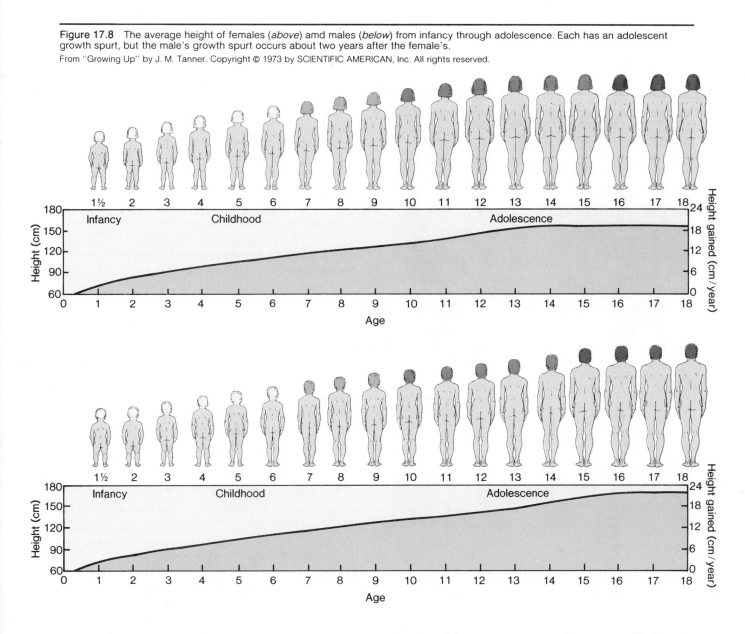

Figure 17.8 The average height of females (*above*) amd males (*below*) from infancy through adolescence. Each has an adolescent growth spurt, but the male's growth spurt occurs about two years after the female's.

From "Growing Up" by J. M. Tanner. Copyright © 1973 by SCIENTIFIC AMERICAN, Inc. All rights reserved.

Adulthood and Aging

Young adults are at their physical peak in muscle strength, reaction time, and sensory perception. The organ systems at this time are best able to respond to altered circumstances in a homeostatic manner. From now on, there will be an almost imperceptible, gradual loss in certain of the body's abilities. Aging (fig. 17.9) encompasses these progressive changes that contribute to an increased risk of infirmity, disease, and death.

Effect of Aging on Body Systems

Most of the organ systems degenerate with age but some to a greater degree than others. For example, a twenty-year-old can hear frequencies as high as 15,000 cycles per second but a seventy-year-old can most likely only hear frequencies as high as 6,000 cycles per second. However, the brain weight is 85 percent and muscle strength is 88 percent that of a twenty-year-old. However, when making these comparisons, we may note that the body has a vast functional reserve so that it can still perform well even when not at 100 percent capacity.

Skin

As aging occurs, the skin becomes thinner and less elastic because the number of elastic fibers decreases and the collagen fibers undergo cross-linking as discussed previously. Also, there is less adipose tissue in

Figure 17.9 Aging is a slow process during which the body undergoes changes that eventually will bring about death even if no marked disease or disorder is present. Although the human life span probably cannot be expanded, it most likely is possible to expand the health span—the length of time the body functions normally.

pumped each minute even though the maximum possible output declines.

Because the middle coat of the arteries contains elastic fibers which are most likely subject to cross-linking, the arteries become more rigid with time. Their size is further reduced by the presence of plaque so that blood pressure readings gradually rise. Such changes are common in individuals living in Western industrialized countries, but not in agricultural societies. As mentioned earlier, diet has been suggested as a way to control degenerative changes in the cardiovascular system (p. 205).

There is reduced blood flow to the liver and this organ does not metabolize drugs as efficiently as before. This means that as a person gets older, less medication is needed to maintain the same level in the bloodstream.

Circulatory problems are often accompanied by respiratory disorders and vice versa. Growing inelasticity of lung tissue means that ventilation is reduced; but since we rarely use the entire vital capacity, these effects will not be noticed unless there is increased demand for oxygen.

There is also less blood supply to the kidneys. The kidneys become smaller and less efficient at filtering off wastes. Salt and water balance are difficult to maintain and the elderly dehydrate faster than younger people. Difficulties involving urination include incontinence and inability to urinate. In men, the prostate gland may enlarge and cut off the urethra.

The loss of teeth, which is often seen in elderly people, is more apt to be the result of long-term neglect than a result of aging. The digestive tract loses tone, and secretion of saliva and gastric juice is reduced, but there is no indication of reduced absorption. Therefore, an adequate diet, rather than vitamin and mineral supplements is recommended. There are common complaints of constipation, increased amount of gas, and heartburn, but gastritis, ulcers, and cancer may also occur.

the subcutaneous layer so that older people are more likely to feel cold. The loss of thickness accounts for sagging and wrinkling of the skin.

Homeostatic adjustment to heat is also limited because there are fewer sweat glands for sweating to occur. There are fewer hair follicles so that the hair on the scalp and extremities thins out. The number of sebaceous glands is reduced and the skin tends to crack.

There is a decrease in the number of melanocytes, making the hair turn gray and the skin become paler. In contrast, some of the remaining pigment cells are larger and pigmented blotches appear in the skin.

Processing and Transporting

Cardiovascular disorders are the leading cause of death. The heart shrinks because there is a reduction of cardiac muscle cell size. This leads to loss of cardiac muscle strength and reduced cardiac output. Still, it is observed that the heart, in the absence of disease, is able to meet the demands of increased activity. It can increase its rate to double or triple the amount of blood

Integration and Coordination

It is often mentioned that while most tissues of the body regularly replace their cells, some at a faster rate than others, the brain and muscles do not. No new nerve or skeletal muscle cells are formed in the adult; however, contrary to previous opinion, recent studies show that few neural cells are lost during the normal aging process. This means that cognitive skills remain unchanged even though there is characteristically a loss in short-term memory. Although the elderly learn more slowly than the young, they can acquire new material and remember it as well as the young. It's noted that when more time is given for the subject to respond, age differences in learning decrease.

Neurons are extremely sensitive to oxygen deficiency and if neuron death does occur, it may not be due to aging itself, but to reduced blood flow in narrowed blood vessels. Specific disorders such as depression, Parkinson's disease, and Alzheimer's disease (p. 125) are sometimes seen, but they are not common. Reaction time, however, does slow and more stimulation is needed for hearing, taste, and smell receptors to function as before. After age fifty, there is a gradual reduction in the ability to hear tones at higher frequencies, and this may make it more difficult to identify individual voices and understand conversation in a group setting. The lens of the eye does not accommodate as well and may develop cataracts. Glaucoma is more likely because there is a reduction in the size of the anterior chamber of the eye.

Loss of skeletal muscle mass is not uncommon, but it can be controlled by a regular exercise program. There is a reduced capacity to do heavy labor, but routine physical work should be no problem. A decrease in the strength of the respiratory muscles contributes to the inability of the lungs to expand as before and reduced muscularity of the urinary bladder contributes to difficulties in urination.

Aging is accompanied by a decline in bone density. Osteoporosis, characterized by a loss of calcium and minerals from bone, is not uncommon, but there is evidence that proper health habits may prevent its occurrence. Arthritis, which restricts the motility of joints, is also seen. In arthritis, ossified spurs develop as the articular cartilage deteriorates, and this causes pain upon movement of the joint.

Weight gain occurs because the metabolic rate decreases and inactivity increases; muscle mass is replaced by stored fat and retained water.

Reproductive System

Females undergo menopause and, thereafter, the level of female sex hormones in the blood falls markedly. The uterus and cervix are reduced in size, and there is a thinning of the walls of the oviducts and vagina. The external genitals become less pronounced. In males, the level of androgens falls gradually over the age span of fifty to ninety, but sperm production continues until death.

Conclusion

We have listed many adverse effects due to aging, but it is important to emphasize that while such effects are seen, they are not a necessary occurrence. The present data about the effects of aging is often based on comparing the characteristics of the elderly to younger age groups. But today's elderly were probably not as aware of the importance of, for example, diet and exercise to general health. It's possible, then, that much of what we attribute to aging is instead due to years of poor health habits.

For example, osteoporosis is common in the elderly. By age sixty-five, one-third of women will have vertebral fractures, and by age eighty-one, one-third of women and one-sixth of men will have suffered a hip fracture. While there is no denying that there is a decline in bone mass as a result of aging, certain extrinsic factors are also important. The occurrence of osteoporosis itself is associated with cigarette smoking, heavy alcohol intake, and inadequate calcium intake. Not only is it possible to eliminate these negative factors by personal choice, it is also possible to add a positive factor. A moderate exercise program has been found to slow down the progressive loss of bone mass.

Rather than collecting data on the average changes observed between different age groups, it might be more useful to note the differences within any particular age group so that extrinsic factors that contribute to a decline and extrinsic factors that promote the health of an organ can be identified. Then, it will be possible for more people to enjoy good health even in their later years.

Summary

I. Embryonic Development
   A. First week. During the first week, the embryo undergoes cleavage, and develops into a hollow ball that contains an inner cell mass (to become the embryo and later the fetus).
   B. Second week. During the second week, the embryo embeds itself in the uterine lining.
   C. Third week. The nervous system, heart, and muscles begin to appear. The embryo is surrounded by two membranes, the outer chorion and the inner amnion.
   D. Fourth to fifth weeks. During the fourth and fifth weeks, these major organs continue to develop. Human features such as the head, arms, and legs begin to appear.
   E. Sixth to eighth weeks. At the end of the embryonic period, all organ systems have been established, and there is a mature and fully functioning placenta. The embryo is only about 2.5 cm.
II. Fetal Development
   A. Third and fourth months. During the third and fourth months, it is obvious that bone

Summary—*continued*

is replacing cartilage in the skeleton. The sex of the individual is now distinguishable.
B. From the fifth to the ninth month, the fetus continues to grow and gain weight. Babies born after six to seven months may survive, but are subject to various illnesses that may have lasting effects or cause an early death.

III. Parturition
A. Stages. During the first stage of birth, the cervix dilates; during the second, the child is born; and during the third, the afterbirth is expelled.
IV. Development after Birth
A. Infancy. Infancy lasts until two years of age, and by this time, the child has developed all sorts of motor skills and is usually able to form sentences.

B. Childhood and adolescence. Puberty marks the division between childhood and adolescence. Females have a growth spurt at about twelve while boys have theirs at fifteen.
C. Adulthood and aging. Aging encompasses progressive changes that contribute to an increased risk of infirmity, disease, and death.

Study Questions

1. List the processes of development.
2. List the events of embryonic development on a weekly basis.
3. List the events of fetal development on a monthly basis.
4. Describe the functioning of the placenta.
5. Describe the three stages of parturition.
6. List and discuss the stages of development after birth.
7. What are the major changes in body systems that have been observed as people age?
8. What is meant by the statement that extrinsic factors affect aging?

Objective Questions

Fill in the blanks.

1. When cells take on a specific structure and function, _____ occurs.
2. The zygote divides as it passes down a uterine tube. This process is called _____ .
3. Once the embryo arrives at the uterus, it begins to _____ itself in the uterine lining.
4. During embryonic development, all major _____ form.
5. Fetal development begins at the end of the _____ month.
6. During development, the nutrient needs of the developing embryo (fetus) are served by the _____ .
7. In most deliveries, the _____ appears before the rest of the body.
8. The time in life when sexual maturity takes place is called _____ .
9. Most deaths are due to a failure of the _____ system.
10. Skin gets (thinner, thicker) as we age. _____

Medical Terminology Reinforcement Exercise

Pronounce and analyze the meaning of the following dissected terms:

1. morphogenesis (mor″fo-jen′ĕ-sis) morpho/genesis
2. neonatologist (ne″o-na-tol′o-jist) neo/nato/log/ist
3. prenatal (pre-na′tal) pre/nat/al
4. postmortem (pōst-mor′tem) post/mort/em
5. degenerative (de-jen′er-a-tiv) de/generat/ive
6. bradykinesia (brad″e-kĭ-ne′se-ah) brady/kines/ia
7. malnutrition (mal″nu-trish′un) mal/nutrit/ion
8. presbycardia (pres″bĭ-kar′de-ah) presby/card/ia
9. gerontophilia (jer″on-to-fil′e-ah) geronto/philia
10. splenogenous (sple-noj′ĕ-nus) spleno/gen/ous
11. pediatrician (pe″de-ah-trish′un) ped/iatri/cian
12. geriatric (jer″e-at′rik) ger/iatr/ic

Genetics

Chapter Outline

Learning Objectives

After you have studied this chapter, you should be able to:

1. Explain the normal chromosome inheritance of humans.
2. Describe the characteristics and cause of the most common abnormal autosomal chromosome inherited condition in humans.
3. Explain how it is possible to detect abnormal autosomal chromosome inheritance in an embryo.
4. Explain how sex is inherited in humans.
5. Describe the characteristics of the common abnormal sex chromosome inherited conditions in humans.
6. Explain dominant, recessive, and sex-linked gene inheritance in humans.
7. Give examples of dominant, recessive, and sex-linked genetic disorders in humans.
8. Explain what is meant by recombinant DNA technology and discuss the possible benefits of biotechnology.

Chromosome Inheritance

Genetics is the study of how biological information is transferred from one generation to another. This information is stored in the **genes,** which are specific parts of chromosomes present in the nucleus of every cell.

A zygote receives 23 pairs of chromosomes when the sex cells unite during fertilization. One of each pair was inherited from the male parent and one from the female parent. Thereafter, due to the process of cell division, each cell in the body contains copies of these chromosomes in the nucleus. It is possible to photograph the nucleus of a cell that is about to divide (the chromosomes are more visible then), cut out the chromosomes, and arrange them according to pairs. The resulting display of chromosomes is called a **karyotype** (fig. 18.1). Twenty-two of these pairs in the karyotype are called the autosomal chromosomes, meaning that they have nothing to do with the sex of the individual, and one pair is called the **sex chromosomes** because they usually determine the sex of the individual. In males, one of the sex chromosomes is larger than the other. The larger of this pair is called the X chromosome and the smaller is called the Y chromosome. Females have two X chromosomes in their karyotype.

Autosomal Chromosome Inheritance

Normally, an individual inherits 22 autosomal chromosomes from the father and 22 autosomal chromosomes from the mother. Sometimes, however, an individual is born with either too many or too few autosomal chromosomes, most likely due to **nondisjunction** of chromosomes during either meiosis I or meiosis II (fig. 18.2).

Down Syndrome

A syndrome is a group or pattern of symptoms that occur together in the same individual due to the presence of an abnormal condition. The most common autosomal abnormality is seen in individuals with **Down syndrome,** who inherit three number-21 chromosomes instead of two (fig. 18.3). Persons with this syndrome are mentally retarded, have an oriental-like fold above

Figure 18.1 Normal karyotype of a male.
From *Antenatal Diagnosis*, HEW, 1979.

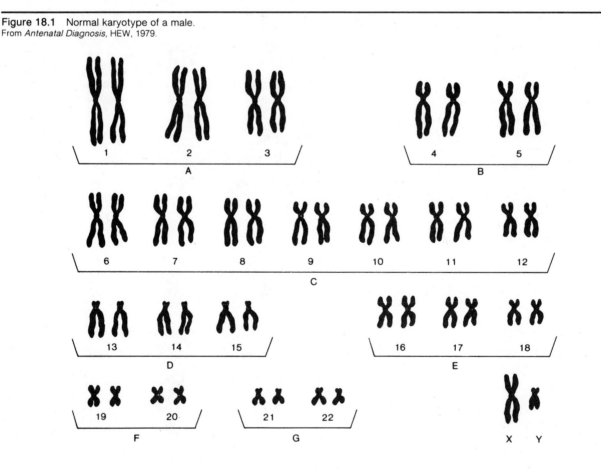

Figure 18.2 Nondisjunction during oogenesis. Nondisjunction can occur during meiosis I or during meiosis II. In either case, the abnormal eggs carry an extra chromosome. Nondisjunction of the number-21 chromosome leads to Down syndrome.

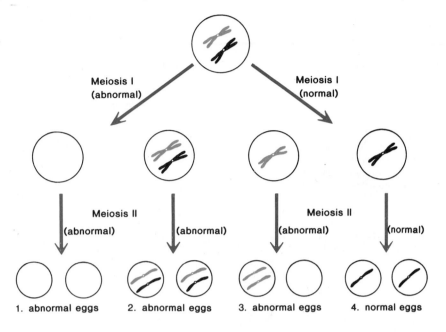

Meiosis I (abnormal)

Meiosis I (normal)

Meiosis II (abnormal)

Meiosis II (normal)

(abnormal)

(abnormal)

1. abnormal eggs

2. abnormal eggs

3. abnormal eggs

4. normal eggs

Figure 18.3 Down syndrome. *a.* Common characteristics include a wide, rounded face and a fold of the upper eyelids. Mental retardation, along with an enlarged tongue, makes it difficult for persons with Down syndrome to learn to speak coherently. *b.* Karyotype of an individual with Down syndrome has an extra number-21 chromosome in the G set. More sophisticated technologies allow investigators to pinpoint the location of specific genes associated with the syndrome. The Gart gene, which leads to a high level of blood purines, may account for the mental retardation seen in persons with Down syndrome.

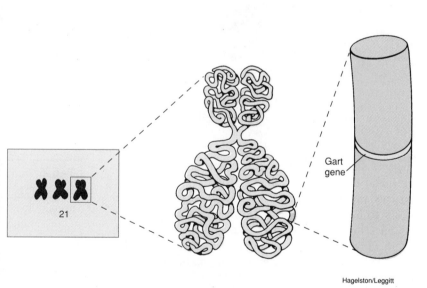

Gart gene

21

Hagelston/Leggitt

a.

b.

More about . . .

Detecting Birth Defects

*I*t is believed that at least one in sixteen newborns has a birth defect, either minor or serious, and the actual percentage may be even higher. Most likely only 20 percent of all birth defects are due to heredity. Those that are can sometimes be detected before birth by subjecting embryonic and/or fetal cells to various tests following chorionic villi sampling or amniocentesis. Chorionic villi sampling (fig. a) allows physicians to collect embryonic cells as early as the fifth week. The doctor inserts a long thin tube through the vagina into the uterus. With the help of ultrasound, which gives a picture of the uterine contents, the tube is placed between the lining of the uterus and the chorion. Then suction is used to remove a sampling of

the chorionic villi cells. Chromosomal analysis and biochemical tests for several different genetic defects can be done immediately on these cells. Physicians have to wait until about the sixteenth week of pregnancy to perform amniocentesis (fig. b). In amniocentesis, a long needle is passed through the abdominal wall to withdraw a small amount of amniotic fluid along with fetal cells. Since there are only a few cells in the amniotic fluid, testing must be delayed for four weeks until cell culture produces enough cells for testing purposes. In fetoscopy, another possible procedure, the physician uses an endoscope to view the fetus so that blood can be withdrawn for prenatal diagnosis.

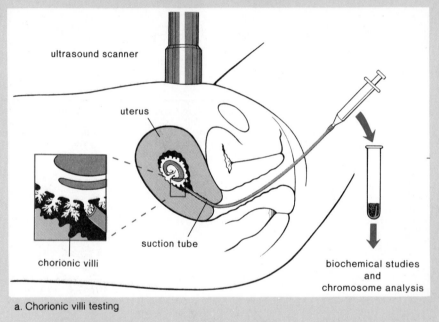

ultrasound scanner

uterus

suction tube

chorionic villi

biochemical studies
and
chromosome analysis

a. Chorionic villi testing

---

the eyes, flattened facial features, unusual palm crease, muscular flaccidity, and short stature. Women who are nearing the end of their reproductive years are more likely to have a child with Down syndrome than are younger women (table 18.1). If a women is concerned about the possibility of having an abnormal child, she can either have chorionic villi sampling or amniocentesis performed as described in the reading above. Following this procedure, a karyotype will reveal whether the child has Down syndrome.

It is known that the genes that cause Down syndrome are located on the bottom third of the number-21 chromosome (fig. 18.3b), and there has been a lot of investigative work to discover the specific genes

Table 18.1	Incidence of Selected Chromosomal Abnormalities
Name	**Frequency/ 100,000 Live Births**
Down syndrome (general)	140
Down syndrome (mothers over age 40)	1,000
Turner syndrome	8
Metafemale	50
Klinefelter syndrome	80
XYY	100

From *Antenatal Diagnosis*, HEW, 1979.

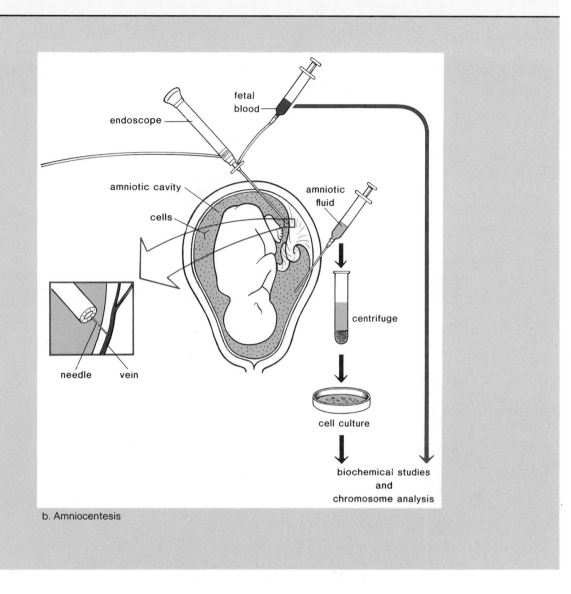

b. Amniocentesis

responsible for the characteristics of the syndrome. Thus far, investigators have discovered several genes that may account for various conditions seen in persons with Down syndrome. For example, they have located genes most likely responsible for the increased tendency toward leukemia, cataracts, accelerated rate of aging, and mental retardation. The latter gene, dubbed the *Gart* gene, uses an increased level of purines in the blood, a finding that is associated with mental retardation. It is hoped that it will someday be possible to find a way to control the expression of the *Gart* gene even before birth so that at least this symptom of Down syndrome will not appear.

Down syndrome is most often due to the inheritance of an extra number-21 chromosome.

Sex Chromosome Inheritance

Due to meiosis during spermatogenesis, a sperm of the male parent carries either an X or Y chromosome. In contrast, an egg always carries an X chromosome; therefore, the sex of the newborn is dependent on whether an X- or Y-bearing sperm fertilizes the egg. This also means that, theoretically, there is a fifty–fifty chance of having a boy or girl (fig. 18.4).

Figure 18.4 In this Punnett Square, the sperm and eggs are shown as carrying only a sex chromosome. Actually, of course, they also carry 22 autosomes. The offspring are either male or female depending on whether they received an X or Y chromosome from the male parent.

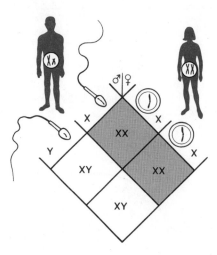

Figure 18.5 *a.* Nondisjunction of sex chromosomes during oogenesis followed by fertilization with normal sperm results in the conditions noted. *b.* Nondisjunction of sex chromosomes during spermatogenesis followed by fertilization of normal eggs results in the conditions noted.

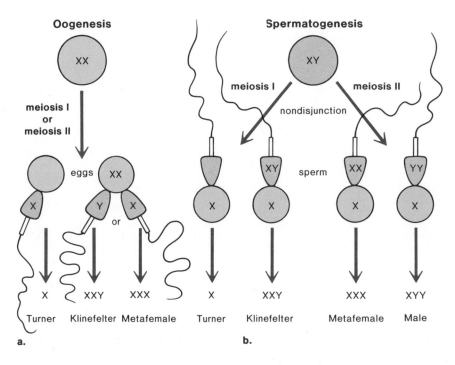

The inheritance of an abnormal sex chromosome (table 18.1) can be due to the occurrence of nondisjunction. Nondisjunction of the sex chromosomes during oogenesis can lead to an egg with either two X chromosomes or no X chromosomes. Nondisjunction of the sex chromosomes during spermatogenesis can result in a sperm that has no sex chromosome, both an X and a Y chromosome, two X chromosomes, or two Y chromosomes. Assuming that the other gamete is normal, the zygote could develop into an individual with one of the conditions noted in figure 18.5.

Sometimes a person inherits an abnormal combination of sex chromosomes due to nondisjunction of these chromosomes during meiosis.

An XO individual with **Turner syndrome** has only one sex chromosome, an X; the O signifies the absence of the second sex chromosome. Because the ovaries never become functional, these females do not undergo puberty or menstruate, and there is a lack of breast development (fig. 18.6b). Generally, these individuals have a stocky build and a webbed neck. They also have difficulty recognizing various spatial patterns.

When an egg having two X chromosomes is fertilized by an X-bearing sperm, a **metafemale** having three X chromosomes results. It might be supposed that the XXX female with forty-seven chromosomes would be especially feminine, but this is not the case. Although there is a tendency toward learning disabilities, most metafemales have no apparent physical abnormalities, and many are fertile and have children with a normal chromosome count.

When an egg having two X chromosomes is fertilized by a Y-bearing sperm, a male with **Klinefelter syndrome** results. This individual is male in general appearance, but the testes are underdeveloped and the breasts may be enlarged (fig. 18.6a). The limbs of these XXY males tend to be longer than average, body hair is sparse, and many have learning disabilities.

XYY males also occur, possibly due to nondisjunction during Meiosis II spermatogenesis. These males are usually taller than average, suffer from persistent acne, and tend to have barely normal intelligence. At one time, it was suggested that these men were likely to be criminally aggressive, but it has been shown that the incidence of such behavior is no greater than that among normal XY males.

Individuals are sometimes born with the sex chromosomes XO (Turner syndrome), XXX (metafemale), XXY (Klinefelter syndrome), and XYY. Individuals with a Y chromosome are always male, no matter how many X chromosomes there may be; however, at least one X chromosome is needed for survival.

Gene Inheritance

The genes are on the chromosomes and ordinarily, an individual receives two genes for each trait (fig 18.7). In simplistic terms, this means that individuals receive two genes for any particular aspect of their anatomy or physiology. One gene can be **dominant** to the other which is called **recessive**.

Occasionally, a person receives a faulty dominant gene and, therefore, is born with an inherited disorder

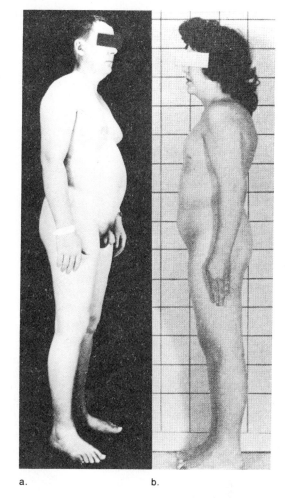

Figure 18.6 Abnormal sex chromosome inheritance. *a.* A male with Klinefelter (XXY) syndrome, which is marked by immature sex organs and development of the breasts. *b.* Female with Turner (XO) syndrome, which includes a bull neck, short stature, and immature sexual features.

a.   b.

even if the other gene is normal. (In these cases, a parent is expected to also have the disorder.) Or, a person can receive two faulty recessive genes and, therefore, be born with an inherited disorder. (In these cases, neither parent need have the disorder since each could have a dominant normal gene.) Commonly inherited dominant and recessive genetic disorders are listed in table 18.2.

The sex chromosomes carry genes just as the autosomal chromosomes do. Some of these genes determine the sex of the individual (that is, whether the individual has testes or ovaries), but most of the genes on the sex chromosomes control traits unrelated to sex characteristics. They are called **sex-linked genes** because they are on the sex chromosomes. A few sex-linked genes are on the Y chromosome, but the most

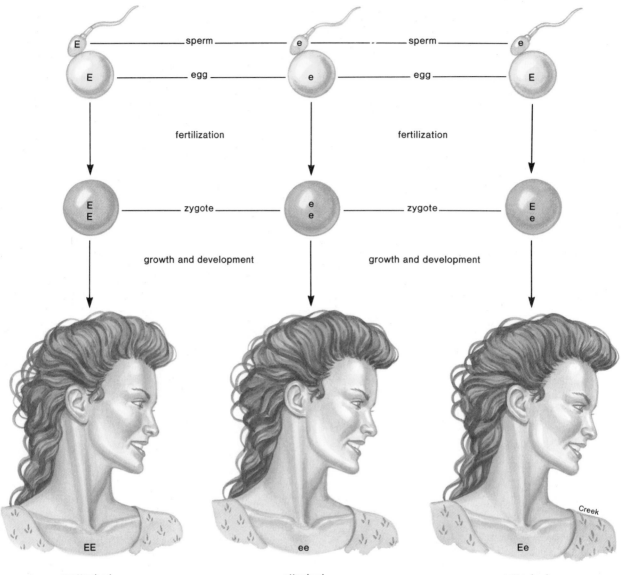

Figure 18.7 Genetic inheritance. Individuals inherit two genes for every characteristic of their anatomy and physiology. This illustration concerns the inheritance of type of earlobe—either unattached (free) or attached. The inheritance of a single dominant gene (E) causes an individual to have unattached earlobes; two recessive genes (ee) cause an individual to have attached earlobes. Notice that each individual receives one gene from the father (by way of a sperm) and one gene from the mother (by way of an egg).

sperm — sperm

egg — egg

fertilization fertilization

zygote — zygote

growth and development growth and development

EE

ee

Ee

unattached
earlobe

attached
earlobe

unattached
earlobe

important ones discovered so far are only on the much larger X chromosome. Since, in a male, the Y chromosome is blank for X-linked genes, a recessive one present on the X chromosome will be expressed. This means that males are more likely to have an X-linked recessive disorder than females. While it might seem that males would inherit a sex-linked disorder from their father, actually, they are more likely to inherit it from their mother. A father always gives his son a Y chromosome and the mother gives the X chromosome. X-linked recessive disorders are listed in table 18.2.

There are several inherited disorders that apparently do not show up unless multiple types of faulty genes are inherited. The inheritance of these conditions, given in table 18.2, cannot be easily explained and are not covered in detail here.

Table 18.2 Gene Disorders

Dominant	Recessive	X-Linked	Multiple Genes
Currently, some 1,489 dominantly inherited disorders have been catalogued. Examples include: • Neurofibromatosis—benign tumors in skin or deeper • Achondroplasia—a form of dwarfism • Chronic simple glaucoma (some forms)—a major cause of blindness if untreated • Huntington disease—progressive nervous system degeneration • Hypercholesterolemia—high blood cholesterol levels, propensity to heart disease • Polydactyly—extra fingers or toes	Among 1,117 recessively inherited disorders catalogued are: • Cystic fibrosis—disorder affecting function of mucous and sweat glands • Galactosemia—inability to metabolize milk sugar • Phenylketonuria—essential liver enzyme deficiency • Sickle-cell anemia—blood disorder primarily affecting blacks • Thalassemia—blood disorder primarily affecting persons of Mediterranean ancestry • Tay-Sachs—lysomal storage disease leading to nervous system destruction	Among 205 catalogued disorders transmitted by a gene or genes on the X chromosome are: • Agammaglobulinemia—lack of immunity to infections • Color blindness—inability to distinguish certain colors • Hemophilia—defect in blood-clotting mechanisms • Muscular dystrophy (some forms)—progressive wasting of muscles • Spinal ataxia (some forms)—spinal cord degeneration	The number of defects due to multifactorial inheritance is unknown. Some that are thought to be multifactorial are: • Cleft lip and/or palate • Clubfoot • Congenital dislocation of the hip • Spina bifida—open spine • Hydrocephalus (with spina bifida)—water on the brain • Pyloric stenosis—narrowed or obstructed opening from stomach into small intestine

Data from the National Foundation/March of Dimes

Biochemical Genetics and Biotechnology

The chemical nature of the genes is known today. They are made up of a chemical called deoxyribonucleic acid (DNA, p. 323). This knowledge has made it possible for researchers to carry out **recombinant DNA** technology. During this procedure, genes from two different sources are recombined into a single unit (fig. 18.8).

For example, a human gene can be placed in a ring of DNA taken from a bacteria, called a plasmid. Now the recombined plasmid can be placed back in a bacterium where it functions normally. Normally, a gene causes a cell, in this case, a bacterium to produce a particular protein. For example, a human gene for insulin can be placed in a plasmid and the bacterium will produce insulin. Further, the gene will be passed on to the offspring of this bacterium and they too will produce insulin. Therefore, products of **biotechnology** can be mass produced today because of recombinant DNA technology.

There are many other possible benefits from the use of biotechnology. For example, hopefully it will be possible to engineer plants that will be resistant to insects or that will grow well without nitrogen fertilizers. And one day, it might be possible to cure human genetic diseases utilizing a form of recombinant DNA technology in which a normal human gene is inserted into the cells of a person with a genetic disorder.

There are some who do not approve of the use of biotechnology to change the genetic inheritance of plants and animals, including humans. They fear that there may be hidden detrimental effects of recombinant DNA technology that are not fully obvious now. The current strategy, then, is to move very slowly in implementing this new technology so that we will have time to thoroughly study all of its possible effects.

Figure 18.8   Recombinant DNA technology. A plasmid is removed from a bacterium, usually *E. coli,* and foreign DNA such as a human gene is incorporated into the plasmid. The recombined plasmid is then reintroduced into a bacterium where it functions normally. When the bacterium reproduces, each of the cloned cells contains a copy of the human gene.

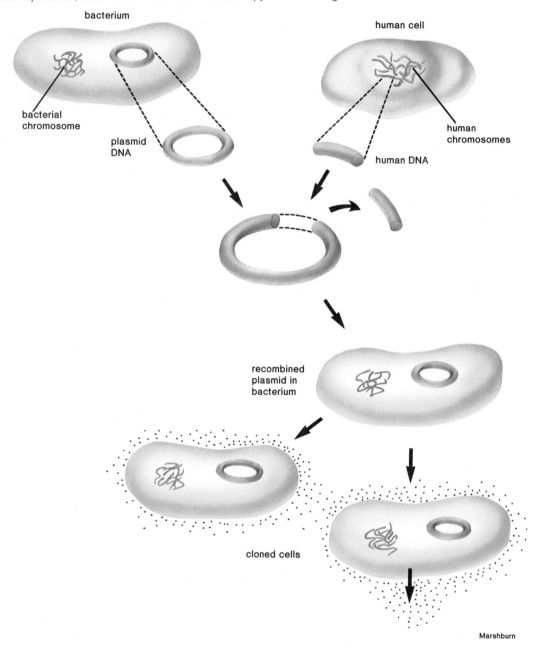

bacterium

human cell

bacterial
chromosome

plasmid
DNA

human
chromosomes

human DNA

recombined
plasmid in
bacterium

cloned cells

Marshburn

Summary

I. Chromosome Inheritance
   A. Normally, humans inherit 22 pairs of autosomal chromosomes and one pair of sex chromosomes. Males are XY and females are XX.
   B. Down syndrome is most often due to the inheritance of an extra number-21 chromosome.
   C. Individuals are sometimes born with the sex chromosomes XO (Turner syndrome), XXX (metafemale), XXY (Klinefelter syndrome), and XYY. Individuals with a Y chromosome are always male; however, at least one X chromosome is needed for survival.

II. Gene Inheritance
   A. Dominant genetic disorders (e.g., neurofibromatosis, Huntington disease) are due to the inheritance of at least one dominant gene.
   B. Recessive genetic disorders (e.g., cystic fibrosis, Tay Sachs) are due to the inheritance of two recessive genes.
   C. Sex-linked recessive disorders (e.g., color blindness, hemophilia) are due to the inheritance of only one recessive gene in males (two in females).
   D. Recombinant DNA technology may eventually make it possible to cure human genetic disorders.

Study Questions

1. What is the normal chromosome inheritance of humans?
2. What are the characteristics and cause of Down syndrome?
3. How is it possible to detect abnormal autosomal chromosome inheritance in an embryo?
4. How is sex inherited in humans?
5. What are the characteristics of the common abnormal sex chromosome inherited conditions in humans?
6. Explain dominant, recessive, and sex-linked gene inheritance in humans.
7. Give examples of dominant, recessive, and sex-linked genetic disorders in humans.
8. Describe how recombinant DNA technology is carried out. What does biotechnology have to do with genetic disorders?

Objective Questions

Fill in the blanks.

1. A person with Down syndrome has inherited _____ number-21 chromosomes.
2. The sex chromosomes of a male are _____ .
3. A person with Klinefelter syndrome has the chromosomes _____ .
4. A karyotype of an embryo's chromosomes can be done following _____ .
5. A dominant genetic disorder only requires the inheritance of _____ (one or two) faulty gene(s).
6. The genes are on the _____ .
7. If a person inherits a recessive genetic disease, then both parents _____ .
8. A cure of genetic disorders might eventually be possible because of _____ .

Medical Terminology Reinforcement Exercise

Pronounce and analyze the meaning of the following dissected terms:

1. neogenesis (ne''o-jen'ĕ-sis) neo/genesis
2. regeneration (re-jen''er-a'shun) re/generat/ion
3. amniocentesis (am''ne-o-sen-te'sis) amnio/centesis
4. fetoscope (fe'to-skōp) feto/scope
5. chromosome (kro'mo-sōm) chromo/some
6. polydysplasia (pol''e-dis-pla'ze-ah) poly/dys/plas/ia
7. congenital (kon-jen'ĭ-tal) con/genit/al
8. hyperplasia (hi''per-pla'ze-ah) hyper/plas/ia
9. atrophy (at'ro-fe) a/trophy
10. agammaglobulinemia (a-gam''ah-glo''bu-li-ne'me-ah) a/gammaglobulin/emia

Appendix A
Chemistry

Inorganic Chemistry

Although inorganic chemistry pertains to nonliving matter, inorganic chemicals are important constituents of all living things. Also, some knowledge of inorganic chemistry is necessary for considering the unique molecules of life.

Atoms

An **atom** is the smallest unit of matter, nondivisible by chemical means. For our purposes, it is satisfactory to think of an atom as having a central *nucleus,* where subatomic particles called *protons* and *neutrons* are located, and *shells,* where *electrons* orbit about the nucleus (fig. A.1). Two important features of protons, neutrons, and electrons are their weight and charge, which are indicated in table A.1.

An **element** is any substance that contains just one type of atom. Figure A.2 gives a simplified Periodic Table of the Elements highlighting the elements that are most common to living things. Notice that in the table each specific atom has a *symbol;* for example,

Figure A.1 Representation of an atom. The nucleus contains protons and neutrons; the shells contain electrons. The first shell is complete with two electrons, and every shell thereafter may contain as many as eight electrons.

p = protons
n = neutrons
● = electrons

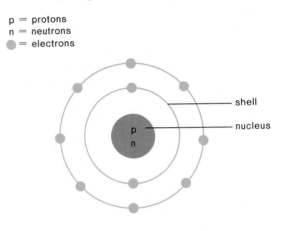

shell
nucleus

Table A.1	Subatomic Particles	
Name	**Charge**	**Weight**
Electron	One negative unit	Almost no weight
Proton	One positive unit	One atomic unit
Neutron	No charge	One atomic unit

Figure A.2 Periodic Table of the Elements (simplified). Each element has an atomic number, atomic symbol, and atomic weight. The elements in dark color are the most common, and those in light color are also common in living things.

I	II	III	IV	V	VI	VII	VIII
1 H hydrogen 1							2 He helium 4
3 Li lithium 7	4 Be beryllium 9	5 B boron 11	6 C carbon 12	7 N nitrogen 14	8 O oxygen 16	9 F fluorine 19	10 Ne neon 20
11 Na sodium 23	12 Mg magnesium 24	13 Al aluminum 27	14 Si silicon 28	15 P phosphorus 31	16 S sulfur 32	17 Cl chlorine 35	18 Ar argon 40
19 K potassium 39	20 Ca calcium 40						

atomic number
atomic symbol
atomic weight

Figure A.3 Carbon atom. The diagram of the atom shows that the number of protons (the atomic number) equals the number of electrons when the atom is electrically neutral. Carbon may also be written in the manner shown below the diagram. The subscript is the atomic number, and the superscript is the weight.

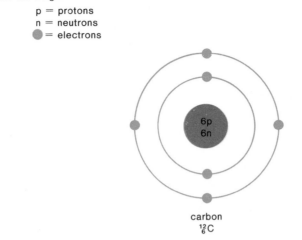

p = protons
n = neutrons
⬤ = electrons

carbon
$^{12}_{6}C$

Figure A.4 Formation of the salt sodium chloride. During this ionic reaction, an electron is transferred from the sodium atom to the chlorine atom. Each resulting ion carries a charge as shown. Most people use the term *salt* to refer only to sodium chloride, but chemists use the term to refer to similar combinations of positive and negative ions.

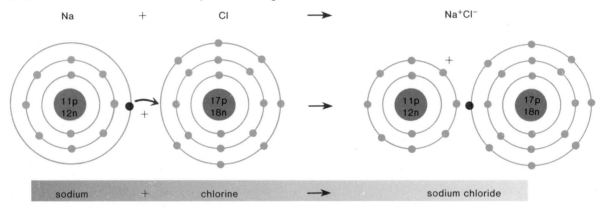

Na + Cl ⟶ Na⁺Cl⁻

sodium + chlorine ⟶ sodium chloride

C = carbon and N = nitrogen. Also, each type of atom has an *atomic number;* for example, carbon is number 6 and nitrogen is number 7. *The atomic number equals the number of protons.* Also, each type of atom has an *atomic weight,* or mass. Carbon has an atomic weight or a mass of 12, and nitrogen has an atomic weight of 14. *The atomic weight equals the number of protons plus the number of neutrons.*

Now, it is possible to diagram a specific *electrically neutral* atom (fig. A.3). In an electrically neutral atom, the number of protons (+) is equal to the number of electrons (−). The first shell of an atom can contain up to two electrons; thereafter, each shell of those atoms in the simplified table (fig. A.2) can contain up to eight electrons.

Reactions between Atoms

Atoms react with one another to form *molecules.* In one type of reaction, there is a transfer of an electron(s) from one atom to another in order to form a molecule.

Such atoms are thereafter called **ions,**[1] and the reaction is called an *ionic reaction.* For example, figure A.4 depicts a reaction between sodium (Na) and chlorine (Cl) in which chlorine takes an electron from sodium. Now the sodium ion (Na⁺) carries a positive charge, and the chlorine ion (Cl⁻) carries a negative charge. Notice that a negative charge indicates that the ion has more electrons (−) than protons (+) and a positive charge indicates that the ion has more protons (+) than electrons (−). Oppositely charged ions are attracted to one another, and this attraction is called an **ionic bond.**

In another type of reaction, atoms form a molecule by sharing electrons. The bond that forms between these is called a **covalent bond.** For example, when oxygen reacts with two hydrogen atoms, water (H_2O) is formed (fig. A.5).

Sometimes the atoms in a covalently bonded molecule share electrons evenly, but in water, the electrons spend more time encircling the larger oxygen than the

1. Ions are also called electrolytes because their charge allows them to carry an electric charge.

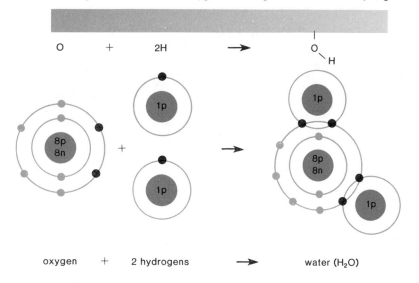

oxygen   +   2 hydrogens   →   water (H₂O)

smaller hydrogens. Therefore, there is a slight positive charge on the hydrogen atoms and a slight negative charge on the oxygen atom. For this reason, water is called a *polar molecule,* and hydrogen bonding occurs between water molecules (fig. A.6). A **hydrogen bond** occurs whenever a partially positive hydrogen is attracted to a partially negative atom. The hydrogen bond is represented by a dotted line in figure A.6 because it is a weak bond that is easily broken.

Dissociation

Polarity also causes water molecules to tend to **dissociate,** or split up, in this manner:

$$H-O-H \longrightarrow H^+ + OH^-$$

The hydrogen ion (H^+) has lost an electron; the hydroxide ion (OH^-) has gained the electron. Because very few molecules actually dissociate, few hydrogen ions and hydroxide ions result.

Acids and Bases

Acids are compounds that dissociate in water and release hydrogen ions. For example, an important inorganic acid is hydrochloric acid (HCl), which dissociates in this manner:

$$HCl \longrightarrow H^+ + Cl^-$$

Dissociation is almost complete, and this acid is called a strong acid. If HCl is added to a beaker of water, the number of hydrogen ions increases.

Bases are compounds that dissociate in water and release hydroxide ions (OH^-). For example, an important inorganic base is sodium hydroxide (NaOH), which dissociates in this manner:

$$NaOH \longrightarrow Na^+ + OH^-$$

Figure A.6 Water molecules are polar; each hydrogen carries a partial positive charge and each oxygen carries a partial negative charge. The polarity of the water molecules brings about hydrogen bonding between the molecules in the manner shown. The dotted lines represent hydrogen bonds (δ = partial).

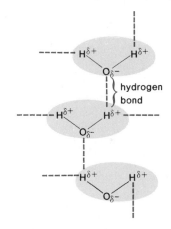

Dissociation is complete, and sodium hydroxide is called a strong base. If NaOH is added to a beaker of water, the number of hydroxide ions increases.

pH

The **pH** scale ranges from 0–14. Any pH value below 7 is acid, with ever increasing acidity toward the lower numbers. Any pH value above 7 is basic (or alkaline), with ever increasing basicity toward the higher numbers. A pH of exactly 7 is neutral. Water has an equal number of H^+ and OH^- ions, and, thus, one of each is formed when water dissociates. The fraction of water molecules that dissociate is 10^{-7} (0.0000001), which is the source of the pH value for neutral solutions. The pH scale was devised to simplify discussion of the hydrogen ion concentration [H^+], without using cumbersome numbers. For example,

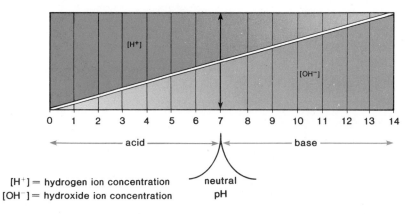

$[H^+]$ = hydrogen ion concentration
$[OH^-]$ = hydroxide ion concentration

neutral pH

a. 1×10^{-6} [H⁺] = pH 6   Each lower pH unit has ten
b. 1×10^{-7} [H⁺] = pH 7   times the amount of H⁺ as
c. 1×10^{-8} [H⁺] = pH 8   the next higher unit.

Of the three values listed here, pH 6 has the greater number of H⁺ ions and is acidic; pH 7 has an equal number of H⁺ and OH⁻ ions and is neutral; pH 8 has the lesser number of H⁺ and is basic. Figure A.7 gives the complete pH scale with proper notations.

All living things need to maintain the hydrogen ion concentration, or pH, at a constant level. For example, the pH of the blood is held constant at about 7.4, or we become ill. The presence of buffers helps keep the pH constant. A **buffer** is a chemical or a combination of chemicals that can take up excess hydrogen ions or excess hydroxide ions. When an acid is added to a buffered solution, a buffer takes up excess hydrogen ions, and when a base is added to a buffered solution, a buffer takes up excess hydroxide ions. Therefore, the pH changes minimally whenever a solution is buffered.

Organic Chemistry

Table A.2 contrasts inorganic compounds with organic compounds. (A *compound* is a substance that contains many copies of the same type of molecule.)

Unit Molecules

The chemistry of carbon accounts for the formation of the very large number of organic compounds we associate with living organisms. Carbon shares electrons with as many as four other atoms. Many times, carbon atoms share with each other to form rings or chains of carbon atoms. These act as a skeleton for the unit molecules found in the life molecules—proteins, carbohydrates, fats, and nucleic acids. Thus the properties of carbon are essential to life as we know it.

Table A.2 Inorganic Versus Organic Chemistry	
Inorganic Compounds	**Organic Compounds**
Usually contain metals and nonmetals	Always contain carbon and hydrogen
Usually ionic bonding	Always covalent bonding
Always contain a small number of atoms	May be quite large with many atoms
Often associated with nonliving elements	Often associated with living organisms

Figure A.8 Synthesis and hydrolysis of an organic polymer. When molecules join together to form the polymer (synthesis), water is released; when the polymer is broken down (hydrolysis), water is added.

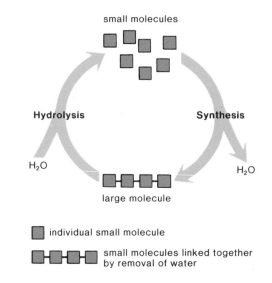

small molecules

Hydrolysis   Synthesis

H_2O H_2O

large molecule

▪ individual small molecule

▭▭▭▭ small molecules linked together by removal of water

Synthesis and Hydrolysis

Figure A.8 diagrammatically illustrates that the large molecules are synthesized or made when small unit molecules join together. A bond that joins two unit

Figure A.9 Each amino acid has the structure shown (R = remainder of the molecule). When two amino acids combine, a peptide bond forms and water is given off. In other words, the water molecule on the right-hand side of the equation is derived from components removed from the amino acids on the left-hand side.

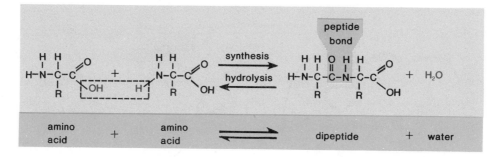

molecules together is created after the removal of H$^+$ from one molecule and OH$^-$ from the next molecule. As water forms, dehydration **synthesis** occurs.

Large molecules are often **polymers,** or chains of unit molecules joined together. They can be broken down in a manner opposite to synthesis: the addition of water leads to the disruption of the bonds linking the unit molecules together. During this process, called **hydrolysis,** one molecule takes on H$^+$ and the next takes on OH$^-$.

Proteins

Functions

Proteins are large, complex macromolecules that sometimes have mainly a structural function. For example, in humans, keratin is a protein that makes up hair and nails, and collagen is a protein found in all types of connective tissue, including ligaments, cartilage, bone, and tendons. The muscles contain proteins that account for their ability to contract.

Some proteins function as **enzymes,** necessary contributors to the chemical workings of the cell and, therefore, of the body. Enzymes are organic catalysts that speed up chemical reactions. They work so quickly that a reaction which might normally take several hours or days takes only a fraction of a second when an enzyme is present.

Structure

The unit molecules found in proteins are called **amino acids.** The name amino acid refers to the fact that the molecule has two functional groups: an *amino group* and an *acid group.*

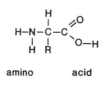

amino   acid

Amino acids differ from one another by their *R groups,* the remainder of the molecule. In amino acids, the R group varies from being a single hydrogen atom to a complicated ring. Because there are about twenty different common amino acids found in the proteins of living things, there are also about twenty different types of R groups.

The bond that joins two amino acids together is called a **peptide bond.** As you can see in figure A.9, when synthesis occurs, the acid group of one amino acid reacts with the amino group of another amino acid and water is given off. A dipeptide contains only two amino acids, but when ten or twenty amino acids have joined together, the resulting chain is called a *polypeptide.* A very long polypeptide of approximately seventy-five amino acids is called a protein. Proteins have three levels of structure. In figure A.10, the final tertiary shape of a common protein is shown at the far left side of the diagram. Within this shape lies the helix of molecules as is apparent when the protein is stretched out. Finally, we see that the helix itself contains a particular sequence of amino acids.

Carbohydrates

Carbohydrates are characterized by the presence of H — C — OH groupings in which the ratio of hydrogen atoms to oxygen atoms is approximately 2:1. Since this ratio is the same as the ratio in water, the meaning of this compound's name, hydrates of carbon, is very appropriate. If the number of carbon atoms in the compound is low (from about three to seven), the carbohydrate is a simple sugar, or monosaccharide. Larger carbohydrates are created by joining together monosaccharides in the manner described in figure A.8 for the synthesis of organic compounds.

Monosaccharides and Disaccharides

As their name implies, *monosaccharides* are simple sugars having only one unit. These compounds are often designated by the number of carbons they contain; for

Figure A.10   Proteins have at least three levels of structure. Primary structure is the order of the amino acids; secondary structure is often a helix; and the tertiary structure is often a twisting and turning of the helix that takes place because of bonding between the R groups.

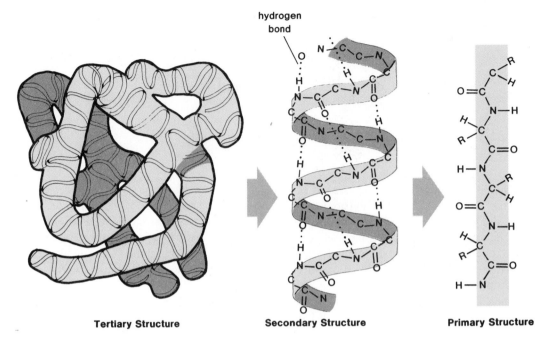

Tertiary Structure   **Secondary Structure**   **Primary Structure**

Figure A.11   Each glucose molecule has the structure shown. When two glucose molecules combine, the disaccharide maltose is formed. During synthesis, a bond forms between the two glucose molecules as a molecule of water is formed. During hydrolysis, the components of water are added as the bond is broken.

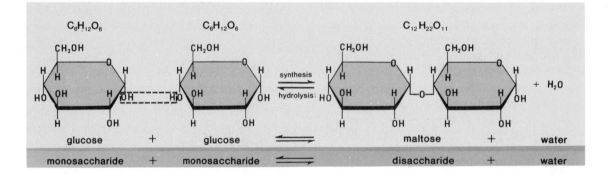

example, pentose sugars have five carbons, and hexose sugars have six carbons. *Glucose* is a six-carbon sugar, with the structural formula shown in figure A.11. Although there are other monosaccharides with the molecular formula $C_6H_{12}O_6$, in this text, we use the molecular formula $C_6H_{12}O_6$ to mean glucose, since glucose is the most common six-carbon monosaccharide found in cells. Cells use glucose as an immediate energy source.

The term *disaccharide* tells us that there are two monosaccharide units joined together in the compound. When two glucose molecules join together, maltose is formed. In figure A.11, the forward direction is a dehydration synthesis and the backward reaction is a hydrolysis. When glucose and another

monosaccharide, fructose, are joined together, the disaccharide called sucrose is formed. *Sucrose* is derived from plants and is commonly used at the table to sweeten foods.

Polysaccharides

A **polysaccharide** is a carbohydrate that contains a large number of monosaccharide molecules. There are three polysaccharides that are common in animals and plants: glycogen, starch, and cellulose. All of these are polymers, or chains, of glucose, just as a necklace might be made up of only one type of bead. Even though all three polysaccharides contain only glucose, they are distinguishable from one another.

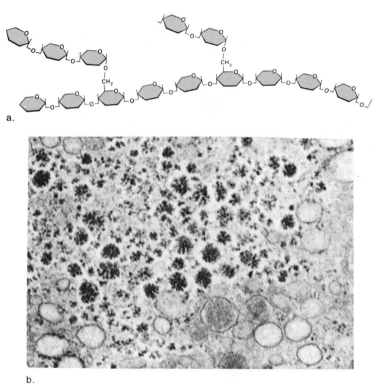

a.

b.

Glycogen, a molecule having many side branches (fig. A.12), is the storage form of glucose in humans. After eating, the liver stores glucose as glycogen; in between eating, the liver releases glucose so that the blood concentration of glucose is always 0.1%.

The polymers *starch* and *cellulose* are found in plants. Plants store glucose as starch, a polymer similar in structure to glycogen except that it has few side branches. Starch is an important source of glucose energy in our diet because it can be hydrolyzed to glucose by digestive enzymes. In cellulose, often called fiber, the glucose units are joined by a slightly different type of linkage compared to that of glycogen and starch. For this reason, we are unable to digest cellulose, and it passes through our digestive tract as roughage. Recently, it has been suggested that the presence of roughage in the diet is necessary to good health and prevention of colon cancer.

Lipids

Many **lipids** are nonpolar and therefore are insoluble in water. This is true of fats, the most familiar lipids, such as lard, butter, and oil, which are used in cooking or at the table. In the body, fats serve as long-term energy sources. Adipose tissue is composed of cells that contain many molecules of fat.

Figure A.13 Fatty acids. *a.* Saturated fatty acids have no double bonds because each carbon is bonded to the maximum of two hydrogen atoms. *b.* Unsaturated fatty acids have double bonds because some carbon atoms are bonded to only one hydrogen.

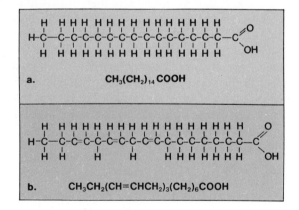

a. $CH_3(CH_2)_{14}COOH$

b. $CH_3CH_2(CH=CHCH_2)_3(CH_2)_6COOH$

Fats (Triglycerides)

A fat contains two types of unit molecules: **glycerol** and **fatty acids.** Each fatty acid has a long chain of carbon atoms, with hydrogens attached, ending in an acid group (fig. A.13). Fatty acids are either *saturated* or *unsaturated.* Saturated fatty acids have no double bonds between the carbon atoms. The carbon chain is saturated, so to speak, with all the hydrogens that can be held. Unsaturated fatty acids have double bonds in

Figure A.14   Synthesis and hydrolysis of a neutral fat. Three fatty acids plus glycerol react to produce a fat molecule and three water molecules. A fat molecule plus three water molecules react to produce three fatty acids and glycerol.

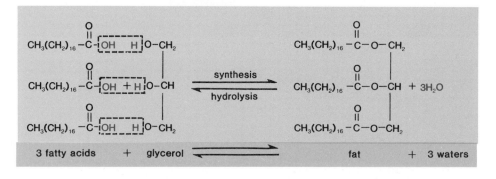

Figure A.15   Fat molecules, being nonpolar, will not disperse in water. An emulsifier contains molecules that have a polar end and nonpolar end. When an emulsifier is added to a beaker containing a layer of nondispersed fat molecules, the nonpolar ends are attracted to the nonpolar fat, and the polar ends are attracted to the water. This causes droplets of fat molecules to become dispersed.

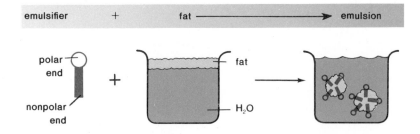

the carbon chain wherever there is only one hydrogen atom per carbon atom. Unsaturated fatty acids are most often found in vegetable oils and account for the liquid nature of these oils. Vegetable oils are hydrogenated to make margarine. Polyunsaturated margarine still contains a large number of unsaturated, or double, bonds.

Glycerol is a compound with three H — C — OH attached by way of the carbon atoms. When fat is formed, by dehydration synthesis, the — OH groups react with the acid portions of three fatty acids so that three molecules of water are formed. The reverse of this reaction represents hydrolysis of the fat molecule into its separate components (fig. A.14).

Emulsifiers

Fats do not mix with water because they are nonpolar. When an emulsifier, a molecule having a polar and nonpolar end, is added to a fat then a fat will mix with water. Figure A.15 shows how an emulsifier positions itself about an oil droplet so that the polar ends project outward. Now the droplet will be soluble in water. This process of causing a fat to disperse in water is called **emulsification,** and it is said that an emulsion has been formed. Emulsification occurs when dirty clothes are washed with soaps and detergents. Also, prior to the digestion of fatty foods, fats are emulsified by bile. Usually a person who has had the gallbladder removed has trouble digesting fatty foods because the gallbladder stores bile for use at the proper time during the digestive process.

Steroids

The **steroids** have a structure that is related to the structure of cholesterol. They are constructed of four fused rings of carbon atoms to which is usually attached a chain of varying length (fig. A.16). Today, there is a great deal of interest in cholesterol because a high blood level is associated with development of coronary heart disease as discussed on page 205. Even so, steroids are very necessary compounds in the body; for example, the sex hormones are steroids. Anabolic steroids, abused by some to increase muscle mass, are a form of testosterone.

Nucleic Acids

Nucleic acids are huge, macromolecular compounds with very specific functions in cells; for example, the genes on the chromosomes within the nucleus of a cell

Like cholesterol (*a*), steroid molecules have four adjacent rings, but their effects on the body largely depend on the type of chain attached and the location indicated. The chain in (*b*) is found in aldosterone, which is involved in the regulation of sodium and water metabolism, while the chain in (*c*) is found in testosterone, the male sex hormone.

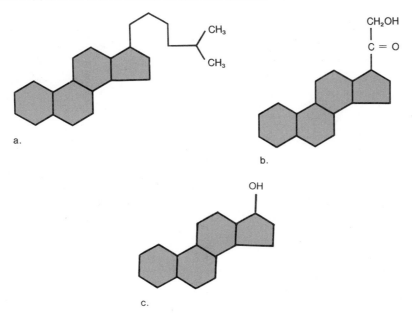

are composed of a nucleic acid called DNA (deoxyribonucleic acid). Another important nucleic acid, **RNA** (ribonucleic acid), works in conjunction with DNA to bring about protein synthesis. RNA is found in the nucleus of a cell and in the cytoplasm. It makes up the ribosomes.

Both DNA and RNA are polymers of nucleotides and, therefore, are chains of nucleotides joined together. Just like the other synthetic reactions we have studied in this section, when these units are joined together to form nucleic acids, water molecules are removed.

Nucleotides

Every **nucleotide** is a molecular complex of three types of unit molecules: phosphoric acid (phosphate), a pentose sugar, and a nitrogen base. In DNA, the sugar is deoxyribose and in RNA, the sugar is ribose, and this difference accounts for their respective names. There are four different types of nucleotides in DNA and RNA. Figure A.17 shows the types of nucleotides that are present in DNA. The base can be the **purines,** adenine or guanine, which have a double ring, or the **pyrimidines,** thymine or cytosine, which have a single ring. These structures are called bases because they have basic characteristics that raise the pH of a solution. RNA differs from DNA in that the base uracil is used in place of the base thymine.

Strands

When nucleotides join together, they form a linear molecule called a strand in which the so-called backbone is made up of phosphate-sugar-phosphate-sugar, with the bases projecting to one side of the backbone. RNA is single stranded (fig. A.18), but DNA is double stranded. The two strands of DNA twist about one another in the form of a **double helix** (fig. A.19). The two strands are held together by hydrogen bonds between purine and pyrimidine bases. Thymine (T) is always paired with adenine (A), and guanine (G) is always paired with cytosine (C). This is called complementary base pairing. If we unwind the DNA helix, it resembles a ladder (fig. A.19). The sides of the ladder are made entirely of phosphate and sugar molecules, and the rungs of the ladder are made only of the *complementary paired bases*. The bases can be in any order, but A is always paired with T, and G is always paired with C, and vice versa. It is now known that each type of gene has its own particular sequence of bases.

ATP

ATP, adenosine triphosphate (fig. A.20), is a very special type of nucleotide. It is composed of the base adenine and the sugar ribose (together called adenosine)

Figure A.17 Nucleotides in DNA. Each nucleotide is composed of phosphate, the sugar deoxyribose, and a base. *a.* The purine bases are adenine and guanine. *b.* The pyrimidine bases are cytosine and thymine.

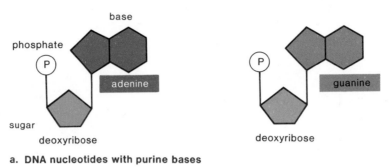

base

phosphate

sugar

deoxyribose

adenine

a. DNA nucleotides with purine bases

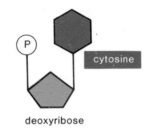

guanine

deoxyribose

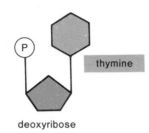

thymine

deoxyribose

cytosine

deoxyribose

b. DNA nucleotides with pyrimidine bases

Table A.3	DNA Structure Compared to RNA Structure	
	DNA	**RNA**
Sugar	Deoxyribose	Ribose
Bases	Adenine, guanine, thymine, cytosine	Adenine, guanine, uracil, cytosine
Strands	Double stranded with base pairing	Single stranded
Helix	Yes	No

Figure A.18 Generalized nucleic acid strand. Nucleic acid polymers contain a chain of nucleotides. Each strand has a backbone made of sugar and phosphate molecules. The bases project to the side.

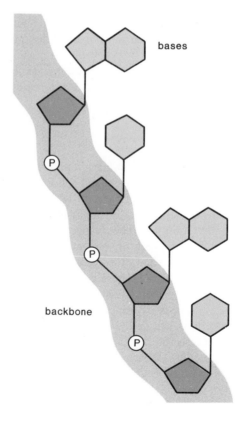

bases

backbone

and three phosphate groups. The wavy lines in the formula for ATP indicate high-energy phosphate bonds; when these bonds are broken, an unusually large amount of energy is released. Because of this property, ATP is the energy currency of cells; when cells "need" something, they "spend" ATP.

ATP is used in body cells for synthetic reactions, active transport, nervous conduction, and muscle contraction. When energy is required for these processes, the end phosphate group is removed from ATP, breaking down the molecule to ADP (adenosine diphosphate) and Ⓟ (phosphate) (fig. A.20). This occurs in both directions; not only is ATP broken down, it is also built up when ADP joins with Ⓟ. Since ATP breakdown is constantly occurring, there is always a ready supply of ADP and Ⓟ to rebuild ATP again.

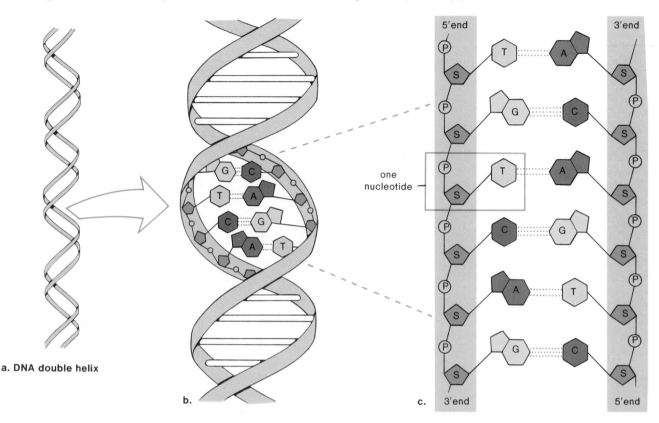

Figure A.19 Overview of DNA structure. *a.* Double helix. *b.* Enlargement of *a* shows twisted ladder. *c.* Ladder configuration. Notice that the uprights are composed of sugar and phosphate molecules and the rungs are complementary paired bases.

a. DNA double helix

b.

c.

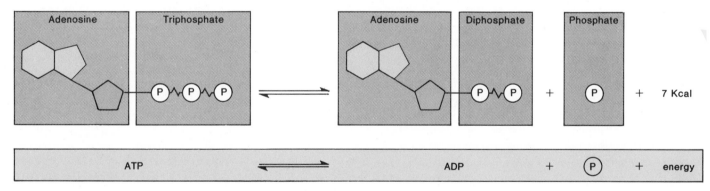

Figure A.20 ATP, the energy molecule in cells, has two high-energy phosphate bonds (indicated in the figure by wavy lines). When cells require energy, the last phosphate bond is broken and a phosphate molecule is released.

Medical Terminology Reinforcement Exercise

Pronounce and analyze the meaning of the following dissected terms:

1. anisotonic (an-i''so-ton'ik) an/iso/ton/ic

2. dehydration (de''hi-dra'shun)—de/hydra/tion

3. hypokalemia (hi''po-ka-le'me-ah) hypo/kal/emia

4. hypovolemia (hi''po-vo-le'me-ah) hypo/vol/emia

5. nonelectrolyte (non''e-lek'tro-līt)—non/electro/lyte

6. hydrolysis (hi-drol'ĭ-sis) hydro/lysis

7. lipometabolism (lip''o-mě-tab'o-lizm) lipo/meta/bol/ism

8. hyperlipoproteinemia (hi''per-lip''o-pro''te-in-e'me-ah) hyper/lipo/protein/emia

9. hyperglycemia (hi''per-gli-se'me-ah) hyper/glyc/emia

10. hypoxemia (hi''pok-se'me-ah) hyp/ox/emia

Appendix B
Further Readings

Berne, R. M., and Levy, M. A. 1981. *Cardiovascular physiology*. 4th ed. St. Louis: C. V. Mosby Co.

Boyd, W., and Sheldon, H. 1980. *Introduction to the study of disease*. 8th ed. Philadelphia: Lea and Febiger.

Burke, S. R. 1975. *Human biology in health and disease*. New York: John Wiley and Sons.

Carlson, B. M. 1988. *Patten's foundations of embryology*. 5th ed. New York: McGraw-Hill.

Clemente, C. D. 1987. *Anatomy: A regional atlas of the human body*. 3d ed. Philadelphia: Lea and Febiger.

Crapo, L. 1985. *Hormones: The messengers of life*. New York: W. H. Freeman and Co.

Crouch, J. E. 1978. *Functional human anatomy*. 3d ed. Philadelphia: Lea and Febiger.

Danforth, D. N., ed. 1982. *Obstetrics and gynecology*. 4th ed. New York: Harper & Row Publishers.

Davenport, H. W. 1982. *Physiology of the digestive tract*. 5th ed. Chicago: Year Book Medical Publishers.

Dewitt, W. 1989. *Human biology: Form, function, and adaptation*. Glenview, IL: Scott, Foresman and Co.

Di Fiore, M., and Schmidt, I. 1981. *Atlas of human histology*. 5th ed. Philadelphia: Lea and Febiger.

Dreamer, D. W. 1981. *Being human*. Philadelphia: Saunders College Publishing.

Fox, E. L., and Mathews, D. K. 1981. *The physiological basis of physical education and athletics*. 3d ed. Philadelphia: Saunders College Publishing.

Gilbert, E., and Hungtington, R. 1978. *An introduction to pathology*. New York: Oxford University Press.

Guyton, A. C. 1979. *Physiology of the human body*. Philadelphia: Saunders College Publishing.

Jensen, M. M., and Wright, D. N. 1985. *Introduction to medical microbiology*. Englewood Cliffs, NJ: Prentice-Hall.

Lankford, T. R. 1979. *Integrated science for health students*. 2d ed. Washington, D.C.: Reston Publishing Co.

Mader, S. S. 1990. *Human reproductive biology*. 2d ed. Dubuque, IA: Wm. C. Brown Publishers.

Nilsson, L. 1985. *The body victorious*. New York: Delacorte Press.

Prescott, D. M. 1988. *Cells*. Boston: Jones and Bartlett Publishers.

Purtilio, D. T. 1978. *A survey of human diseases*. Reading, MA: Addison-Wesley.

Stine, G. J. 1989. *The new human genetics*. Dubuque, IA: Wm. C. Brown Publishers.

Thompson, R. F. 1985. *The brain: An introduction to neuroscience*. San Francisco: W. H. Freeman and Co.

Wantz, M. S., and Gay, J. E. 1981. *The aging process: A health perspective*. Cambridge, MA: Winthrop.

Glossary

A

abdomen (ab-do'men) Portion of the body between the diaphragm and the pelvis.

abduction (ab-duk'shun) Movement of a body part away from the midline.

accommodation (ah-kom''o-da'-shun) Adjustment of the lens for close vision.

acetabulum (as''e-tab'u-lum) A socket in the lateral surface of the hipbone (os coxae) into which the head of the femur articulates.

acetylcholine (as''e-til-ko'len) See ACh.

acetylcholinesterase (as''e-til-ko''lin-es'ter-as) An enzyme in the membrane of postsynaptic cells that breaks down ACh; AChE. This enzymatic reaction inactivates the neurotransmitter.

ACh (acetylcholine) A neurotransmitter substance secreted at the ends of many neurons; responsible for the transmission of a nerve impulse across a synaptic cleft.

acid (as'id) A solution in which pH is less than 7; a substance that contributes or liberates hydrogen ions in a solution.

acromegaly (ak''ro-meg'ah-le) Condition resulting from an increase in growth hormone production after adult height has been achieved.

acrosome (ak'ro-som) Covering on the tip of a sperm cell's nucleus believed to contain enzymes necessary for fertilization.

ACTH Adrenocorticotropic hormone secreted by the anterior lobe of the pituitary gland that stimulates activity in the adrenal cortex.

actin (ak'tin) One of the two major proteins of muscle; makes up thin filaments in myofibrils of muscle cells. *See* myosin.

action potential (ak'shun po-ten'shal) The change in potential propagated along the membrane of a neuron, the nerve impulse.

active immunity (ak'tiv i-mu'ni-te) Immunity involving sensitization, in which antibody production is stimulated by exposure to an antigen.

active site (ak'tiv sit) The region on the surface of an enzyme where the substrate binds and where the reaction occurs.

active transport (ak'tiv trans'port) Transfer of a substance into or out of a cell against a concentration gradient by a process that requires a carrier and expenditure of energy.

Addison's disease (ad'i-sonz di-zez') A condition resulting from a deficiency of adrenal cortex hormones.

adduction (ah-duk'shun) Movement of a body part toward the midline.

adenosine triphosphate (ah-den'o-sen tri-fos'fat) *See* ATP.

ADH Antidiuretic hormone released from the posterior lobe of the pituitary gland that enhances the conservation of water by the kidneys; sometimes called vasopressin.

adrenalin (ah-dren'ah-lin) A hormone produced by the adrenal medulla that stimulates "fight or flight" reactions. Also called epinephrine.

adrenocorticotropic hormone (ah-dre''no-kor''te-ko-trop'ik hor'mon) *See* ACTH.

aerobic cellular respiration (a-er-o'bik sel'u-lar res''pi-ra'shun) The breakdown of carbohydrate in mitochondria that utilizes oxygen and gives off carbon dioxide results in ATP buildup.

afterbirth (af'ter-berth'') The placenta that is expelled after the birth of a child.

agglutination (ah-gloo''ti-na'shun) Clumping of cells, particularly in reference to red blood cells involved in an antigen–antibody reaction.

agranulocytes (ah-gran'u-lo-sits'') White blood cells that do not contain distinctive granules, i.e. agranular leukocytes.

AIDS (adz) Acquired immune deficiency syndrome, a disease caused by a retrovirus and transmitted via body fluids; characterized by failure of the immune system.

albumin (al-bu'min) A water-soluble protein, produced in the liver, that is the major component of the plasma proteins.

aldosterone (al''do-ster'on) A hormone, secreted by the adrenal cortex, that functions in regulating sodium and potassium excretion by the kidneys.

allantois (ah-lan'to-is) An extraembryonic membrane that serves as a source of blood vessels for the umbilical cord.

allele (ah-lel') An alternative form of a gene that occurs at a given chromosomal site (locus).

all-or-none law (awl' or nun' law) Muscle fibers either contract maximally or not at all, and neurons either conduct a nerve impulse completely or not at all.

alveoli (al-ve'o-li) Saclike structures that are the air sacs of a lung.

amino acid (ah-me'no as'id) A unit of protein that takes its name from the fact that it contains an amino group (NH_2) and an acid group (COOH).

ammonia (ah-mo'ne-ah) NH_3, a nitrogenous waste product resulting from deamination of amino acids.

amnion (am'ne-on) One of the extraembryonic membranes; a fluid-filled sac around the embryo.

ampulla (am-pul'ah) An expansion at the end of each semicircular canal that contains receptors for dynamic equilibrium.

amylase (am'il-as) A starch-digesting enzyme secreted by the salivary glands (salivary amylase) and the pancreas (pancreatic amylase).

anatomical position (an''ah-tom'e-kal) An erect body stance with the eyes directed forward, the arms at the sides, and the palms of the hands facing forward.

anatomy (ah-nat'o-me) Branch of science dealing with the form and structure of body parts.

anemia (ah-ne'me-ah) A condition characterized by a deficiency of red blood cells or hemoglobin.

angina pectoris (an'ji-nah pec'tor-is) A condition characterized by thoracic pain resulting from occluded coronary arteries and preceding a heart attack.

antagonist (an-tag'o-nist) A muscle that acts in opposition to a "prime mover," or an agonist.

anterior (an-te're-or) Pertaining to the front; the opposite of posterior.

anterior pituitary (an-te're-or pi-tu'i-tar''e) The front lobe of the pituitary gland.

antibody (an'ti-bod''e) A protein produced in response to the presence of some foreign substance in the blood or tissues.

antibody-mediated immunity (an'ti-bod''e me'de-āted i-mu'ni-te) Resistance to disease-causing agents resulting from the production of specific antibodies by B lymphocytes; humoral immunity.

antidiuretic hormone (an''ti-di''u-ret'ik hōr'mōn) See ADH.

antigen (an'ti-jen) A foreign substance, usually a protein, that stimulates the immune system to produce antibodies.

anus (a'nus) Outlet of the digestive tube.

aorta (a-or'tah) Major systemic artery that receives blood from the left ventricle.

appendicular (ap''en-dik'u-lar) Pertaining to the upper limbs (arms) and lower limbs (legs).

appendicular skeleton (ap''en-dik'u-lar skel'e-ton) Part of the skeleton forming the upper limbs, shoulder girdle, lower limbs, and hip girdle.

appendix (ah-pen'diks) A small, tubular appendage that extends outward from the cecum of the large intestine.

aqueous humor (a'kwe-us hu'mor) Watery fluid that fills the anterior cavity of the eye.

arachnoid (ah-rak'noid) The weblike middle covering (one of the three meninges) of the central nervous system.

ARAS Ascending reticular activating system by which the thalamus is connected to various parts of the brain and composed of the diffuse thalamic projection system and the reticular formation.

arrector pili muscle (ah-rek'tor pil'i-mus'l) Smooth muscle in the skin associated with a hair follicle.

arterial duct (ar-te're-al dukt) Fetal connection between the pulmonary artery and the aorta, ductus arteriosus.

arteriole (ar-te're-ōl) A branch from an artery that leads into a capillary.

artery (ar'ter-e) A vessel that takes blood away from the heart; characteristically possessing thick elastic walls.

articular cartilage (ar-tik'u-lar kar'ti-lij) A hyaline cartilaginous covering over the articulating surface of the bones of synovial joints.

articulation (ar-tik''u-la'shun) The joining together of parts at a joint.

ascending tracts (ah-send'ing trakts) Groups of nerve fibers in the spinal cord that transmit sensory impulses upward to the brain.

aster (as'ter) Short rays of microtubules that appear at the ends of the spindle apparatus in animal cells during cell division.

atherosclerosis (ath''er-o-skle-ro'sis) Condition in which fatty substances accumulate abnormally on the inner linings of arteries.

ascending reticular activating system (ah-send'ing re-tik'u-lar ak'ti-vāt''ing sis'tem) See ARAS.

atom (at'om) Smallest unit of matter.

ATP Adenosine triphosphate, the molecule used by cells when energy is needed.

atria (a'tre-ah) Chambers; particularly the upper chambers of the heart that lie above the ventricles (*sing.* atrium).

atrioventricular (a''tre-o-ven-trik'u-lar) A structure in the heart that pertains to both the atria and ventricles; for example, an atrioventricular valve is located between an atrium and a ventricle.

atrioventricular node (a''tre-o-ven-trik'u-lar nōd) See AV node.

auditory canal (aw'di-to''re kah-nal') A tube in the outer ear that leads to the tympanic membrane.

autonomic nervous system (aw''to-nom'ik) The sympathetic and parasympathetic portions of the nervous system that function to control the actions of the visceral organs and skin.

autosomal chromosome (aw''to-so'mal kro'mo-sōm) A chromosome other than a sex chromosome.

AV node (a-ve nōd) A small region of neuromuscular tissue located near the septum of the heart that transmits impulses from the SA node to the ventricular walls.

axial skeleton (ak'se-al skel'e-ton) Portion of the skeleton that supports and protects the organs of the head, neck, and trunk.

axon (ak'son) Process of a neuron that conducts nerve impulses away from the cell body.

B

ball-and-socket joint The most freely movable type of joint (e.g., the shoulder or hip joint).

basal metabolic rate (BMR) (ba'sal met''ah-bol'ik) The rate of metabolism (expressed as oxygen consumption or heat production) under resting or basal conditions (fourteen to eighteen hours after eating).

base (bās) A solution in which pH is more than 7; a substance that contributes or liberates hydroxide ions in a solution; alkaline; opposite of acidic.

basement membrane (bās'ment mem'brān) Thin interior non-cellular surface of epithelium that attaches to connective tissue.

bile (bīl) A secretion of the liver that is temporarily stored in the gallbladder before being released into the small intestine where it emulsifies fat.

bilirubin (bil''i-roo'bin) Bile pigment derived from the breakdown of the heme portion of hemoglobin.

blastocyst (blas'to-sist) An early stage of embryonic development that consists of a hollow ball of cells.

blind spot (blīnd spot) Area where the optic nerve passes through the retina and where vision is not possible due to the lack of rods and cones.

blood (blud) Connective tissue composed of cells separated by plasma.

B lymphocyte (bē lim'fo-sīt) Type of lymphocyte that is responsible for antibody mediated immunity.

bone (bōn) Connective tissue having a hard matrix of calcium salts deposited around protein fibers.

Bowman's capsule (bo'manz kap'sul) A double-walled cup that surrounds the glomerulus at the beginning of the kidney tubule.

bradykinins (brad''e-ki'nins) Short polypeptides that stimulate vasodilation and other cardiovascular changes.

breathing (brēth'ing) Entrance and exit of air into and out of the lungs.

Broca's area (bro'kahz a're-ah) Region of the frontal lobe that coordinates complex muscular actions of the mouth, tongue, and larynx, making speech possible.

bronchi (brong'ki) The two major divisions of the trachea, leading to the lungs.

bronchiole (brong'ke-ōl) The smaller air passages in the lungs.

buffer (buf'er) A substance or compound that prevents large changes in the pH of a solution.

bursa (bur'sah) A saclike, fluid-filled structure, lined with synovial membrane, that occurs near a joint.

C

calcitonin (kal''si-to'nin) Hormone secreted by the thyroid gland that helps to regulate the level of blood calcium.

calorie (kal'o-re) The amount of heat required to raise one kilogram of water one degree centigrade.

capillaires (kap'i-ler''es) Microscopic vessels located in the tissues connecting arterioles to venules; molecules either exit or enter the blood through the thin walls of the capillaries.

carbohydrate (kar″bo-hi′drāt) Organic compounds with the general formula $(CH_2O)_n$ including sugars and glycogen.

cardiac (kar′de-ak) Of or pertaining to the heart.

cardiac muscle (kar′de-ak mus′el) Heart muscle (myocardium) consisting of striated muscle cells that interlock.

caries (kar′ēz) Tooth decay caused by bacterial activity.

carpals (kar′pals) Bones of the wrist.

carrier (kar′e-er) A molecule that combines with a substance and actively transports it through the cell membrane.

cartilage (kar′ti-lij) A connective tissue, usually part of the skeleton, which is composed of cells in a flexible matrix.

CCK Cholecystokinin, a hormone secreted by the small intestine that stimulates the release of pancreatic juice from the pancreas and bile from the gallbladder.

cell (sel) The structural and functional unit of an organism; the smallest structure capable of performing all the functions necessary for life.

cell body (sel bod′e) Portion of a nerve cell that includes a cytoplasmic mass and a nucleus, and from which the nerve fibers extend.

cell-mediated immunity (i-mu′ni-te) Immunological defense provided by killer T cells, which destroy cells infected with viruses, foreign cells, and cancer cells.

cell membrane (sel mem′brān) A membrane that surrounds the cytoplasm of cells and regulates the passage of molecules into and out of the cell.

central canal (sen′tral kah-nal′) Tube within the spinal cord that is continuous with the ventricle of the brain and contains cerebrospinal fluid.

central nervous system (sen′tral ner′vus sis′tum) See CNS.

centriole (sen′tre-ōl) A short, cylindrical organelle that contains microtubules in a 9 + 0 pattern and is associated with the formation of the spindle during cell division.

cerebellum (ser′ĕ-bel′um) The part of the brain that controls muscular coordination.

cerebral hemisphere (ser′ĕ-bral hem′i-sfēr) One of the large, paired structures that together constitute the cerebrum of the brain.

cerebrospinal fluid (ser″ĕ-bro-spi′nal floo′id) Fluid found within ventricles of the brain and surrounding the CNS in association with the meninges.

cerebrovascular accident (or stroke) (strōk) Condition resulting when an arteriole bursts or becomes blocked by an embolism.

cerebrum (ser′ĕ-brum) The main portion of the vertebrate brain that is responsible for consciousness.

cholecystokinin (ko″le-sis″to-ki′nin) See CCK.

cholesterol (ko-les′ter-ol) A lipid produced by body cells that is used in the synthesis of steroid hormones and is excreted into the bile.

chorion (ko′re-on) An extraembryonic membrane that forms an outer covering around the embryo and contributes to the formation of the placenta.

chorionic villi (ko″re-on′ik vil′i) Projections from the chorion that appear during implantation and that in one area contribute to the development of the placenta.

choroid (ko′roid) The vascular, pigmented middle layer of the wall of the eye.

chromatids (kro′mah-tidz) The two identical parts of a chromosome following replication of DNA.

chromatin (kro′mah-tin) Threadlike network in the nucleus that condenses to become the chromosomes just before cell division.

chromosomes (kro′mo-sōmz) Rod-shaped bodies in the nucleus, particularly during cell division, that contain the hereditary units or genes.

ciliary muscle (sil′e-er″e mus′el) A muscle that controls the curvature of the lens of the eye.

circle of Willis (sir′kl uv wil′is) An arterial ring located on the ventral surface of the brain.

circumcision (ser″kum-sizh′un) Removal of the foreskin of the penis.

circumduction (ser″kum-duk′shun) A conelike movement of a body part, such that the distal end moves in a circle while the proximal portion remains relatively stable.

cleavage (klēv′ij) Cell division of the fertilized egg that is unaccompanied by growth so that numerous small cells result.

clone (klōn) DNA fragments from an external source that have been reproduced by E. coli.; also, asexually produced organisms having the same genetic makeup.

clotting (klot′ing) Process of blood coagulation, usually when injury occurs.

CNS The central nervous system; the brain and spinal cord.

cochlea (kok′le-ah) Portion of the inner ear that contains the receptors of hearing.

codon (ko′don) The sequence of three nucleotide bases in mRNA that specifies a given amino acid and binds through complementary base pairing with an anticodon in RNA.

collecting duct (ko-lekt′ing dukt) A tube that receives urine from several distal convoluted tubules.

colon (ko′lon) The large intestine.

colostrum (ko-los′trum) The first secretion of the mammary glands following the birth of an infant.

columnar epithelium (ko-lum′nar ep″i-the le-um) Pillar-shaped cells usually having the nuclei near the bottom of each cell and found lining the digestive tract, for example.

compact bone (kom-pakt′ bōn) Hard bone consisting of Haversian systems cemented together.

complement (kom′ple-ment) A group of proteins in plasma that aid the general defense of the body by destroying bacteria.

cones (kōns) Color receptors located in the retina of the eye.

connective tissue (ko-nek′tiv tish′u) A type of tissue, characterized by cells separated by a matrix, that often contains fibers.

coronary arteries (kor′ŏ-na-re ar′ter-ēz) Arteries that supply blood to the wall of the heart.

corpus callosum (kor′pus kah-lo′sum) A mass of white matter within the brain, composed of nerve fibers connecting the right and left cerebral hemisphere.

corpus luteum (kor′pus lut′e-um) Structure that forms from the tissues of a ruptured ovarian follicle and functions to secrete female hormones.

cortex (kor′teks) The outer layer of an organ such as the convoluted cerebrum, adrenal gland, or kidney.

cortisol (kor′ti-sol) A glucocorticoid secreted by the adrenal cortex.

covalent bond (ko′va-lent bond) Chemical bond created by the sharing of electrons between atoms.

cranial (kra′ne-al) Pertaining to the cranium.

cranial nerve (kra′ne-al nerv) Nerve that arises from the brain.

creatine phosphate (kre′ah-tin fos′fāt) A compound unique to muscles that contains a high-energy phosphate bond and is used to regenerate a supply of ATP.

creatinine (kre-at′i-nin) Excretion product from creatine phosphate breakdown.

cretinism (kre′tin-izm) A condition resulting from a lack of thyroid hormone in an infant.

crista (kris′ta) A crest, such as the crista galli, extending superiorly from the cribiform plate.

cuboidal epithelium (ku-boi′dal ep″i-the′le-um) Cube-shaped cells found lining the kidney tubules.

Cushing's syndrome (koosh′ingz sin′drōm) A condition characterized by thin arms and legs, and a ''moon face,'' and accompanied by high blood glucose and sodium levels due to hypersecretion of cortical hormones.

cutaneous (ku-ta′ne-us) Pertaining to the skin.

cytokinesis (si″to-ki-ne′sis) Division of the cytoplasm of a cell.

cytoplasm (si′to-plazm) The ground substance of cells located between the nucleus and the cell membrane.

cytoskeleton (si″to-skel′e-ton) Microfilaments and microtubules in the cytoplasm that help maintain the shape of the cell.

D

deamination (de-am″i-na′shun) Removal of an amino group ($-NH_2$) from an amino acid or other organic compound.

dendrite (den′drit) Process of a neuron, typically branched, that conducts nerve impulses toward the cell body.

deoxyribonucleic acid (de-ok″se-ri′bo-nu-kle′ik as′id) See DNA.

depolarization (de-po″lar-i-za′shun) A loss in polarization as when the nerve impulse occurs.

dermis (der′mis) The thick skin layer that lies beneath the epidermis.

descending tracts (de-send′ing trakts) Groups of nerve fibers that carry nerve impulses downward from the brain through the spinal cord.

diabetes insipidus (di″ah-be′tez in-sip′i-dus) Condition characterized by an abnormally large production of urine due to a deficiency of antidiuretic hormone.

diabetes mellitus (di″ah-be′tez me-li′tus) Condition characterized by a high blood glucose level and the appearance of glucose in the urine due to a deficiency of insulin.

diaphragm (di′ah-fram) A sheet of muscle that separates the thoracic cavity from the abdominal cavity. Also, a birth control device inserted in front of the cervix in females.

diaphysis (di-af′i-sis) The shaft of a long bone.

diastole (di-as′to-le) Relaxation of heart chambers.

diastolic pressure (di-a-stol′ik presh′ur) Arterial blood pressure during the diastolic phase of the cardiac cycle.

differentiation (dif′er-en″she-a′shun) The process by which a cell becomes specialized for a particular function.

diffusion (di-fu′zhun) The movement of molecules from an area of greater concentration to an area of lesser concentration.

dissociation (dis-so″she-a′shun) The breaking of a chemical bond such that ions are released.

distal (dis′tal) Further from the midline or origin; opposite of proximal.

distal convoluted tubule (dis′tal kon′vo-lūt-ed tu′bul) Highly coiled region of a nephron that is distant from Bowman's capsule.

DNA A nucleic acid; the genetic material found in the nucleus of a cell.

dominant gene (dom′i-nant jēn) Hereditary factor that expresses itself even when there is only one copy in the genotype.

dorsal (dor′sal) Pertaining to the back or posterior portion of a body part; the opposite of ventral.

dorsal root ganglion (dor′sal root gang′gle-on) Mass of sensory neuron cell bodies located in the dorsal root of a spinal nerve.

double helix (du′b′l he′liks) A double spiral often used to describe the three-dimensional shape of DNA.

Down syndrome (down sin′drōm) Human congenital disorder associated with an extra twenty-third chromosome.

duodenum (du″o-de′num) The first portion of the small intestine into which ducts from the gallbladder and pancreas enter.

dura mater (du′rah ma′ter) Tough outer layer of the meninges.

dynamic equilibrium (di-nam′ik e″kwi-lib′re-um) The maintenance of balance when the head and body are suddenly moved or rotated.

E

ECG See electrocardiogram.

ectoderm (ek′to-derm) The outer germ layer of the embryonic gastrula; it gives rise to the skin and nervous system.

ectopic pregnancy (ek-top′ik preg′nan-se) Implantation and development of a fertilized egg outside the uterus.

edema (e-de′mah) Swelling due to tissue fluid accumulation in the intercellular spaces.

EEG Electroencephalogram, a graphic recording of the brain's electrical activity.

effector (e-fek′tor) A structure that allows a response to environmental stimuli such as the muscles and glands.

EKG See electrocardiogram.

elastic cartilage (e-las′tik kar′ti-lij) Cartilage composed of elastic fibers allowing greater flexibility.

electrocardiogram (e-lek″tro-kar′de-o-gram″) A recording of the electrical activity that accompanies the cardiac cycle; ECG or EKG.

electroencephalogram (e-lek″tro-en-sef′ah-lo-gram″) See EEG.

element (el′e-ment) The simplest of substances consisting of only one type of atom; i.e., carbon, hydrogen, oxygen.

embolus (em′bo-lus) A moving blood clot that is carried through the bloodstream.

embryo (em′bre-o) The organism in its early stages of development; first week to two months.

emulsification (e-mul″si-fi′ka′shun) The act of dispersing one liquid in another.

endocrine gland (en′do-krin gland) A gland that secretes hormones directly into the blood or body fluids.

endocytosis (en″do-si-to′sis) A process in which extracellular material is enclosed within a vesicle and taken into the cell. Phagocytosis and pinocytosis are forms of endocytosis.

endoderm (en′do-derm) An inner layer of cells that line the primitive gut of the gastrula. It becomes the lining of the digestive tract and associated organs.

endometrium (en″do-me′tre-um) The lining of the uterus that becomes thickened and vascular during the uterine cycle.

endoplasmic reticulum (ER) (en-do-plaz′mic re-tik′u-lum) A complex system of tubules, vesicles, and sacs in cells; sometimes having attached ribosomes.

enzyme (en′zim) A protein catalyst that speeds up a specific reaction or a specific type of reaction.

epidermis (ep″i-der′mis) The outer layer of cells of an organism.

epididymis (ep″i-did′i mis) Coiled tubules next to the testes where sperm mature and may be stored for a short time.

epiglottis (ep″i-glot′is) A structure that covers the glottis during the process of swallowing.

epimysium (ep″i-mis′e-um) A fibrous, outer sheath of connective tissue surrounding a skeletal muscle.

epinephrine (ep''i-nef'rin) *See* adrenalin.

epiphyseal disk (ep''i-fiz'e-al disk) Cartilaginous layer within the epiphysis of a long bone that functions as a growing region.

epiphysis (e-pif'i-sis) The end segment of a long bone, separated from the diaphysis early in life by an epiphyseal plate, but later becoming part of the larger bone.

epithelial tissue (ep''i-the'le-al tish'u) A type of tissue that lines cavities and covers the external surface of the body.

erection (e-rek'shun) erect penis prepared for copulation.

erythrocyte (e-rith'ro-sīt) Non-nucleated, hemoglobin containing blood cells capable of carrying oxygen; the red blood cell.

esophagus (e-sof'ah-gus) A tube that transports food from the mouth to the stomach.

eustachian tube (u-sta'ke-an tūb) An air tube that connects the pharynx to the middle ear.

eversion (e-ver'zhun) A movement of the foot in which the sole is turned outward.

excretion (ek-skre'shun) Removal of metabolic wastes.

exocrine gland (ek'so-krin gland) Secreting externally; particular glands with ducts whose secretions are deposited into cavities, such as salivary glands.

exocytosis (eks''o-si-to'sis) A process in which an intracellular vesicle fuses with the cell membrane so that the vesicle's contents are released outside the cell.

exophthalmic goiter (ek''sof-thal'mik goi'ter) An enlargement of the thyroid gland accompanied by an abnormal protrusion of the eyes.

expiration (eks''pi-ra'shun) Process of expelling air from the lungs; exhalation.

extension (ek-sten'shun) Movement by which the angle between parts at a joint is increased.

external auditory meatus (aw'di-to''re me-a'tus) An opening through the temporal bone that connects with the tympanum and the middle ear chamber, and through which sound vibrations pass.

external respiration (eks-ter'nal res''pi-ra'shun) Exchange between alveoli and blood of oxygen and carbon dioxide.

extraembryonic membranes (eks''trah-em''bre-on'ik mem'branz) Membranes that are not a part of the embryo, but are necessary to the continued existence and health of the embryo.

F

fascia (fash'e-ah) A tough sheet of fibrous tissue binding the skin to underlying muscles, or supporting and separating muscles.

fascicle A small bundle of muscle fibers.

fat An organic molecule that the body uses for long-term energy storage.

fatigue (fah-tēg') Muscle relaxation in the presence of stimulation due to energy reserve depletion.

feces (fe'sēz) Indigestible wastes expelled from the digestive tract; excrement.

femur (fe'mur) The thighbone found in the upper leg.

fetal position (fe'tal po-zish'un) Curled position of the body in which the head touches the knees; the position assumed by the fetus in the womb.

fetus (fe'tus) Human development in its later stages following the embryonic stages; three months to term.

fibrin (fi'brin) Insoluble protein threads formed from fibrinogen during blood clotting.

fibrinogen (fi-brin'o-jen) Plasma protein that is converted into fibrin threads during blood coagulation.

fibrocartilage (fi''bro-kar'ti-lij) Cartilage with a matrix of strong collagenous fibers.

fibrous connective tissue (fi'brus ko-nek'tiv tish'u) Tissue composed mainly of closely packed collagenous fibers and found in tendons and ligaments.

fibula (fib'u-lah) A long slender bone located on the lateral side of the tibia.

filtrate (fil'trāt) The filtered portion of blood that is contained within Bowman's capsule of a kidney tubule.

fimbria (fim'bre-ah) Fingerlike extensions from the oviduct near the ovary.

fissure (fish'ur) A narrow cleft separating parts, such as the lobes of the cerebrum.

flagella (flah-jel'ah) Slender, long processes used for locomotion, for example by sperm.

flexion (flek'shun) Blending at a joint so that the angle between bones is decreased.

focusing (fo'kus-ing) Manner by which light rays are bent by the cornea and lens, creating an image on the retina.

follicle (fol'i-kl) A structure in the ovary that produces the egg and particularly the female sex hormone, estrogen.

follicle-stimulating hormone (fol'i-kl stim'u-lat''ting hor'mon) *See* FSH.

fontanel (fon''tah-nel') Membranous region located between certain cranial bones in the skull of a fetus or infant.

foramen (fo-ra'men) An opening, usually in a bone or membrane (plural, *foramina*).

foramen magnum (fo-ra'men mag'num) Opening in the occipital bone of the skull through which the spinal cord passes.

foreskin (fōr'skin) Skin covering the glans penis in uncircumcised males.

formed element (form'd el'e-ment) A cellular constituent of blood.

fovea centralis (fo've-ah sen-tral'is) Region of the retina, consisting of densely packed cones, which is responsible for the greatest visual acuity.

fracture (frak'tūr) A break in a bone.

frontal (frun'tal) Pertaining to the region of the forehead.

frontal lobe (frun'tal lōb) Area of the cerebrum responsible for voluntary movements and higher intellectual processes.

FSH Follicle stimulating hormone secreted by the anterior pituitary gland that stimulates the development of an ovarian follicle in a female or the production of sperm cells in a male.

G

gallbladder (gawl'blad-er) Saclike organ associated with the liver that stores and concentrates bile.

gallstones (gawl'stonz) Precipitated crystals of cholesterol or calcium carbonate formed from bile within the gallbladder or bile duct.

gamete (gam'et) A sex cell (egg or sperm) that joins in fertilization to form a zygote.

ganglion (gang'gle-on) A collection of neuron cell bodies outside the central nervous system.

gastric (gas'trik) Of or pertaining to the stomach.

gastric gland (gas'trik gland) Gland within the stomach wall that secretes gastric juice.

gastrin (gas'trin) A hormone secreted by stomach cells that regulates the release of pepsin by the stomach wall.

gene (jēn) A unit of heredity located on a chromosome.

genotype (je'no-tip) The genetic makeup of any individual.

germ layers (jerm la'ers) Primary tissues of an embryo (ectoderm, mesoderm, endoderm) that give rise to the major tissue systems of the adult animal.

GH *See* growth hormone.

glial cells (gli′al selz) Supporting cells within the brain and spinal cord that perform functions other than transmission of nerve impulse.

globin (glo′bin) The protein portion of a hemoglobin molecule.

glomerulus (glo-mer′u-lus) A cluster; for example, the cluster of capillaries surrounded by Bowman's capsule in a kidney tubule.

glottis (glot′is) Slitlike opening between the vocal cords.

glucagon (gloo′kah-gon) Hormone secreted by the pancreatic islets of Langerhans that causes the release of glucose from glycogen.

glucose (gloo′kōs) Blood sugar which is broken down in cells to acquire energy for ATP production.

glycogen (gli′ko-jen) A polysaccharide that is the principal storage compound for sugar in animals.

Golgi apparatus (gol′ge ap″ah-ra′tus) An organelle that consists of concentrically folded membranes and functions in the packaging and secretion of cellular products.

gonad (go′nad) An organ that produces sex cells; the ovary, which produces eggs, and the testis, which produces sperm.

gonadotropic (gon″ah-do-trōp′ik) A type of hormone that regulates the activity of the ovaries and testes; principally FSH and LH (ICSH).

Graafian follicle (graf′e-an fol′li-k′l) Mature follicle within the ovaries which houses a developing egg.

granulocytes (gran′u-lo-sīts) White blood cells that contain distinctive granules, i.e. granular leukocytes.

groin (groin) Region of the body between the abdomen and thighs.

growth (grōth) Increase in the number of cells and/or the size of these cells.

growth hormone (groth hor′mon) A hormone released by the anterior lobe of the pituitary gland that promotes the growth of the organism; GH or somatotropin.

gyrus (ji′rus) A convoluted elevation or ridge.

H

hair follicle (hār fol′i-kl) Tubelike depression in the skin in which a hair develops.

hard palate (hard pal′at) Anterior portion of the roof of the mouth which contains several bones.

HCG hormone Human chorionic gonadotropic hormone produced by the placenta that helps maintain pregnancy and is the basis for the pregnancy test.

head (hed) An enlargement on the end of a bone.

heart (hart) Muscular organ located in thoracic cavity responsible for maintenance of blood circulation.

heart attack (hart ah-tak′) Condition resulting when circulation in the coronary arteries is blocked; also myocardial infarction.

hematocrit (he-mat′o-krit) The volume percentage of red blood cells within a sample of whole blood.

heme (hēm) The iron-containing portion of a hemoglobin molecule.

hemoglobin (he″mo-glo′bin) Pigment of red blood cells responsible for the transport of oxygen.

hepatic portal system (he-pat′ik por′tal sis′tem) Portal system that begins at the villi of the small intestine and ends at the liver.

hepatic portal vein (he-pat′ik por′tal vān) Vein leading to the liver formed by the merging blood vessels of the small intestine.

heterozygous (het″er-o-zi′gus) Having two different alleles (as *Aa*) for a given trait.

hinge joint A type of joint characterized by a convex surface of one bone fitting into a concave surface of another so that movement is confined to one place, such as in the knee or interphalangeal joint.

histamine (his′tah-min) Substance produced by basophil-derived mast cells in connective tissue which causes capillaries to dilate. It also causes many of the symptoms of allergy.

HIV Virus responsible for AIDs; human immunodeficiency virus.

homeostasis (ho″me-o-sta′sis) The constancy of conditions, particularly the environment of the body's cells: constant temperature, blood pressure, pH, and other body conditions.

homozygous (ho″mo-zi′gus) Having identical alleles (as *AA* or *aa*) for a given trait; pure breeding.

human chorionic gonadotropic hormone (hu′man ko″re-on′ik go-nad″o-trōp′ik hor′mōn) *See* HCG.

humerus (hu′mer-us) A heavy bone that extends from the scapula to the elbow.

hyaline cartilage (hi′ah-lin kar′ti-lij) Cartilage composed of very fine collagenous fibers and matrix of a clear milk-glass appearance.

hybridoma (hi-brid-o′mah) Fused lymphocyte and cancer cell used in the manufacture of monoclonal antibodies.

hypertension (hi″per-ten′shun) Elevated blood pressure, particularly the diastolic pressure.

hypertonic solution (hi″per-ton′ik so-lu′shun) One that has a greater concentration of solute, a lesser concentration of water than the cell.

hypothalamus (hi″po-thal′ah-mus) A region of the brain; the floor of the third ventricle that helps maintain homeostasis.

hypotonic solution (hi″po-ton′ik so-lu′shun) One that has a lesser concentration of solute, a greater concentration of water than the cell.

I

ilium (il′e-um) One of the bones of a coxal bone or hipbone.

immune complex (i-mūn′ kom′pleks) The product of an antigen-antibody reaction.

implantation (im″plan-ta′shun) The attachment and penetration of the embryo to the lining (endometrium) of the uterus.

impotency (im′po-ten″se) Failure of the penis to achieve erection.

induction (in-duk′shun) In development, the ability of one body part to influence the development of another part.

inferior (in-fēr′e-or) Situated below something else; pertaining to the lower surface of a part.

inflammatory reaction (in-flam′ah-to″re re-ak′shun) A tissue response to injury that is characterized by dilation of blood vessels and an accumulation of fluid in the affected region.

inner cell mass (in′er sel mas) The portion of a blastocyst that will develop into the embryo and fetus.

inner ear (in′er ēr) Portion of the ear consisting of a vestibule, semicircular canals, and the cochlea where balance is maintained and sound is transmitted.

innervation (in″er-va′shun) Stimulation of muscle fiber to contract by a motor axon.

insertion (in-ser′shun) The end of a muscle that is attached to a movable part.

inspiration (in″spi-ra′shun) The act of breathing in; inhalation.

insulin (in′su-lin) A hormone produced by the pancreas that regulates glucose storage in liver and glucose uptake by cells.

integration (in″te-gra′shun) The summing up of negative and positive stimuli within a dendrite.

integument (in-teg′u-ment) Pertaining to the skin.

interferon (in″ter-fēr′on) A protein formed by a cell infected with a virus that can increase the resistance of other cells to the virus.

internal respiration (in-ter'nal res''pi-ra'shun) Exchange between blood and tissue fluid of oxygen and carbon dioxide.

interneuron (in''ter-nu'ron) A neuron that is found within the central nervous system and takes nerve impulses from one portion of the system to another.

interphase (in'ter-faz) Interval between successive cell divisions, during which the chromosomes are extended and the cell is metabolically active.

interstitial cells (in''ter-stish'al selz) Hormone-secreting cells located between the seminiferous tubules of the testes.

inversion (in-ver'zhun) A movement of the foot in which the sole is turned inward.

ionic bond (i-on'ik bond) A chemical attraction between a positive and negative ion.

ischemia (is-ke'me-ah) A deficiency of blood in a body part.

islets of Langerhans (i'lets uv lahng'er-hanz) Distinctive groups of cells within the pancreas that secrete insulin and glucagon.

isometric contraction (i''so-met'rik kon-trak'shun) Muscular contraction in which the muscle fails to shorten.

isotonic solution (i''so-ton'ik so-lu'shun) One that contains the same concentration of solutes and water as does the cell.

J

joint (joint) The union of two or more bones; an articulation.

K

karyotype (kar'e-o-tip) The arrangement of all the chromosomes from a nucleus by pairs in a fixed order.

keratin (ker'ah-tin) An insoluble protein present in the epidermis and in epidermal derivatives such as hair and nails.

kidneys (kid'nez) Organs in the urinary system which form, concentrate, and excrete urine.

Klinefelters' syndrome (klin'fel-terz sin'drom) A condition caused by the inheritance of XXY chromosomes.

L

labium (la'be-um) A fleshy border or liplike fold of skin, as in the labia majora and labia minora of the female genitalia.

lacteal (lak'te-al) A lymph vessel in a villus of the wall of the small intestine.

lactogenic hormone (lak''to-jen'ik hor'mon) see LTH.

lacuna (lah-ku'nah) A small pit or hollow cavity, as in bone or cartilage, where a cell or cells are located.

lanugo (lah-nu'go) Short, fine hair that is present during the later portion of fetal development.

larynx (lar'ingks) Structure that contains the vocal cords; voice box.

lateral (lat'er-al) Pertaining to the side.

lens (lenz) A clear membranelike structure found in the eye behind the iris. The lens brings objects into focus.

leukemia (lu-ke'me-ah) Form of cancer characterized by uncontrolled production of leukocytes in red bone marrow.

leukocyte (lu'ko-sit) Refers to several types of colorless, nucleated blood cells which, among other functions, resist infection; white blood cells.

leukocytosis (lu''ko-si-to'sis) An abnormally large increase in the number of white blood cells.

leukopenia (lu''ko-pe'ne-ah) An abnormally low number of leukocytes in the blood.

LH Hormone produced by the anterior pituitary gland that stimulates the development of the corpus luteum in females and the production of testosterone in males.

ligament (lig'ah-ment) A strong connective tissue that joins bone to bone.

limbic system (lim'bik sis'tem) A system involving many different centers of the brain that is concerned with visceral functioning and emotional responses.

lipase (li'pas) An enzyme secreted by the pancreas that digests or breaks down fats.

lipid (lip'id) A group of organic compounds that are insoluble in water; notably fats, oils, and steroids.

loose connective tissue (loos ko-nek'tiv tish'u) Tissue composed mainly of fibroblasts that are separated by collagen and elastin fibers, and found beneath epithelium.

LTH A hormone secreted by the anterior pituitary that stimulates the production of milk from the mammary glands.

lumen (lu'men) The cavity inside any tubular structure, such as the lumen of the gut.

luteinizing hormone (lu'te-in-iz''ing hor'mon) See LH.

lymph (limf) Fluid having the same composition as tissue fluid; carried in lymph vessels.

lymphatic system (lim-fat'ik sis'tem) Vascular system which takes up excess tissue fluid and transports it to the bloodstream.

lymph node (limf nod) A mass of lymphoid tissue located along the course of a lymphatic vessel.

lymphocyte (lim'fo-sit) A type of white blood cell characterized by agranular cytoplasm. Lymphocytes usually constitute about 20 to 25 percent of the white cell count.

lymphokines (lim'fo-kinz) Chemicals secreted by T lymphocytes that have the ability to affect the characteristics of monocytes.

lysosome (li'so-som) An organelle in which digestion takes place due to the action of powerful hydrolytic enzymes.

M

macrophage (mak'ro-faj) An enlarged monocyte that ingests foreign material and cellular debris.

major histocompatibility protein (ma'jor his''to-kom-pat''i-bil'i-te pro'te-in) See MHC protein.

marrow (mar'o) Connective tissue that occupies the spaces within bones.

matrix (ma'triks) The secreted basic material or medium of biological structures, such as the matrix of cartilage or bone.

medial (me'de-al) Toward or near the midline.

medulla oblongata (me-dul'ah ob''long-gah'tah) The lowest portion of the brain that is concerned with the control of internal organs.

medullary cavity (med'u-lar''e kav'i-te) Cavity within the diaphysis of a long bone occupied by yellow marrow.

meiosis (mi-o'sis) Type of cell division in which the daughter cells have 23 chromosomes; occurs during spermatogenesis and oogenesis.

melanin (mel'ah-nin) A pigment found in the skin and hair of humans that is responsible for their coloration.

melanocyte (mel'ah-no-sit'') Melanin-producing cell.

memory cells (mem'o-re selz) Cells derived from B lymphocytes that remain within the body for some time and account for the presence of active immunity.

meninges (me-nin'jez) Protective membranous coverings about the brain and spinal cord (singular, meninx).

meniscus (me-nis'kus) A piece of fibrocartilage that separates the surfaces of bones in the knee; (plural, menisci).

menopause (men′o-pawz) Termination of the ovarian and uterine cycles in older women.

menstruation (men″stroo a′shun) Loss of blood and tissue from the uterus at the beginning of a female uterine cycle.

mesentery (mes′en-ter″e) A fold of peritoneal membrane that attaches an abdominal organ to the abdominal wall.

mesoderm (mes′o-derm) The middle germ layer of an animal embryo that gives rise to the muscles, connective tissue, and circulatory system.

metabolism (me-tab′o-lizm) All of the chemical changes that occur within cells considered together.

metacarpals (met″ah-kar′pals) Bones of the hand between the wrist and finger bones.

metatarsal (met″ah-tar′sal) Bones found in the foot between the ankle and the toes.

microfilament (mi″kro-fil′ah-ment) Tiny rod of protein that occurs in cytoplasm and functions in causing various cellular movements.

microtubule (mi″kro-tu′bul) An organelle composed of thirteen rows of globular proteins; found in multiple units in several other cell organelles such as the centriole, cilia, and flagella.

midbrain (mid′brān) A small region of the brain stem located between the forebrain and the hindbrain.

middle ear (mid″l ēr) Portion of the ear consisting of the tympanic membrane, the oval and round windows, and the ossicles where sound is amplified.

mitochondrion (mi″to-kon′dre-on) An organelle in which cellular respiration produces the energy molecule, ATP.

mitosis (mi-to′sis) Cell division by means of which two daughter cells receive 46 chromosomes; occurs during growth and repair.

mixed nerves (mikst nervs) Nerves that contain both the long dendrites of sensory neurons and the long axons of motor neurons.

monoclonal antibodies (mon″-o-klōn′al an′ti-bod′ēz) Antibodies of one type that are produced by cells derived from a lymphocyte that has fused with a cancer cell.

mononucleosis (mon″o-nu″kle-o′sis) Viral disease characterized by the presence of an increase in atypical lymphocytes in the blood.

morphogenesis (mor″fo-jen′i-sis) The establishment of shape and structure in an organism.

morula (mor′u-lah) An early stage in development in which the embryo consists of a mass of cells, often spherical.

motor nerve (mo′tor nerv) Nerve containing only the long axons of motor neurons.

motor neuron (mo′tor nu′ron) A neuron that takes nerve impulses from the central nervous system to an effector.

mucous membrane (mu′kus mem′bran) Membrane that lines a cavity or tube with an opening to the outside; also called mucosa.

muscle action potential (mus′el ak′shun po-ten′shal) An electrochemical change due to increased sarcolemma permeability that is propagated down the T system and results in muscle contraction.

muscle spindle (mus′el spin′dul) Modified skeletal muscle fiber that can respond to changes in muscle length.

muscularis (mus″ku-la′ris) A muscular layer or tunic of an organ, composed of smooth muscle tissue.

muscular tissue (mus″ku-lar tish′u) A major type of tissue that is adapted to contract. The three kinds of muscle are cardiac, smooth, and skeletal.

mutagen (mu′tah-jen) An agent, such as a chemical, that increases the rate of mutations.

mutation (mu-ta′shun) A chromosomal or genetic change that is inherited either by daughter cells following mitosis or by an organism following reproduction.

myelin sheath (mi′e-lin) The fatty cell membranes of Schwann cells that cover long neuron fibers and give them a white, glistening appearance.

myocardial infarction (mi″o-kar′de-al in-fark′shun) Damage to the myocardium due to an interruption in blood supply; a heart attack.

myocardium (mi″o-kar′de-um) Heart (cardiac) muscle consisting of striated muscle cells that interlock.

myofibrils (mi″o-fi′brilz) The contractile portions of muscle fibers.

myogram (mi′o-gram) A recording of a muscular contraction.

myosin (mi′o-sin) The thick filament in myofibrils made of protein and capable of breaking down ATP. *See* actin.

myxedema (mik″se-de′mah) A condition resulting from a deficiency of thyroid hormone in an adult.

N

NA Noradrenalin, an excitatory neurotransmitter active in the peripheral and central nervous systems; norepinephrine.

negative feedback (neg′ah-tiv fēd′bak) Mechanism that is activated by a surplus imbalance and acts to correct it by stopping the process that brought about the surplus.

nephron (nef′ron) The anatomical and functional unit of the kidney; kidney tubule.

nerve (nerv) A bundle of long nerve fibers that run to and/or from the central nervous system.

nerve impulse (nerv im′puls) An electrochemical change due to increased membrane permeability that is propagated along a neuron from the dendrite to the axon following excitation.

neural tube (nu′ral tūb) During development, a tube formed from ectoderm located just above the notochord; the spinal cord.

neurolemmal sheath (nu″ro-lem′al) Sheath on the outside of some nerve fibers due to the presence of Schwann cells.

neuromuscular junction (nu″ro-mus′ku-lar jungk′shun) A junction that occurs between a neuron and a muscle fiber.

neuron (nu′ron) Nerve cell that characteristically has three parts: dendrite, cell body, and axon.

neurotransmitter substance (nu″ro-trans mit′er sub′stans) A chemical made at the ends of axons that is responsible for transmission across a synapse.

neutrophil (nu′tro-fil) A type of phagocytic white blood cell, normally constituting about 60 to 70 percent of the white blood cell count.

node of Ranvier (rah-ve-a′) A gap in the myelin sheath of a nerve fiber.

nondisjunction (non″dis junk′shun) The failure of the chromosomes (or chromatids) to separate during meiosis.

noradrenalin (nor″ah-dren′ah-lin) *See* NA.

nuclear envelope (nu′kle-ar en′ve-lōp) Double-layered membrane enclosing the nucleus.

nucleic acid (nu-kle′ik as′id) A large organic molecule found in the nucleus (DNA and RNA) and cytoplasm (RNA).

nucleolus (nu-kle′o-lus) An organelle found inside the nucleus; composed largely of RNA for ribosome formation; (*pl.* nucleoli).

nucleus (nu'kle-us) A large organelle containing the chromosomes and acting as a control center for the cell.

O

occipital (ok-sip'i-tal) Pertaining to the inferior, dorsal portion of the head.

occipital lobe (ok-sip'i-tal lōb) Area of the cerebrum responsible for vision, visual images, and other sensory experiences.

olfactory cells (ol-fak'to-re selz) The cells located high in the nasal cavity that bear receptor sites on cilia for various chemicals and whose stimulation results in smell.

oncogene (ong'ko-jēn) A gene that contributes to the transformation of a cell into a cancerous cell.

oogenesis (o''o-jen'e-sis) Production of eggs in females by the process of meiosis and maturation.

optic nerve (op'tik nerv) Nerve composed of the ganglion cell fibers that form the innermost layer of the retina.

orbit (or'bit) The eye socket.

organ (or'gan) A structure consisting of a group of tissues that performs a specialized function.

organelle (or''gah-nel') A part of a cell that performs a specialized function.

organism (or'gah-nizm) An individual living thing.

organ of Corti (or'gan uv kor'ti) The organ that contains the hearing receptors in the inner ear.

orgasm (or'gazm) Physical and emotional climax during sexual intercourse; results in ejaculation in the male.

origin (or'i-jin) End of a muscle that is attached to a relatively immovable part.

osmosis (oz-mo'sis) The movement of water from an area of greater concentration of water to an area of lesser concentration of water across the cell membrane.

osmotic pressure (oz-mot'ik presh'ur) Pressure generated by the osmotic flow of water.

ossicles (os'i-k'lz) The tiny bones found in the middle ear: hammer, anvil, and stirrup.

osteoblast (os'te-o-blast'') A bone-forming cell.

osteoclast (os'te-o-klast'') A cell that causes the erosion of bone.

osteocyte (os'te-o-sīt) A mature bone cell.

otoliths (o'to-liths) Granules that lie above and whose movement stimulates ciliated cells in the utricle and saccule.

outer ear (out'er ēr) Portion of the ear consisting of the pinna and the auditory canal.

oval opening (o'val o'pen-ing) An opening between the two atria in the fetal heart; also foramen ovale.

oval window (o'val win'do) Membrane covered opening between the stapes and the inner ear.

ovarian cycle (o-va're-an si'k'l) Monthly occurring changes in the ovary that affect the level of sex hormones in the blood of females.

ovaries (o'var-ez) The female gonads, the organs that produce eggs, and estrogen and progesterone.

ovulation (o''vu-la'shun) The discharge of a mature egg from the follicle within the ovary.

oxygen debt (ok'si-jen det) The amount of oxygen needed to metabolize lactic acid that accumulates during vigorous exercise.

oxyhemoglobin (ok''se-he''mo-glo'bin) Hemoglobin bound to oxygen in a loose, reversible way.

oxytocin (ok''se-to'sin) Hormone released by posterior pituitary that causes contraction of uterus and milk letdown.

P

pacemaker (pās'māk-er) A small region of neuromuscular tissue that initiates the heartbeat; also SA node.

palate (pal'at) The roof of the mouth.

pancreas (pan'kre-as) An endocrine organ located near the stomach that secretes digestive enzymes into the duodenum and produces hormones, notably insulin.

parasympathetic nervous system (par''ah-sim''pa-thet'ik ner'vus sis'tem) A portion of the autonomic nervous system that usually promotes those activities associated with a normal state.

parathyroid hormone (par''ah-thi'roid hor'mōn) See PTH.

parietal lobe (pah-ri'e-tal lōb) Area of the cerebrum responsible for sensations involving temperature, touch, pressure, pain, and speech.

parturition (par''tu-rish'un) The processes that lead to and include the birth of a human, and the expulsion of the extraembryonic membranes through the terminal portion of the female reproductive tract.

pectoral (pek'tor-al) Pertaining to the thorax.

pectoral girdle (pek'tor-al ger'dl) Portion of the skeleton that provides support and attachment for the upper limbs.

pelvic girdle (pel'vik ger'dl) Portion of the skeleton to which the lower limbs are attached.

pelvis (pel'vis) Bony ring formed by the sacrum and coxal bones; also a hollow chamber in the kidney that lies inside the medulla and receives freshly prepared urine from the collecting ducts.

penis (pe'nis) Male copulatory organ.

pepsin (pep'sin) A protein-digesting enzyme secreted by gastric glands.

peptide bond (pep'tīd bond) The bond that joins two amino acids.

pericardium (per''i-kar'de-um) A protective serous membrane that surrounds the heart.

perimysium (per''i-mis'e-um) Fascia or connective tissue surrounding a bundle (fascicle) of muscle fibers.

periodontitis (per''e-o-don-ti'tis) Inflammation of the gums.

periosteum (per''e-os'te-um) A fibrous connective tissue covering the surface of bone.

peripheral nervous system (pe-rif'er-al) See PNS.

peristalsis (per''i-stal'sis) A rhythmical contraction that serves to move the contents along in tubular organs such as the digestive tract.

peritoneum (per''i-to-ne'um) A serous membrane that lines the abdominal cavity and encloses the abdominal viscera.

peritubular capillary (per''i-tu'bu-lar kap'i-lar''e) Capillary that surrounds a nephron and functions in reabsorption during urine formation.

pH (pe āch) A measure of the hydrogen ion concentration; any pH below 7 is acid and any pH above 7 is basic.

phagocytosis (fag''o-si-to'sis) The taking in of bacteria and/or debris by engulfing; cell eating.

phalanges (fah-lan'jēz) Bones of the fingers and thumb in the hand and in the toes of the foot (*sing.* phalanx).

pharynx (far'ingks) A common passageway (throat) for both food intake and air movement.

phenotype (fe'no-tīp) The outward appearance of an organism caused by the genotype and environmental influences.

physiograph (fiz'e-o-graf) Instrument used to record a myogram.

physiology (fiz''e-ol'o-je) The branch of science dealing with the study of body functions.

pia mater (pi'ah ma'ter) The innermost meningeal layer that is in direct contact with the brain and spinal cord.

pinna (pin'nah) Outer, funnellike structure of the ear that picks up sound waves.

placenta (plah-sen'tah) A structure formed from the chorion and uterine tissue through which nutrient and waste exchange occur for the embryo and later the fetus.

plaque (plak) An accumulation of soft masses of fatty material, particularly cholesterol, beneath the inner linings of arteries.

plasma (plaz'mah) The liquid portion of blood.

plasma cell (plaz'mah sel) A cell derived from a B lymphocyte that is specialized to mass produce antibodies.

plasmid (plaz'mid) A circular DNA segment that is present in bacterial cells, but is not part of the bacterial chromosome.

platelet (plāt'let) Cell-like disks formed from fragmentation of megakaryocytes that initiate blood clotting.

pleural membranes (ploor'al mem'brānz) Serous membranes that enclose the lungs.

PNS Peripheral nervous system. The nerves and ganglia of the nervous system that lie outside of the brain and spinal cord.

polar bodies (po'lar bod'ēz) Nonfunctioning daughter cells that have little cytoplasm and are formed during oogenesis.

polycythemia (pol''e-si-the'me-ah) An excessive concentration of red blood cells.

polysome (pol'e-sōm) A cluster of ribosomes all bringing about the synthesis of the same protein.

pons (ponz) A portion of the brain stem above the medulla oblongata and below the midbrain.

posterior (pos-tēr'e-or) Toward the back; opposite of anterior.

posterior pituitary (pos-tēr'e-or pi-tu'i-tār-e) Portion of the pituitary gland connected by a stalk to the hypothalamus.

postganglionic axon (pōst''gang-gle-on'ik ak'son) In the autonomic nervous system, the axon that leaves, rather than goes to, a ganglion.

postsynaptic membrane (pōst''si-nap'tik mem'brān) A membrane that is part of a synapse and receives a neurotransmitter substance.

preganglionic axon (pre''gang-gle-on'ik ak'son) In the autonomic nervous system, the axon that goes to, rather than leaves, a ganglion.

pressure filtration (presh'ur fil-tra'shun) Process by which small molecules leave a capillary due to blood pressure.

presynaptic membrane (pre''si-nap'tik mem'brān) A membrane that is part of a synapse and releases a neurotransmitter substance.

prime mover (prīm' mov'r) The muscle most directly responsible for a particular movement; an agonist.

process (pros'es) A prominent projection on a bone.

prolactin (pro-lak'tin) See lactogenic hormone.

proprioceptor (pro''pre-o-sep'tor) Sensory receptor that assists the brain in knowing the position of the limbs.

prostaglandins (pros''tah-glan'dinz) Hormones that have various and powerful effects often within the cells that produce them.

prostate gland (pros'tāt gland) A gland in males that is located about the urethra at the base of the bladder; contributes to the seminal fluid.

protein (pro'te-in) A macromolecule composed of amino acids.

prothrombin (pro-throm'bin) Plasma protein made by the liver that must be present in blood before clotting can occur.

proximal (prok'si-mal) Closer to the midline or origin; opposite of distal.

proximal convoluted tubule (prok'si-mal kon'vo-lūt-ed tu'būl) Highly coiled region of a nephron near Bowman's capsule.

pseudostratified (su''do-strat'if-fīd) The appearance of layering in some epithelial cells when actually each cell touches a base line and true layers do not exist.

PTH Parathyroid hormone secreted by the parathyroid glands that raises blood calcium level primarily by stimulating reabsorption of bone.

pulmonary (pul'mo-ner''e) Referring to the lungs.

pulmonary artery (pul'mo-ner''e ar'ter-e) A blood vessel that takes blood away from heart to the lungs.

pulmonary circuit (pul'mo-ner''e) The blood vessels that take deoxygenated blood to and oxygenated blood away from the lungs.

pulmonary vein (pul'mo-ner''e vān) A blood vessel that takes blood away from lungs to the heart.

pulse (puls) Vibration felt in arterial walls due to expansion of the aorta following ventricle contraction.

Purkinje fibers (pur-kin'je fi'berz) Specialized muscle fibers that conduct the cardiac impulse from A–V bundle into the ventricular walls.

pus (pus) Thick yellowish fluid composed of dead phagocytes, dead tissue, and bacteria.

R

radius (ra'de us) An elongated bone located on the thumb side of the lower arm.

receptor (re-sep'tor) A sense organ specialized to receive information from the environment. Also a structure found in the membrane of cells that combines with a specific chemical in a lock and key manner.

recessive gene (re-ses'iv jēn) Hereditary factor that only expresses itself when two copies are present in the genotype.

recombinant DNA (re-kom'bi-nant) DNA having genes from two different sources.

rectum (rek'tum) The terminal portion of the intestine.

red blood cell (red blud sel) See erythrocyte.

red marrow (red mar'o) Blood cell-forming tissue located in spaces within certain bones.

reduced hemoglobin (re-dust' he-mo-glo'bin) Hemoglobin that has released its oxygen.

reflex (re'fleks) An inborn autonomic response to a stimulus that is dependent on the existence of fixed neural pathways.

REM Rapid eye movement; a stage in sleep that is characterized by eye movements and dreaming.

renal cortex (rēn'al kor'teks) The outer portion of the kidney, primarily vascular.

renal medulla (rēn'al me-dul'ah) The inner portion of the kidney, including the renal pyramids.

renal pelvis (rēn'al pel'vis) The inner cavity of the kidney formed by the expanded ureter and into which the collecting ducts open.

replication (re''pli-ka'shun) The duplication of DNA; occurs when the cell is not dividing.

repolarization (re-po''lar-i-za'shun) The recovery of a neuron's polarity to the resting potential after it ceases transmitting impulses.

respiration (res''pi-ra'shun) Cellular process by which air enters the body, gases are exchanged with tissue fluid and ATP is produced in cells.

resting potential (rest'ing po-ten'shal) The voltage recorded from inside a neuron when it is not conducting nerve impulses.

retina (ret'i-nah) The innermost layer of the eyeball that contains the rods and cones.

retinene (ret'i-nēn) Substance used in the production of rhodopsin.

Rh factor (ar'āch fak'tor) A type of antigen on the red blood cells.

rhodopsin (ro-dop'sin) Visual purple, a pigment found in the rods of one type of receptor in the retina of the eye.

ribonucleic acid (ri''bo-nu-kle'ik as'id) See RNA.

ribosomes (ri'bo-somz) Minute particles, found attached to endoplasmic reticulum or loose in the cytoplasm, that are the site of protein synthesis.

ribs (ribz) Bones hinged to the vertebral column and sternum which, along with muscle, define the top and sides of the chest cavity.

RNA A nucleic acid that helps DNA in the synthesis of proteins.

rods (rodz) Dim-light receptors in the retina of the eye that detect motion but no color.

rotation (ro-ta'shun) The movement of a bone around its own longitudinal axis.

round window (rownd win'do) A membrane-covered opening between the inner ear and the middle ear.

S

saccule (sak'ul) A saclike cavity of the inner ear that contains receptors for static equilibrium.

sagittal (saj'i-tal) A plane or section that divides a structure into right and left portions.

salivary gland (sal'i-ver-e gland) A gland associated with the mouth that secretes saliva.

SA node (es a nod) Small region of neuromuscular tissue that initiates the heartbeat. Also called the pacemaker.

sarcolemma (sar''ko-lem'ah) The membrane that surrounds striated muscle cells, corresponding to the cell membrane of other cells.

sarcoplasmic reticulum (sar''ko-plaz'mik re-tik'u-lum) Membranous network of channels and tubules within a muscle fiber, corresponding to the endoplasmic reticulum of other cells.

Schwann cell (shwahn sel) Cell that surrounds a fiber of a peripheral nerve and forms the neurolemmal sheath and myelin.

sclera (skle'rah) White fibrous outer layer of the eyeball.

scotopsin (sko-top'sin) Protein portion of rhodopsin.

scrotal sacs (skro'tal saks) The sac that contains the testes.

scrotum (skro'tum) A pouch of skin that encloses the testes.

sebaceous gland (se-ba'shus gland) Gland of the skin that secretes sebum.

sebum (se'bum) Oily secretion of the sebaceous glands.

secretin (se-kre'tin) Hormone secreted by the small intestine that stimulates the release of pancreatic juice from the pancreas.

selective reabsorption (se-lek'tiv re''ab-sorp'shun) One of the processes involved in the formation of urine; involves the greater reabsorption of nutrient molecules compared to waste molecules from the contents of the kidney tubule into the blood.

semen (se'men) The sperm-containing secretion of males; seminal fluid plus sperm.

semicircular canals (sem''e-ser'ku-lar kah-nal') Tubular structures within the inner ear with ampullae that contain the receptors responsible for the sense of dynamic equilibrium.

semilunar valves (sem''e-lu'nar val'vz) Valves resembling half moons located between the ventricles and their attached vessels.

seminal fluid (sem'i-nal floo'id) Fluid produced by various glands situated along the male reproductive tract.

seminal vesicles (sem'i-nal ves'i-k'l) Convoluted saclike structures attached to vas deferens near the base of the bladder in males, contribute to seminal fluid.

seminiferous tubules (sem''i-nif'er-us tu'bulz) Highly coiled ducts within the male testes that produce and transport sperm.

sensory nerve (sen'so-re nerv) Nerve containing only sensory neuron dendrites.

sensory neuron (sen'so-re nu'ron) A neuron that takes the nerve impulse to the central nervous system; afferent neuron.

septum (sep'tum) Partition or wall such as the septum in the heart, which divides the right half from the left half.

serous membrane (se'rus mem'bran) Membrane that covers organs and lines a cavity without an opening to the outside of the body; also called serosa.

serum (se'rum) Light-yellow liquid left after clotting of the blood.

sex chromosome (seks kro'mo-som) A chromosome responsible for the development of characteristics associated with maleness or femaleness; and X or Y chromosome.

sex-linked genes (seks-linkt' jenz) Genes found on the sex chromosomes that control traits other than sexual traits.

simple goiter (sim'p'l goi'ter) Condition in which an enlarged thyroid produces low levels of thyroxin.

sinus (si'nus) A cavity, as the sinuses in the human skull.

skeletal muscle (skel'e-tal mus'el) The contractile tissue that comprises the muscles attached to the skeleton; also called striated muscle.

skin (skin) Organ system covering the body, which serves in sensory, protective, excretory, and temperature regulating capacities.

sliding filament theory (slid'ing fil'ah-ment the'o-re) The movement of actin in relation to myosin in explaining the mechanics of muscle contraction.

smooth muscle (smooth mus'el) The contractile tissue that comprises the muscles found in the walls of internal organs.

soft palate (soft pal'at) Entirely muscular posterior portion of the roof of the mouth.

somatic nervous system (so-mat'ik ner'vus sis'tem) That portion of the PNS containing motor neurons that control skeletal muscles.

somatotropin (so''mah-to-tro'pin) See growth hormone.

spermatogenesis (sper''mah-to-jen'e-sis) Production of sperm in males by the process of meiosis and maturation.

sphincter (sfingk'ter) A muscle that surrounds a tube, and closes or opens the tube by contracting and relaxing.

spinal (spi'nal) Pertaining to the spinal cord or to the vertebral canal.

spinal nerve (spi'nal nerv) Nerve that arises from the spinal cord.

spindle (spin'd'l) An apparatus composed of microtubules to which the chromosomes are attached during cell division.

spleen (splen) A large, glandular organ located in the upper left region of the abdomen that stores and purifies blood.

spongy bone (spun'je bon) Bone found at the ends of long bones that consists of bars and plates separated by irregular spaces.

squamous epithelium (skwa'mus ep''i-the'le-um) Flat cells found lining the lungs and blood vessels.

stereoscopic vision (ste''re-o-skop'ik vizh'un) The product of two eyes and both cerebral hemispheres functioning together so that depth perception results.

sternum (ster'num) The breastbone to which the ribs are ventrally attached.

steroid (ste'roid) Lipid soluble, biologically active molecules having four interlocking rings; examples are cholesterol, progesterone, testosterone.

stratified (strat'i-fīd) Layered, as in stratified epithelium, which contains several layers of cells.

stretch receptors (strech re-sep'torz) Muscle fibers which, upon stimulation, cause muscle spindles to increase the rate at which they send impulses to the CNS.

striated (stri'āt-ed) Having bands; cardiac and skeletal muscle are striated with bands of light and dark.

subcutaneous (sub''ku-ta'ne-us) A tissue beneath the dermis that tends to contain fat cells.

submucosa (sub''mu-ko'sah) A layer of supportive connective tissue that underlies a mucous membrane.

summation (sum-ma'shun) Ever greater contraction of a muscle due to constant stimulation that does not allow complete relaxation to occur.

superfemale (su''per-fe'māl) A female that has three X chromosomes.

superficial (soo''per-fish'al) Near the surface.

superior Toward the upper part of a structure or toward the head.

suture (su'chur) A type of immovable joint articulation found between bones of the skull.

sweat gland (swet gland) A skin gland that secretes a fluid substance for evaporative cooling.

sympathetic nervous system (sim''pah-thet'ik ner'vus sis'tem) That part of the autonomic nervous system that generally causes effects associated with emergency situations.

symphysis (sim'fī-sis) A slightly movable joint between bones separated by a pad of fibrocartilage.

synapse (sin'aps) The region between two nerve cells where the nerve impulse is transmitted from one to the other; usually from axon to dendrite.

synapsis (si-nap'sis) The attracting and pairing of homologous chromosomes during meiosis.

synaptic cleft (si-nap'tik kleft) Small gap between presynaptic and postsynaptic membranes of a synapse.

synaptic ending (si-nap'tik end'ing) The knob at the end of an axon in a synapse.

synergist (sin'er-jist) A muscle that assists the action of the prime mover.

synovial fluid (si-no've-al floo'id) Fluid secreted by the synovial membrane.

synovial joint (si-no've-al joint) A freely movable joint.

synovial membrane (si-no've-al mem'brān) Membrane that forms the inner lining of the capsule of a freely movable joint.

synthesis (sin'the-sis) To build up, such as the combining together of two small molecules to form a larger molecule.

syphillis (sif'i-lis) Chronic, sexually transmitted disease caused by a spirochete bacterium.

systemic circuit (sis-tem'ik) That part of the circulatory system that serves body parts other than the gas-exchanging surfaces in the lungs.

systole (sis'to-le) Contraction of the heart chambers, particularly the left ventricle.

systolic pressure (sis-tol'ik presh'ur) Arterial blood pressure during the systolic phase of the cardiac cycle.

T

tarsal (tahr'sal) A bone of the ankle in humans.

taste bud (tāst bud) Organ containing the receptors associated with the sense of taste.

tectorial membrane (tek-to're-al mem'brān) A membrane in the organ of Corti that lies above and makes contact with the receptor cells for hearing.

temporal lobe (tem'po-ral lōb) Area of the cerebrum responsible for hearing and smelling, the interpretation of sensory experience and memory.

tendon (ten'don) A tissue that connects muscle to bone.

testes (tes'tez) The male gonads, the organs that produce sperm and testosterone.

testosterone (tes-tos'te-rōn) The most potent of the androgens.

tetanus (tet'ah-nus) Sustained muscle contraction without relaxation.

tetany (tet'ah-ne) Severe twitching caused by involuntary contraction of the skeletal muscles due to a lack of calcium.

thalamus (thal'ah-mus) A mass of gray matter located at the base of the cerebrum in the wall of the third ventricle; receives sensory information and selectively passes it to the cerebrum.

thoracic (tho-ras'ik) Pertaining to the chest.

thrombin (throm'bin) The enzyme derived from prothrombin that converts fibrinogen to fibrin threads during blood clotting.

thrombus (throm'bus) A blood clot that remains in the blood vessel where it formed.

thymus (thi'mus) A lymphatic organ that lies in the neck and chest area, and is absolutely necessary to the development of immunity.

thyroid stimulating hormone (thi'roid stim'u-la't''ing hor'mōn) See TSH.

thyroxin (thi-rok'sin) The hormone produced by the thyroid that speeds up the metabolic rate.

tibia (tib'e-ah) The shinbone found in the lower leg.

tidal volume (tid'al vol'ūm) Amount of air that enters the lungs during a normal, quiet inspiration.

tight junction (tīt junk'shun) A zipperlike junction between two cells that prevents passage of molecules.

tissue (tish'u) A group of similar cells that performs a specialized function.

tissue fluid (tish'u floo'id) Fluid found about tissue cells containing molecules that enter from or exit to the capillaries.

T lymphocyte (lim'fo-sit) A lymphocyte called killer T cells that interacts directly with antigen-bearing cells and is responsible for cell-mediated immunity, or is a helper T cell that stimulates other immune cells.

tone (tōn) The continuous partial contraction of muscle; also the quality of a sound.

trachea (tra'ke-ah) The windpipe that serves as a passageway for air.

tract (trakt) A bundle of neurons forming a transmission pathway through the brain and spinal cord.

trait (trāt) Specific term for a distinguishing feature studied in heredity.

transcription (trans-krip'shun) The process that results in the production of a strand of mRNA that is complementary to a segment of DNA.

trophoblast (trof'o-blast) The outer membrane that surrounds the human embryo and, when thickened by a layer of mesoderm, becomes the chorion.

trunk The thorax and abdomen together.

trypsin (trip'sin) A protein-digesting enzyme secreted by the pancreas.

TSH Thyroid stimulating hormone that causes the thyroid to produce thyroxin.

tubal ligation (tu'bal li-ga'shun) Cutting the oviducts in females as a birth control measure.

tubular excretion (tu'bu-lar esk-kre'shun) Process occurring in the distal convoluted tubule during which substances are added to urine.

Turner's syndrome (tur'nerz sin'drōm) A condition caused by the inheritance of a single X chromosome.

twitch (twitch) A brief muscular contraction followed by relaxation.

tympanic membrane (tim-pan'ik mem'brān) A membrane located between the external and middle ear; the eardrum.

U

ulna (ul'nah) An elongated bone found within the lower arm.

umbilical cord (um-bil'i-kal kord) Cord through which blood vessels pass, connecting the fetus to the placenta.

urea (u-re'ah) Primary nitrogenous waste of mammals.

ureters (u-re'ters) Tubes that take urine from the kidneys to the bladder.

urethra (u-re'thrah) Tube that takes urine from bladder to outside of the body.

uric acid (u'rik as'id) Waste product of nucleotide breakdown.

urinalysis (u''ri-nal'i-sis) A medical procedure in which the composition of a patient's urine is determined.

urinary bladder (u'ri-ner''e blad'der) An organ where urine is stored before being discharged by way of the urethra.

uterine cycle (u'ter-in si'kl) The female reproductive cycle that is characterized by regularly occurring changes in the uterine lining.

uterus (u'ter-us) The organ in females in which the fetus develops.

utricle (u'tre-k'l) A saclike cavity of the inner ear that contains receptors for static equilibrium.

V

vaccine (vak'sēn) Treated antigens that can promote active immunity when administered.

vagina (vah-ji'nah) Copulatory organ in females.

valve (valv) A structure that opens and closes, insuring one-way flow; common to vessels such as the systemic veins and the lymphatic veins, and to the heart.

varicose veins (var'i-kōs vānz) Abnormally swollen and enlarged veins, especially in the legs.

vas deferens (vas def'er enz) Tube connecting epididymis to the urethra; sperm duct.

vasectomy (vah-sek'to-me) Cutting of the vas deferens in males as a birth control measure.

vein (vān) A blood vessel that takes blood to the heart.

vena cava (ve'nah ka'vah) One of two large veins that convey deoxygenated blood to the right atrium of the heart; (plural, *venae cavae*).

venous duct (ve'nus dukt) Fetal connection between the umbilical vein and the inferior vena cava; ductus venosus.

ventilation (ven''ti-la'shun) Breathing; the process of moving air into and out of the lungs.

ventral (ven'tral) Toward the front or belly surface; the opposite of *dorsal*.

ventricle (ven'tri-k'l) A cavity in an organ such as the ventricles of the heart or the ventricles of the brain.

venule (ven'ul) Type of blood vessel that takes blood from capillaries to veins.

vernix caseosa (ver'niks ka''se-o'sah) Cheeselike substance covering the skin of the fetus.

vertebral column (ver'te-bral kol'um) The backbone of vertebrates composed of individual bones called vertebrae.

vestibule (ves'ti-būl) A space or cavity at the entrance of a canal such as the cavity that lies between the semicircular canals and the cochlea.

villi (vil'i) Fingerlike projections that line the small intestine and function in absorption.

visceral (vis'er-al) Pertaining to the contents of a body cavity.

vital capacity (vi'tal kah-pas'i-te) The maximum amount of air a person can exhale after taking the deepest breath possible.

vitreous humor (vit're-us hu'mor) The substance that occupies the posterior cavity of the eye.

vocal cords (vo'kal kordz) Folds of tissue within the larynx that create vocal sounds when they vibrate.

vulva (vul'vah) The external genitalia of the female that lie near the opening of the vagina.

W

white blood cell (wit blud sel) *See* leukocyte.

X

X-linked (eks'linked) Refers to inherited genes or traits carried by the female (X) chromosome.

Y

yellow marrow (yel'o mar'o) Fat storage tissue found in the medullary cavities within certain bones.

yolk sac (yōk sak) An extraembryonic membrane that serves as the first site for blood cell formation.

Z

zygote (zi'gōt) Cell formed by the union of the sperm and egg, the product of fertilization.

Credits

Photographs

Table of Contents
p. v: © Manfred Kage/Peter Arnold, Inc.; p. vi (left): © Edwin Reschke; p. vi (right): © Manfred Kage/Peter Arnold, Inc.; p. vii: © David Gifford/Science Photo Library/Photo Researchers, Inc.; p. viii: © David Scharf/Peter Arnold, Inc.

Part Openers
One: © Manfred Kage/Peter Arnold, Inc.; Two: © Edwin Reschke; Three: © Manfred Kage/Peter Arnold, Inc.; Four: © David Gifford/Science Photo Library/Photo Researchers, Inc.; Five: © David Scharf/Peter Arnold, Inc.

Chapter 1
1.1c: Edwin Reschke; 1.7a: © Dr. Sheril D. Burton; 1.8a: © Dr. Sheril D. Burton

Chapter 2
2.3: © Dr. Stephen Wolfe; 2.4a: © Dr. Don Fawcett/Photo Researchers, Inc.; 2.7a: © Lennart Nilsson; 2.8a: © Keith Porter; 2.9: © Dr. Jean Revel; 2.10: © John Walsh/Photo Researchers, Inc.

Chapter 3
3.1a, 3.1B1: © Edwin Reschke; 3.1C1: © Manfred Kage/Peter Arnold, Inc.; 3.3a, 3.4, 3.5, 3.6b, 3.7, 3.9a, 3.10a, 3.11a, 3.12: © Edwin Reschke

Chapter 4
4.2 all: © American Institute of Applied Science; 4.3: © James M. Clayton; 4.4b: © CNRI/Photo Researchers, Inc.; 4.6: © George P. Bogumill; 4.9: © Ira Wyman/Sygma

Chapter 8
8.8b: © Thomas Sims

Chapter 9
9.5: © Bettina Cirone/Photo Researchers, Inc; 9.6 both: © Dr. Charles Blake/"The Pituitary Gland"/Carolina Reader #118, pg. 11; 9.8: © Lester V. Bergman and Associates; 9.10: From *Clinical Endocrinology and Its Physiological Basis* by Arthur Grollman, 1964. Used by permission of J. B. Lippincott Co.; 9.11: © Lester V. Bergman and Associates; 9.15: © From *Clinical Endocrinology and Its Physiological Basis* by Arthur Grollman, 1964. Used by permission of J. B. Lippincott Co.; 9.16: © F. A. David Co., Philadelphia and R. H. Kampmeier

Chapter 11
11.2: © CNRI/Photo Researchers, Inc.; 11.3: © Bill Longcore/Photo Researchers, Inc.; 11.5a: From *The Morphology of Human Blood Cells* by L. W. Diggs, M.D., Dorothy Sturm and Ann Bell © 1984 Abbott Laboratories; 11.5c: © Dr. Elienne de Harven and Ms. Nina Lampen, Sloan-Kettering Institute, New York; 11.10a: © Stuart I. Fox

Chapter 12
12.5b: © Igaku Shoin, Ltd

Chapter 13
13.11: © Dr. Kirk Ziegler; 13.4b: © Guy Gillette/Science Source/Photo Researchers, Inc.

Chapter 14
14.7: © John Watney Photo Library; 14.11b: © Dan McCoy/Rainbow; 14.13: © American Lung Association/Patricia Delaney; 14.14: © CNRI/Science Photo Library/Photo Researchers, Inc.

Chapter 15
15.8: From Bloom, W. and Fawcett, D. W. A., *Text-Book of Histology*, Philadelphia, W.B. Saunders Company, 1966, photo by Ruth Bulger

Chapter 17
17.1 both: © Claude Edelmann, Petit Format et Guigorz from *First Days of Life* / Black Star; 17.4a: © Lennart Nilsson

Chapter 18
18.3a: © Jill Cannefax/EKM-Nepenthe; 18.6a: © F. A. Davis Company, Philadelphia and R. H. Kampmeier; 18.6a: From M. Bartolos and T. A. Baramski, *Medical Cytogenetics,* Williams and Wilkins, Baltimore, MD, 1967

Appendix A
Figure A12b: © Northern Illinois University

Illustrators

Chris Creek: 18.7
Ruth Krabach: 5.8a, b, c; 5.17a, b, c
Rob Margulies: 10.10; 15.7
Nancy Marshburn: 12.10; 18.8
Precision Graphics: 6.3; 6.5; 13.9
Rolin Graphics: 12.17
Mike Schenk: 1.3; 1.5

Line Art

Chapter 1
Figure 1.2: From Kent M. Van De Graaff and Stuart Ira Fox, *Concepts of Human Anatomy and Physiology*, 2d ed. Copyright © 1989 Wm. C. Brown Publishers, Dubuque, Iowa. All Rights Reserved. Reprinted by permission.
Figure 1.4: From John W. Hole, Jr., *Essentials of Human Anatomy and Physiology,* 3d ed. Copyright © 1989 Wm. C. Brown Publishers, Dubuque, Iowa. All Rights Reserved. Reprinted by permission.
Figure 1.6: From Kent M. Van De Graaff, *Human Anatomy,* 2d ed. Copyright © 1988 Wm. C. Brown Publishers, Dubuque, Iowa. All Rights Reserved. Reprinted by permission.
Figure 1.7 bottom: From Kent M. Van De Graaff, *Human Anatomy,* 2d ed. Copyright © 1988 Wm. C. Brown Publishers, Dubuque, Iowa. All Rights Reserved. Reprinted by permission.
Figure 1.8 bottom: From Kent M. Van De Graaff, *Human Anatomy,* 2d ed. Copyright © 1988 Wm. C. Brown Publishers, Dubuque, Iowa. All Rights Reserved. Reprinted by permission.
Figure 1.9: From Kent M. Van De Graaff and Stuart Ira Fox, *Concepts of Human Anatomy and Physiology,* 2d ed. Copyright © 1989 Wm. C. Brown Publishers, Dubuque, Iowa. All Rights Reserved. Reprinted by permission.
Figure 1.10: From John W. Hole, Jr., *Human Anatomy and Physiology,* 4th ed. Copyright © 1987 Wm. C. Brown Publishers, Dubuque, Iowa. All Rights Reserved. Reprinted by permission.

Chapter 3
Figure 3.6a: From Kent M. Van De Graaff, *Human Anatomy,* 2d ed. Copyright © 1988 Wm. C. Brown Publishers, Dubuque, Iowa. All Rights Reserved. Reprinted by permission.
Figure 3.9b: From Kent M. Van De Graaff, *Human Anatomy,* 2d ed. Copyright © 1988 Wm. C. Brown Publishers, Dubuque, Iowa. All Rights Reserved. Reprinted by permission.
Figure 3.11b: From Kent M. Van De Graaff, *Human Anatomy,* 2d ed. Copyright © 1988 Wm. C. Brown Publishers, Dubuque, Iowa. All Rights Reserved. Reprinted by permission.
Figure 3.13: From Kent M. Van De Graaff, *Human Anatomy,* 2d ed. Copyright © 1988 Wm. C. Brown Publishers, Dubuque, Iowa. All Rights Reserved. Reprinted by permission.

Figure 3.14a, b: From Kent M. Van De Graaff and Stuart Ira Fox, *Concepts of Human Anatomy and Physiology*, 2d ed. Copyright © 1989 Wm. C. Brown Publishers, Dubuque, Iowa. All Rights Reserved. Reprinted by permission.

Reference Figures

Figure 1: From John W. Hole, Jr., *Human Anatomy and Physiology*, 5th ed. Copyright © 1990 Wm. C. Brown Publishers, Dubuque, Iowa. All Rights Reserved. Reprinted with permission.

Figure 2: From John W. Hole, Jr., *Human Anatomy and Physiology*, 5th ed. Copyright © 1990 Wm. C. Brown Publishers, Dubuque, Iowa. All Rights Reserved. Reprinted with permission.

Figure 3: From John W. Hole, Jr., *Human Anatomy and Physiology*, 5th ed. Copyright © 1990 Wm. C. Brown Publishers, Dubuque, Iowa. All Rights Reserved. Reprinted with permission.

Figure 4: From John W. Hole, Jr., *Human Anatomy and Physiology*, 5th ed. Copyright © 1990 Wm. C. Brown Publishers, Dubuque, Iowa. All Rights Reserved. Reprinted with permission.

Figure 5: From John W. Hole, Jr., *Human Anatomy and Physiology*, 5th ed. Copyright © 1990 Wm. C. Brown Publishers, Dubuque, Iowa. All Rights Reserved. Reprinted with permission.

Figure 6: From John W. Hole, Jr., *Human Anatomy and Physiology*, 5th ed. Copyright © 1990 Wm. C. Brown Publishers, Dubuque, Iowa. All Rights Reserved. Reprinted with permission.

Figure 7: From John W. Hole, Jr., *Human Anatomy and Physiology*, 5th ed. Copyright © 1990 Wm. C. Brown Publishers, Dubuque, Iowa. All Rights Reserved. Reprinted with permission.

Chapter 4

Figure 4.4a: From John W. Hole, Jr., *Essentials of Human Anatomy and Physiology*, 3d ed. Copyright © 1989 Wm. C. Brown Publishers, Dubuque, Iowa. All Rights Reserved. Reprinted by permission.

Figure 4.5: From John W. Hole, Jr., *Human Anatomy and Physiology*, 5th ed. Copyright © 1990 Wm. C. Brown Publishers, Dubuque, Iowa. All Rights Reserved. Reprinted with permission.

Figure 4.8: From John W. Hole, Jr., *Human Anatomy and Physiology*, 4th ed. Copyright © 1987 Wm. C. Brown Publishers, Dubuque, Iowa. All Rights Reserved. Reprinted by permission.

Chapter 5

Figure 5.1: From John W. Hole, Jr., *Human Anatomy and Physiology*, 5th ed. Copyright © 1990 Wm. C. Brown Publishers, Dubuque, Iowa. All Rights Reserved. Reprinted with permission.

Figure 5.2: From John W. Hole, Jr., *Human Anatomy and Physiology*, 5th ed. Copyright © 1990 Wm. C. Brown Publishers, Dubuque, Iowa. All Rights Reserved. Reprinted with permission.

Figure 5.3: From Kent M. Van De Graaff, *Human Anatomy*, 2d ed. Copyright © 1988 Wm. C. Brown Publishers, Dubuque, Iowa. All Rights Reserved. Reprinted by permission.

Figure 5.4: From Kent M. Van De Graaff, *Human Anatomy*, 2d ed. Copyright © 1988 Wm. C. Brown Publishers, Dubuque, Iowa. All Rights Reserved. Reprinted by permission.

Figure 5.5a, b: From Kent M. Van De Graaff, *Human Anatomy*, 2d ed. Copyright © 1988 Wm. C. Brown Publishers, Dubuque, Iowa. All Rights Reserved. Reprinted by permission.

Figure 5.6a, b: From Kent M. Van De Graaff, *Human Anatomy*, 2d ed. Copyright © 1988 Wm. C. Brown Publishers, Dubuque, Iowa. All Rights Reserved. Reprinted by permission.

Figure 5.7: From John W. Hole, Jr., *Human Anatomy and Physiology*, 5th ed. Copyright © 1990 Wm. C. Brown Publishers, Dubuque, Iowa. All Rights Reserved. Reprinted with permission.

Figure 5.9: From John W. Hole, Jr., *Human Anatomy and Physiology*, 4th ed. Copyright © 1987 Wm. C. Brown Publishers, Dubuque, Iowa. All Rights Reserved. Reprinted by permission.

Figure 5.11: From John W. Hole, Jr., *Human Anatomy and Physiology*, 5th ed. Copyright © 1990 Wm. C. Brown Publishers, Dubuque, Iowa. All Rights Reserved. Reprinted with permission.

Figure 5.12: From John W. Hole, Jr., *Human Anatomy and Physiology*, 5th ed. Copyright © 1990 Wm. C. Brown Publishers, Dubuque, Iowa. All Rights Reserved. Reprinted with permission.

Figure 5.13: From John W. Hole, Jr., *Human Anatomy and Physiology*, 5th ed. Copyright © 1990 Wm. C. Brown Publishers, Dubuque, Iowa. All Rights Reserved. Reprinted with permission.

Figure 5.14: From John W. Hole, Jr., *Human Anatomy and Physiology*, 5th ed. Copyright © 1990 Wm. C. Brown Publishers, Dubuque, Iowa. All Rights Reserved. Reprinted with permission.

Figure 5.15: From John W. Hole, Jr., *Human Anatomy and Physiology*, 5th ed. Copyright © 1990 Wm. C. Brown Publishers, Dubuque, Iowa. All Rights Reserved. Reprinted with permission.

Figure 5.18: From John W. Hole, Jr., *Human Anatomy and Physiology*, 5th ed. Copyright © 1990 Wm. C. Brown Publishers, Dubuque, Iowa. All Rights Reserved. Reprinted with permission.

Figure 5.19: From John W. Hole, Jr., *Human Anatomy and Physiology*, 5th ed. Copyright © 1990 Wm. C. Brown Publishers, Dubuque, Iowa. All Rights Reserved. Reprinted with permission.

Figure 5.20: From John W. Hole, Jr., *Human Anatomy and Physiology*, 5th ed. Copyright © 1990 Wm. C. Brown Publishers, Dubuque, Iowa. All Rights Reserved. Reprinted with permission.

Figure 5.21: From John W. Hole, Jr., *Human Anatomy and Physiology*, 4th ed. Copyright © 1987 Wm. C. Brown Publishers, Dubuque, Iowa. All Rights Reserved. Reprinted by permission.

Figure 5.22a: From Kent M. Van De Graaff, *Human Anatomy*, 2d ed. Copyright © 1988 Wm. C. Brown Publishers, Dubuque, Iowa. All Rights Reserved. Reprinted by permission.

Figure 5.22b: From John W. Hole, Jr., *Human Anatomy and Physiology*, 5th ed. Copyright © 1990 Wm. C. Brown Publishers, Dubuque, Iowa. All Rights Reserved. Reprinted with permission.

Figure 5.23: From John W. Hole, Jr., *Human Anatomy and Physiology*, 5th ed. Copyright © 1990 Wm. C. Brown Publishers, Dubuque, Iowa. All Rights Reserved. Reprinted with permission.

Chapter 6

Figure 6.1: From John W. Hole, Jr., *Human Anatomy and Physiology*, 5th ed. Copyright © 1990 Wm. C. Brown Publishers, Dubuque, Iowa. All Rights Reserved. Reprinted with permission.

Figure 6.4: From Bernard Katz, *Nerve, Muscle, & Synapse*. Copyright © 1966 McGraw-Hill, Inc., New York, NY. Reprinted by permission of McGraw-Hill, Inc.

Figure 6.9: From John W. Hole, Jr., *Human Anatomy and Physiology*, 4th ed. Copyright © 1987 Wm. C. Brown Publishers, Dubuque, Iowa. All Rights Reserved. Reprinted by permission.

Figure 6.10: From John W. Hole, Jr., *Human Anatomy and Physiology*, 4th ed. Copyright © 1987 Wm. C. Brown Publishers, Dubuque, Iowa. All Rights Reserved. Reprinted by permission.

Figure 6.11: From Kent M. Van De Graaff, *Human Anatomy*, 2d ed. Copyright © 1988 Wm. C. Brown Publishers, Dubuque, Iowa. All Rights Reserved. Reprinted by permission.

Figure 6.12: From Kent M. Van De Graaff, *Human Anatomy*, 2d ed. Copyright © 1988 Wm. C. Brown Publishers, Dubuque, Iowa. All Rights Reserved. Reprinted by permission.

Figure 6.13: From John W. Hole, Jr., *Human Anatomy and Physiology*, 4th ed. Copyright © 1987 Wm. C. Brown Publishers, Dubuque, Iowa. All Rights Reserved. Reprinted by permission.

Figure 6.14: From Kent M. Van De Graaff and Stuart Ira Fox, *Concepts of Human Anatomy and Physiology*, 2d ed. Copyright © 1989 Wm. C. Brown Publishers, Dubuque, Iowa. All Rights Reserved. Reprinted by permission.

Figure 6.15: From Kent M. Van De Graaff, *Human Anatomy*, 2d ed. Copyright © 1988 Wm. C. Brown Publishers, Dubuque, Iowa. All Rights Reserved. Reprinted by permission.

Figure 6.16: From Kent M. Van De Graaff, *Human Anatomy*, 2d ed. Copyright © 1988 Wm. C. Brown Publishers, Dubuque, Iowa. All Rights Reserved. Reprinted by permission.

Figure 6.17: From Kent M. Van De Graaff and Stuart Ira Fox, *Concepts of Human Anatomy and Physiology*, 2d ed. Copyright © 1989 Wm. C. Brown Publishers, Dubuque, Iowa. All Rights Reserved. Reprinted by permission.

Figure 6.18: From Kent M. Van De Graaff and Stuart Ira Fox, *Concepts of Human Anatomy and Physiology*, 2d ed. Copyright © 1989 Wm. C. Brown Publishers, Dubuque, Iowa. All Rights Reserved. Reprinted by permission.

Chapter 7

Figure 7.2: Reprinted with permission of Macmillan Publishing Company from *Anatomy and Physiology*, 17th edition, by Miller, Drakontides and Leavell. Copyright © 1977 by Macmillan Publishing Company.

Figure 7.3: From Stuart Ira Fox, *Human Physiology*, 3d ed. Copyright © 1990 Wm. C. Brown Publishers, Dubuque, Iowa. All Rights Reserved. Reprinted by permission.

Figure 7.4: From John W. Hole, Jr., *Human Anatomy and Physiology*, 3d ed. Copyright © 1984 Wm. C. Brown Publishers, Dubuque, Iowa. All Rights Reserved. Reprinted by permission.

Figure 7.6a, b: From John W. Hole, Jr., *Human Anatomy and Physiology*, 5th ed. Copyright © 1990 Wm. C. Brown Publishers, Dubuque, Iowa. All Rights Reserved. Reprinted with permission.

Figure 7.7: From Stuart Ira Fox, *Human Physiology*, 3d ed. Copyright © 1990 Wm. C. Brown Publishers, Dubuque, Iowa. All Rights Reserved. Reprinted by permission.

Figure 7.10b: From Kent M. Van De Graaff, *Human Anatomy*. Copyright © 1988 Wm. C. Brown Publishers, Dubuque, Iowa. All Rights Reserved. Reprinted by permission.

Figure 7.11: From John W. Hole, Jr., *Human Anatomy and Physiology*, 5th ed. Copyright © 1990 Wm. C. Brown Publishers, Dubuque, Iowa. All Rights Reserved. Reprinted with permission.

Figure 7.12: From Kent M. Van De Graaff, *Human Anatomy*, 2d ed. Copyright © 1988 Wm. C. Brown Publishers, Dubuque, Iowa. All Rights Reserved. Reprinted by permission.

Figure 7.13: From John W. Hole, Jr., *Human Anatomy and Physiology*, 5th ed. Copyright © 1990 Wm. C. Brown Publishers, Dubuque, Iowa. All Rights Reserved. Reprinted with permission.

Figure 7.14: From John W. Hole, Jr., *Human Anatomy and Physiology*, 5th ed. Copyright © 1990 Wm. C. Brown Publishers, Dubuque, Iowa. All Rights Reserved. Reprinted with permission.

Figure 7.15: From T. L. Peele, *Neuroanatomical Basis for Clinical Neurology*, 3d ed. Copyright © 1977 McGraw-Hill, Inc., New York, NY. Reprinted by permission of McGraw-Hill, Inc.

Figure 7.20: From John W. Hole, Jr., *Human Anatomy and Physiology*, 5th ed. Copyright © 1990 Wm. C. Brown Publishers, Dubuque, Iowa. All Rights Reserved. Reprinted with permission.

Figure 7.22: From John W. Hole, Jr., *Human Anatomy and Physiology*, 5th ed. Copyright © 1990 Wm. C. Brown Publishers, Dubuque, Iowa. All Rights Reserved. Reprinted with permission.

Chapter 8

Figure 8.1b: From John W. Hole, Jr., *Human Anatomy and Physiology*, 5th ed. Copyright © 1990 Wm. C. Brown Publishers, Dubuque, Iowa. All Rights Reserved. Reprinted with permission.

Figure 8.2a, b: From John W. Hole, Jr., *Human Anatomy and Physiology*, 5th ed. Copyright © 1990 Wm. C. Brown Publishers, Dubuque, Iowa. All Rights Reserved. Reprinted with permission.

Figure 8.4: From John W. Hole, Jr., *Human Anatomy and Physiology*, 5th ed. Copyright © 1990 Wm. C. Brown Publishers, Dubuque, Iowa. All Rights Reserved. Reprinted with permission.

Figure 8.5: From John W. Hole, Jr., *Human Anatomy and Physiology*, 5th ed. Copyright © 1990 Wm. C. Brown Publishers, Dubuque, Iowa. All Rights Reserved. Reprinted with permission.

Figure 8.6: From John W. Hole, Jr., *Human Anatomy and Physiology*, 5th ed. Copyright © 1990 Wm. C. Brown Publishers, Dubuque, Iowa. All Rights Reserved. Reprinted with permission.

Figure 8.8a, b: Modified from Crouch, J. E.: *Functional Human Anatomy*, 4th edition, Philadelphia, Lea & Febiger, 1984.

Figure 8.12a: From John W. Hole, Jr., *Human Anatomy and Physiology*, 5th ed. Copyright © 1990 Wm. C. Brown Publishers, Dubuque, Iowa. All Rights Reserved. Reprinted with permission.

Figure 8.13a, b: From John W. Hole, Jr., *Human Anatomy and Physiology*, 5th ed. Copyright © 1990 Wm. C. Brown Publishers, Dubuque, Iowa. All Rights Reserved. Reprinted with permission.

Figure 8.14: From John W. Hole, Jr., *Human Anatomy and Physiology*, 4th ed. Copyright © 1987 Wm. C. Brown Publishers, Dubuque, Iowa. All Rights Reserved. Reprinted with permission.

Chapter 9

Figure 9.1: From John W. Hole, Jr., *Human Anatomy and Physiology*, 5th ed. Copyright © 1990 Wm. C. Brown Publishers, Dubuque, Iowa. All Rights Reserved. Reprinted by permission.

Figure 9.2: From Kent M. Van De Graaff and Stuart Ira Fox, *Concepts of Human Anatomy and Physiology*, 2d ed. Copyright © 1989 Wm. C. Brown Publishers, Dubuque, Iowa. All Rights Reserved. Reprinted by permission.

Figure 9.4: From Kent M. Van De Graaff and Stuart Ira Fox, *Concepts of Human Anatomy and Physiology*, 2d ed. Copyright © 1989 Wm. C. Brown Publishers, Dubuque, Iowa. All Rights Reserved. Reprinted by permission.

Figure 9.13a: From John W. Hole, Jr., *Human Anatomy and Physiology*, 5th ed. Copyright © 1990 Wm. C. Brown Publishers, Dubuque, Iowa. All Rights Reserved. Reprinted with permission.

Figure 9.17: From Kent M. Van De Graaff, *Human Anatomy*, 2d ed. Copyright © 1988 Wm. C. Brown Publishers, Dubuque, Iowa. All Rights Reserved. Reprinted by permission.

Chapter 10

Figure 10.3: From Kent M. Van De Graaff, *Human Anatomy*, 2d ed. Copyright © 1988 Wm. C. Brown Publishers, Dubuque, Iowa. All Rights Reserved. Reprinted by permission.

Figure 10.4: From John W. Hole, Jr., *Human Anatomy and Physiology*, 5th ed. Copyright © 1990 Wm. C. Brown Publishers, Dubuque, Iowa. All Rights Reserved. Reprinted with permission.

Figure 10.6: From John W. Hole, Jr., *Human Anatomy and Physiology*, 5th ed. Copyright © 1990 Wm. C. Brown Publishers, Dubuque, Iowa. All Rights Reserved. Reprinted with permission.

Figure 10.7: From John W. Hole, Jr., *Human Anatomy and Physiology*, 5th ed. Copyright © 1990 Wm. C. Brown Publishers, Dubuque, Iowa. All Rights Reserved. Reprinted with permission.

Figure 10.9: From Kent M. Van De Graaff and Stuart Ira Fox, *Concepts of Human Anatomy and Physiology*, 2d ed. Copyright © 1989 Wm. C. Brown Publishers, Dubuque, Iowa. All Rights Reserved. Reprinted by permission.

Figure 10.10: From John W. Hole, Jr., *Essentials of Human Anatomy and Physiology*, 3d ed. Copyright © 1989 Wm. C. Brown Publishers, Dubuque, Iowa. All Rights Reserved. Reprinted by permission.

Chapter 11

Figure 11.4: From John W. Hole, Jr., *Human Anatomy and Physiology*, 5th ed. Copyright © 1990 Wm. C. Brown Publishers, Dubuque, Iowa. All Rights Reserved. Reprinted with permission.

Figure 11.8: From John W. Hole, Jr., *Human Anatomy and Physiology*, 5th ed. Copyright © 1990 Wm. C. Brown Publishers, Dubuque, Iowa. All Rights Reserved. Reprinted with permission.

Chapter 12

Figure 12.1: From Kent M. Van De Graaff and Stuart Ira Fox, *Concepts of Human Anatomy and Physiology*, 2d ed. Copyright © 1989 Wm. C. Brown Publishers, Dubuque, Iowa. All Rights Reserved. Reprinted by permission.

Figure 12.2: From Kent M. Van De Graaff, *Human Anatomy*, 2d ed. Copyright © 1988 Wm. C. Brown Publishers, Dubuque, Iowa. All Rights Reserved. Reprinted by permission.

Figure 12.4: From J. E. Crouch: *Functional Human Anatomy*, 2nd edition. Copyright © 1985. Lea & Febiger.

Figure 12.5a: From John W. Hole, Jr., *Human Anatomy and Physiology*, 5th ed. Copyright © 1990 Wm. C. Brown Publishers, Dubuque, Iowa. All Rights Reserved. Reprinted by permission.

Figure 12.6a: From John W. Hole, Jr., *Human Anatomy and Physiology*, 5th ed. Copyright © 1990 Wm. C. Brown Publishers, Dubuque, Iowa. All Rights Reserved. Reprinted with permission.

Figure 12.8: From Kent M. Van De Graaff, *Human Anatomy*, 2d ed. Copyright © 1988 Wm. C. Brown Publishers, Dubuque, Iowa. All Rights Reserved. Reprinted by permission.

Index

Greater trochanter, 90
Greater tubercle, 87
Great saphenous vein, 54, 55
Greek, and medical terminology, 1
Greenstick fracture, 92
Groin region, 17–18
Groups, blood, 206, 207
Growth
    embryonic, 291
    skeletal, 75
Growth hormone (GH), 166, 167, 168
Guanine, 323, 324
Gums, 181, 183
Gyri, 134

H

Haemophilus influenza, 241
Hair, 64
    follicle, 63, 64, 65, 66
    shaft, 63, 65, 66
Hamate, 88
Hammer, 157, 158, 159
Hamstring muscle group, 108, 113, 114, 116
Hand, 88, 112
Hard palate, 82, 181, 183, 184
Haversian canal, 46
Haversian system, 73, 75
HCG. See Human chorionic gonadotropic hormone (HCG)
HCl. See Hydrochloric acid (HCl)
HCT. See Hematocrit (HCT)
Head
    body regions, 17–18
    bone, 77
    of femur, 90
    of fibula, 115
    of humerus, 86
    muscles, 108–9, 109–10, 115
    of radius, 87
Healing, wound, 69, 70
Hearing, 159–60
Heart, 11, 57, 58, 211–16
    anatomy, 224
    attack, 205
    beat, 213
    blood path, 212–13, 217
    blood supply, 223
    and cardiac conduction system, 215
    cardiac cycle, 213–14
    disease, 205, 256, 257
    heartbeat control, 215–16
    internal view, 212
    sounds, 214
    valves, 214–15
Heat exhaustion, 68
Heat stroke, 68
Heelbone. See Calcaneus
Height, 299
Heimlich maneuver, 247, 248
Helix, 323
Helper T cells, 239, 240
Hematocrit (HCT), 197
Hematoma. See Blood
Heme, 198
Hemispheres, cerebral, 133–34
Hemodialysis, 269
Hemoglobin, 179, 198
Hemolysis, 36, 37
Hemolytic anemia, 198

Hemophilia, 311
Hemorrhoids, 189, 218
Henle, loops of. See Loops of Henle
Hepatic portal system, 223, 225
Hepatic portal vein, 190
Hepatitis, 192
Hinge joint, 95, 96
Hipbones, 88–90
His, bundle of. See Bundle of His
Histamine, 235, 236, 240
Hodgkin's disease, 234
Homeostasis, 24–25, 202
Horizontal plane, 16–17
Hormones, 197
    ACTH. See Adrenocorticotropic hormone (ACTH)
    ADH. See Antidiuretic hormone (ADH)
    of endocrine glands, 165
    and female reproductive system, 282–85
    female sex, 284–85
    FSH. See Follicle-stimulating hormone (FSH)
    GH. See Growth hormone (GH)
    GnRH. See Gonadotropic releasing hormone (GnRH)
    gonadotrophic, 165
    HCG. See Human chorionic gonadotropic hormone (HCG)
    ICSH. See Interstitial cell-stimulating hormone (ICSH)
    lactogenic, 284–85
    LH. See Leuteinizing hormone (LH)
    and male reproductive system, 278–79
    MSH. See Melanocyte-stimulating hormone (MSH)
    PTH. See Parathyroid hormone (PTH)
    sex, 165
    TSH. See Thyroid-stimulating hormone (TSH)
Horns, dorsal and ventral, 129
Hot, 67
Human chorionic gonadotropic hormone (HCG), 284, 292
Human organism, reference figures, 53–60
Human organization, 11–72
Human torso, 54–60
Humerus, 78, 86–87, 106, 112
Humors, of eye, 152, 153
Huntington disease, 125–26, 311
Hyaline cartilage, 45, 46
Hydrocephalus, 129, 311
Hydrochloric acid (HCl), 185, 190
Hydrogen bond, 317
Hydrogen ion concentration, 202
Hydrolysis, and synthesis, 318–19, 322
Hyoid bone, 58, 184
Hypercholesterolemia, 311
Hyperextension, 104
Hypertension, 226
Hyperthermia, 68–69
Hypertonic solution, 36, 37
Hypertrophy, 104
Hypochondriac region, 21
Hypogastric region, 21
Hypoglossal canal, 79
Hypoglossal nerves, 140, 141
Hypothalamus, 130, 133, 137, 164, 165, 166, 167, 169, 279, 283
Hypothermia, 69
Hypotonic solution, 36, 37
H zone, 99, 101

I

I bands, 99, 101
ICSH. See Interstitial cell-stimulating hormone (ICSH)
Identical twins, 292
IgA, 239
IgD, 239
IgE, 239, 240
IgG, 238, 239, 240
IgM, 239
Ileum, 58, 186, 187
Iliac artery, 58, 59, 220
Iliac crest, 89
Iliac region, 21
Iliac spine, 54, 55
Iliacus muscle, 60, 113
Iliac vein, 59
Iliopsoas muscle, 107, 112, 113, 116
Ilium, 88, 89
Immovable bones, 91
Immune system, 24
Immunity, 199, 235–42
    active and passive, 241–42
    antibody-mediated, 237–39
    cell-mediated, 239–40
Immunizations, 241
Immunoglobulins, 199, 238, 239, 240
Immunotherapy, 241–42
Impetigo, 69
Implantation, 284, 292
Impotency, 277
Impulse, nerve, 124–25
Incisive foramen, 81
Incisors, 181, 183
Incus, 157, 158, 159
Infancy, 297–98
Infarction, 205
Infections
    and deafness, 160–61
    respiratory, 254–56
    spinal cord, 132
    See also specific infections and Diseases and Disorders
Inferior, 15–16
Inferior mesenteric artery, 59
Inferior nasal concha, 79, 80, 82
Inferior oblique muscle, 150
Inferior rectus muscle, 149, 151
Inferior vena cava, 59, 60, 212, 219, 220
Infertility, 287–88
Inflammatory response, 199, 235–36
Influenza, 241, 254–55
Infraorbital foramen, 80
Inguinal canal, 55, 277
Inguinal region, 17–18, 21
Inheritance
    chromosome, 304–9
    gene, 309–12
    sex chromosome, 307–9
Inhibin, 279
Inner ear, 157, 158
Innervation, 102
Inorganic chemistry, 315–18
Insertion, 105, 106, 115–16
Inspiration, 245, 250–51, 252
Inspiratory reserve volume, 252
Insulin, 165, 174, 176
Insulin shock, 176
Integration, 20–21, 119–78, 301
Integument, 62
Integumentary system, 19, 22, 62
Intercalated disks, 48, 49
Intercostal muscle, 55

Vertical plane, 16–17
Vesicles, 102
Vessels
    blood, 216–18
    lymphatic, 232
Vestibule, 157, 158, 160, 282
Vestibulocochlear nerves, 140, 141
Villi, 181, 187
Viscera
    abdominal, 56
    thoracic, 57
Visceral, 15–16
Visceral membranes, 250
Visceral pericardial membrane, 211
Visual field, 154
Vital capacity, 252, 253
Vitamins, 67, 154, 170, 188, 193, 197
Vitreous humor, 152, 153
Vocal cords, 247
Voice box. *See* Larynx
Volume, blood, 267–68
Vomer bones, 79, 80, 81, 82

W

Wastes, 197
Water, 24, 317
WBCs. *See* White blood cells (WBCs)
White blood cell count, 199
White blood cells (WBCs), 47, 197,
    198–201
Whiteheads, 65
White matter, 119, 127, 128, 129
Whole skeletal muscle, 99
Whooping cough, 241
Willis, circle of. *See* Circle of Willis
Windows, oval and round, 157, 158, 160
Windpipe. *See* Trachea
Withdrawal. *See* Coitus interruptus
Womb. *See* Uterus
Words, root. *See* Medical terminology
Wound healing, 69, 70
Wrist region, 17–18

X

Xiphoid process, 56, 84, 85
X rays, 132
XYY, 306, 308, 309

Y

Yellow marrow, 75, 76, 77
    See also Blood

Z

Z lines, 99, 101, 103
Zygomatic arch, 81
Zygomatic bones, 80, 81, 82, 149
Zygomatic process, 80, 82
Zygomaticus muscle, 107, 109, 115

Anatomy and physiology is the study of the structure and function of the vital organs, such as the heart. The heart is muscular and has four chambers: two thin-walled atria above and two thick-walled ventricles below. Their contraction, evidenced by the electrocardiogram (ECG) shown in green, moves the blood through the heart and into the attached blood vessels. The tall spike of the ECG is caused by ventricular contraction; the smaller spike to the left of the tall one is caused by atrial contraction, while the smaller spike to the right of the tall one is caused by ventricular recovery.

ISBN 0-697-07856-6

90000

9 780697 078568

"Recycled"

WCB

Wm. C. Brown Publishers